*Tai-Kai Ng*

**Introduction to Classical and Quantum Field Theory**

## *Related Titles*

Gasiorowicz, S.

**Quantum Physics**

2003
ISBN: 978-0-471-05700-0

Zettili, N.

**Quantum Mechanics**

**Concepts and Applications**

2001
ISBN: 978-0-471-48944-3

Jackson, J. D.

**Classical Electrodynamics**

1999
ISBN: 978-0-471-30932-1

Huang, K.

**Quantum Field Theory**

**From Operators to Path Integrals**

1998
ISBN: 978-0-471-14120-4

Cohen-Tannoudji, C., Diu, B., Laloe, F.

**Quantum Mechanics**

**Volume 1**

1977
ISBN: 978-0-471-16433-3

Cohen-Tannoudji, C., Diu, B., Laloe, F.

**Quantum Mechanics**

**Volume 2**

1977
ISBN: 978-0-471-16435-7

*Tai-Kai Ng*

# Introduction to Classical and Quantum Field Theory

WILEY-VCH

WILEY-VCH Verlag GmbH & Co. KGaA

**The Author**

***Prof. Tai-Kai Ng***
Hong Kong University
Department of Physics
Clear Water Bay Road
Hong Kong
Hongkong



**Library of Congress Card No.:**
applied for

**British Library Cataloguing-in-Publication Data**
A catalogue record for this book is available from the British Library.

**Bibliographic information published by the Deutsche Nationalbibliothek**
The Deutsche Nationalbibliothek lists this publication in the Deutsche Nationalbibliografie; detailed bibliographic data are available on the Internet at http://dnb.d-nb.de.



**Composition** Thomson Digital, Noida, India
**Printing** Betz Druck GmbH, Darmstadt
**Bookbinding** Litges & Dopf GmbH, Heppenheim

Printed in the Federal Republic of Germany
Printed on acid-free paper

**ISBN:** 978-3-527-40726-2

## Acknowledgements

The book is a result of exchanges and collaborations between me and my colleagues, post-docs. and students. I want to mention in particular Prof. P.A. Lee, Prof. N. Nagaosa, Dr. Yi Zhou, Mr. Ching-Kit Chan, Mr. Cheung Chan and Mr. Zhengxin Liu.

I am also grateful to my parents and my wife for their support and encouragement during the course of my work.

*Introduction to Classical and Quantum Field Theory*. Tai-Kai Ng

ISBN: 978-3-527-40726-2

# Contents



*Introduction to Classical and Quantum Field Theory.* Tai-Kai Ng

ISBN: 978-3-527-40726-2

# Introduction to Classical and Quantum Field Theory

## Content Outline

Many excellent textbooks summarizing modern discoveries in classical and quantum field theory have appeared in the past ten years. However, these textbooks typically assume that students have already some familiarity with basic concepts and techniques in statistical/quantum field theory like the Landau theory of phase transitions, perturbative diagrammatic approaches, superfluidity and Fermi liquid theory, etc. A textbook which introduces and explains these concepts and shows how the concepts are applied in modern quantum (and classical) field theory is missing. With the aid of several basic examples, this textbook tries to fill this gap by explaining to students at an introductory level why and how these concepts and techniques enter (quantum) field theory.

The course is focused mainly on application of quantum/classical field theory techniques to condensed matter and statistical physics systems and is divided into three parts. The first part (Part I) covers fundamental physics and mathematics background that students need in order to enter the field and the second part (Part II) introduces students to more advanced and modern concepts/techniques in quantum and classical field theory. The third part (Part III) discusses how these concepts/techniques are applied in a few important examples. The limitation of standard quantum field theory techniques in facing real physics problems will be discussed.

The first part of the book (Chapters 1–4) is introductory and its content has appeared in many 'older' textbooks (e.g. *Quantum Field Theory* by Itzykson and Zuber, *Many Particle Physics* by Mahan) in one form or another. The only difference is that these concepts and techniques are introduced with modern context and terminology in mind here.

The second part of the book introduces to students more advanced concepts and techniques in classical and quantum field theory. The fundamentals of perturbation theory are introduced in Chapter 5 and the relation between statistics, Berry phase, and gauge theory is discussed in Chapter 6. The important concepts of effective field theory, phases, and phase transitions are introduced in Chapter 7, followed by a chapter introducing the more advanced topics of solitons and instantons (Chapter 8).

*Introduction to Classical and Quantum Field Theory.* Tai-Kai Ng

ISBN: 978-3-527-40726-2

Let us emphasize that the topics in Part II are gigantic topics in themselves, and special textbooks have been written on these topics. The purpose of Part II is to introduce these topics to students who have no prior exposure to the subjects hoping that it can help the students when studying more advanced textbooks such as *Quantum Phase Transitions* by Sachdev or when the students are studying papers involving vortices and skyrmions.

The third part of the book demonstrates how the techniques and concepts developed in previous chapters are applied to a few important topics in quantum field theory, including simple boson and fermion liquids, superconductivity, and U (1) lattice gauge theories. The limitation of the 'standard' perturbation theory techniques when facing real physics problem is pointed out. These chapters can be considered as introductory materials for more advanced textbooks such as *Quantum Field Theory in Condensed Matter Physics* by Nagaosa or *Theory of Quantum Many-body Systems* by Wen.

Readers who have no prior exposure to quantum field theory are recommended to read first Chapters 1–6. Afterward the readers may like to go to Chapters 9 and 10 to have a feeling of how the techniques developed in the first six chapters are applied to real-life problems. The readers may then go back to the more advanced concepts and techniques discussed in Chapters 7 and 8, which are necessary to understand the more advanced topics in Chapters 11 and 12. The readers may have to go back and forth between Part II and Part III of the book a few times before fully understanding why these advanced concepts and techniques are needed when facing real-life problems.

# Part One

*Introduction to Classical and Quantum Field Theory*. Tai-Kai Ng

ISBN: 978-3-527-40726-2

# 1
# Introduction

## 1.1
## What is a Field Theory?

Field theory deals with objects which form a *medium*, for example objects like water, air, solids, magnets, electromagnetic fields, or even our universe. A common feature of these objects is that they are all composed of a large number of microscopic constituents. The goal of field theory is to study the *thermodynamics* and *dynamical* properties of these media.

### 1.1.1
### Mathematical Description

To describe these objects, we have to introduce variables that are functions of space and time. These variables can be real or complex scalars, vectors, tensors, and so on. For example, to describe the motion of air (or fluid), we have to introduce the variables $\rho(\vec{x}, t)$ and $\vec{j}(\vec{x}, t)$, which are the mass and current density of air (or fluid) molecules, respectively. To describe magnets, we introduce the magnetization density $\vec{M}(\vec{x}, t)$, which is a vector variable describing the magnitude and direction of magnetization at point $\vec{x}$ inside the magnet at time $t$.

**Question**
How about vacuum (or our universe)? Can you think of any *continuous* physical quantities that can describe different states of the vacuum?

**Answer**
Yes, electromagnetic and gravitational fields, density of stars (galaxies), and so on.

In physics, the properties of the media are characterized by two important ingredients, the thermodynamic and dynamical properties of the media. Examples of thermodynamic properties include

– How density and pressure of air change with temperature, volume, and so on.

*Introduction to Classical and Quantum Field Theory.* Tai-Kai Ng

ISBN: 978-3-527-40726-2

– How magnetization of a magnet changes with temperature, external magnetic field, and so on.

An important concept coming from thermodynamics is *phases and phase transitions* – that the same physical medium can behave in qualitatively different ways (different phases) under different conditions, and that the medium can evolve from one phase to another when some thermodynamic parameters change. For example, water can evolve from the solid to the liquid phase when the temperature rises, and the physical properties of the two phases are qualitatively different. We shall see later that we can ask similar questions even for our vacuum. For example, how will the dynamics of an electromagnetic field change if our vacuum evolves to a different phase?

In studying dynamics we typically ask the question of how the variables describing a medium will evolve with time with an initial condition or under external, time-dependent force. These are the situations we encounter when we want to investigate a medium with our experimental probes.

**Question**

What do we need to study the thermodynamics of a field theory?

According to Gibbs, the (equilibrium) thermodynamics of any physical system is determined by the *partition function*

$$Z = \int e^{-\frac{H[\phi]}{k_B T}} D\phi,$$

where $H$ is the Hamiltonian (energy) function, $T$ is the temperature, and $\phi_{i,j,\ldots}(\vec{x})$ are the variables characterizing the medium. The integral sign $\int \ldots D\phi$ is a shorthand notation meaning that we have to sum over all configurations of the field $\phi$. Note that to evaluate the partition function we need to know how the energy of the system depends on the field configurations $H[\phi(\vec{x})]$. In quantum systems the situation is complicated by the fact that different variables may not commute with each other. The simplest way to get out of this trouble is to replace the sum over all configurations $\phi$ by a sum over all eigenstates $\varphi_n$. We shall discuss concrete examples later in the book.

**Question**

Can you give us some examples of $H[\phi(\vec{x})]$?

$$\text{Ising model for magnet: } H[S_i] = -J \sum_{\langle i \cdot j \rangle} S_i S_j \quad (S_i = \pm 1).$$

$$\text{(Classical) Heisenberg model: } H[\vec{S}_i] = -J \sum_{\langle i \cdot j \rangle} \vec{S}_i \cdot \vec{S}_j \quad (\vec{S} = \text{unit vector}).$$

*Can you think of other examples?*

The computation of the partition function of a classical field system is in general a very complicated problem. We shall look into this subject more carefully in the second part of this chapter. Let us discuss the question of dynamics first.

**Question**
What is required to understand the dynamics of a field theory?

In classical field theory we need to solve equations of motion of the fields typically of the form $\frac{\partial}{\partial t}\phi(\vec{x},t) = F\left[\phi(\vec{x}',t')\right],\ldots,$ and so on. A well-known example of equations of motion of a classical field theory are the Maxwell equations

$$\nabla\cdot\vec{E} = \frac{\rho}{\varepsilon_0},\quad \nabla\cdot\vec{B} = 0,\quad \nabla\times\vec{B} = \mu_0\vec{J} + \frac{1}{c^2}\frac{\partial\vec{E}}{\partial t},\quad \nabla\times\vec{E} + \frac{\partial\vec{B}}{\partial t} = 0.$$

Another well-known example is the Navier–Stokes equation. We also have equations for elastic waves in solids, surface waves in water, and so on. Equations of motion in quantum field theories have a different meaning because the field variables become operators. We shall explore the quantum situation in later chapters.

Summarizing, for classical field theory we need mathematical tools to (i) evaluate the partition function and (ii) solve the equations of motion. Both of these tasks can be achieved rigorously only if the system is *linear*. In general it is impossible to accomplish these two tasks and approximate methods have to be used. We shall review the basic mathematical techniques needed for the above problems in the following.

## 1.2 Basic Mathematical Tools in (Classical) Field Theory

### 1.2.1 Solution of Field Equations of Motion

First, we shall consider linear systems where the above mathematical tasks can be solved exactly in principle.

For linear equations of motion, we can write the equations in a general form

$$\widehat{L}\left(\frac{\partial}{\partial t},\nabla\right)\psi(\vec{x},t) = 0, \tag{1.1a}$$

where $\widehat{L}$ is in general an $N\times N$ Hermitian matrix operator and $\psi$ is an $N$-dimensional vector function. An example is the Maxwell equation where $\psi\to\vec{E},\vec{B}$ are field vectors each with three components. Many equations of motion involving vectorial fields can be written in matrix form.

Examples of $\widehat{L}$:

$$\widehat{L} = \frac{\partial^2}{\partial t^2} - c^2\nabla^2 \text{ (wave equation)},\quad \widehat{L} = i\frac{\partial}{\partial t} - \widehat{H} \text{ (Schrödinger equation)}.$$

Note that the above equations can be viewed as eigenvalue equations with zero eigenvalue, that is

$$\widehat{L}\left(\frac{\partial}{\partial t},\nabla\right)\psi_0(\vec{x},t) = (0)\cdot\psi_0(\vec{x},t). \tag{1.1b}$$

Very often, we also need to solve the equations with an external driving force. In this case we have equations of the form

$$\hat{L}\left(\frac{\partial}{\partial t}, \nabla\right)\psi(\vec{x}, t) = f(\vec{x}, t), \tag{1.2}$$

where $f(\vec{x}, t)$ is a vector function with the same dimension as $\psi$.

A general strategy to solve this kind of problem is the eigenfunction expansion method plus Green's functions. We shall treat space and time on equal footings in the following general discussion.

#### 1.2.1.1 Eigenfunction Expansion Method

In the theory of differential equations a very important and useful result is the completeness and orthonormality of the set of eigenfunctions that satisfies an eigenvalue equation $H\psi_n = \lambda_n\psi_n$, where $H$ is any Hermitian differential operator operating in space and time and the $\lambda_n$ are the corresponding eigenvalues.

It can be proved that the eigenstates $\psi_n$ form a complete set, that is any (continuous, differentiable) function which is defined in the (Hilbert) space where $H$ operates can be written as

$$f(\vec{x}t) = \sum_n a_n\psi_n(\vec{x}t), \tag{1.3a}$$

where $a_n$ are complex, numerical coefficients that can be determined from another important property of the $\psi$: the orthonormality of eigenstates with different eigenvalues, that is

$$\int_V \psi_n^*(\vec{x}t)\psi_m(\vec{x}t)\mathrm{d}^d x\,\mathrm{d}t = \delta_{nm}. \tag{1.3b}$$

With Equation (1.3a) it can be shown easily that

$$a_n = \int_V f(\vec{x}t)\psi_n^*(\vec{x}t)\mathrm{d}^d x\,\mathrm{d}t. \tag{1.3c}$$

**Degeneracy** In many cases the eigenstates are degenerate, that is there is more than one eigenstate with the same eigenvalue. In this case let $\{\{\psi'^{(n)}_m\}\}$ be a complete set of eigenstates with eigenvalue $\lambda_n$; then, a result from linear algebra tells us that we can always construct a set of orthonormal states $\{\psi_m^{(n)}\}$ from $\{\psi_m'^{(n)}\}$ and from completeness any function $g_n(\vec{x}t)$ that satisfies $\hat{L}g_n(\vec{x}t) = \lambda_n g_n(\vec{x}t)$ can be written as

$$g_n(\vec{x}t) = \sum_m a_m^{(n)}\psi_m^{(n)}(\vec{x}t), \tag{1.4a}$$

where

$$\begin{aligned}&\int_V \psi_m^{(n)^*}(\vec{x}t)\psi_{m'}^{(n)}(\vec{x}t)\mathrm{d}^d x\,\mathrm{d}t = \delta_{mm'} \quad \text{and}\\ &a_m^{(n)} = \int_V g_n(\vec{x}t)\psi_m^{(n)^*}(\vec{x}t)\mathrm{d}^d x\,\mathrm{d}t.\end{aligned} \tag{1.4b}$$

One should keep in mind that the set $\{\psi_m^{(n)}\}$ is not unique. A different set can be constructed if a different procedure is used to construct the orthonormal set. However, Equation (1.4(a), (b)) remains correct; the only difference is that the coefficients $a_m^{(n)} = \int_V f(\vec{x}t)\psi_m^{(n)^*}(\vec{x}t)\mathrm{d}^d x\,\mathrm{d}t$ will become different when the set $\{\psi_m^{(n)}\}$ changes.

**Homework Q1**

Consider the solutions of the free-particle Schrödinger equation in three dimensions,

$$-\frac{\hbar^2}{2m}\nabla^2\varphi_\varepsilon = \varepsilon\varphi_\varepsilon,$$

with eigenvalue $\varepsilon > 0$. Construct two different orthonormal sets of eigenstates $\{\varphi_m^{(1)}\}$, $\{\varphi_m^{(2)}\}$ with the same eigenvalue $\varepsilon$. Expand the function $\cos(\vec{k}.\vec{x})$ $\left(\frac{\hbar^2\vec{k}^2}{2m} = \varepsilon\right)$ in the two sets of eigenstates and evaluate the corresponding coefficients $a_m^{(n)} = \int_V \cos(\vec{k}.\vec{x})\varphi_m^{(n)^*}(\vec{x})\mathrm{d}^d x$.

The above result can be applied readily to Equation (1.1b) and the general solution of the equation can be written as

$$\psi_0(\vec{x}t) = \sum_m a_m^{(0)}\psi_m^{(0)}(\vec{x}t),$$

where $\{\psi_m^{(0)}\}$ is a chosen orthonormal set of solutions. Note that the $a_m^{(0)}$ are completely arbitrary and an infinite number of possible solutions exists in general. The coefficients $a_m^{(0)}$ are fixed by initial and boundary conditions, which determine the solution uniquely.

#### 1.2.1.2 Eigenfunction Expansions for Green's Functions

For any linear differential operator $\widehat{L}$, we can introduce a corresponding Green's function through

$$\widehat{L}G(\vec{x}t, \vec{x}'t') = \delta(\vec{x}-\vec{x}')\delta(t-t')\widehat{I}, \tag{1.5a}$$

where $\widehat{I}$ is an $N \times N$-dimensional identity matrix.

With the Green's function determined, the solution of the equation $\widehat{L}(\frac{\partial}{\partial t}, \nabla)\psi(\vec{x}, t) = f(\vec{x}, t)$ can be written as

$$\psi(\vec{x}, t) = \psi_0(\vec{x}, t) + \int \mathrm{d}^d x'\mathrm{d}t' G(\vec{x}t, \vec{x}'t')f(\vec{x}', t'), \tag{1.5b}$$

as can be seen by direct substitution.

Note that, in the language of linear algebra, the Green's function is nothing but the inverse operator of $\widehat{L}$, that is $\widehat{G} = \widehat{L}^{-1}$ and $\widehat{L}\widehat{G} = \widehat{I}$. A formal solution of the Green's function can be found using a general method based on completeness and orthogonality.

Let $\psi_n(\vec{x}t)$ be the complete set of eigenstates for the linear operator $L$, that is $\widehat{L}\psi_n(\vec{x}t) = \lambda_n\psi_n(\vec{x}t)$. To obtain the Green's function defined by

$$\widehat{L}\left(\frac{\partial}{\partial t}, \nabla_x\right)G(\vec{x}t, \vec{x}'t') = \delta(\vec{x}-\vec{x}')\delta(t-t'),$$

we expand $G(\vec{x}t,\vec{x}'t')=\sum_n a_n(\vec{x}'t')\psi_n(\vec{x}t)$ and, from completeness, $\delta(\vec{x}-\vec{x}')\delta(t-t')=\sum_n \psi_n^*(\vec{x}'t')\psi_n(\vec{x}t)$. Therefore,

$$\hat{L}\left(\frac{\partial}{\partial t},\nabla_x\right)G(\vec{x}t,\vec{x}'t')=\sum_n \lambda_n a_n(\vec{x}'t')\psi_n(\vec{x}t)=\sum_n \psi_n^*(\vec{x}'t')\psi_n(\vec{x}t),$$

which leads to $a_n(\vec{x}'t')=\frac{\psi_n^*(\vec{x}'t')}{\lambda_n}$

or

$$G(\vec{x}t,\vec{x}'t')=\sum_n \frac{\psi_n^*(\vec{x}'t')\psi_n(\vec{x}t)}{\lambda_n}. \tag{1.6}$$

Therefore, the Green's function can be determined if we can evaluate the above sum over all eigenstates. This is in general a very difficult task and alternative methods are often used when we want to evaluate the Green's function explicitly. However, the expansion form (1.6) for the Green's function is very useful for formal manipulations, as we shall see later.

**Homework Q2**

Determine the Green's function of the differential operator $a\frac{\partial}{\partial t}+D\frac{\partial^2}{\partial x^2}$ in one dimension. $D$ is a real number and $a$ can be either real or imaginary. Consider both situations.

(Hint: you have to fix the boundary condition to determine the Green's function uniquely. There are different possible boundary conditions. Choose one which you think is physically meaningful.)

#### 1.2.1.3 A Variant of the Above Method: Initial Condition Problem

Note that time and space are treated on equal footings in the above general discussion. However, in facing real problems we often have to treat space and time separately. This is often reflected in the initial condition problem: what is $\psi(\vec{x},t)$ if we know $\psi(\vec{x},0)$ (plus some time-derivative terms)?

To illustrate this, we shall first consider the Schrödinger equation

$$\hat{L}=i\frac{\partial}{\partial t}-\hat{H}(\vec{x}). \tag{1.7a}$$

The key feature of the Schrödinger equation is that the operator $\hat{L}$ can be written in a separable form $\hat{L}=L_0(t)+\hat{H}(\vec{x})$ where the time and space coordinates enter in different pieces in $\hat{L}$. For this particular form the eigenstates can be written as

$$\psi_n(\vec{x}t)=\phi_n(\vec{x})\chi_m(t), \tag{1.7b}$$

where

$$\hat{H}\phi_n(\vec{x})=\lambda_n\phi_n(\vec{x}),\ L_0\chi_m(t)=-\omega_m\chi_m(t), \tag{1.7c}$$

and

$$\hat{L}\psi_n(\vec{x}t) = (\lambda_n - \omega_m)\psi_n(\vec{x}t). \tag{1.7d}$$

Moreover, the eigenstates $\phi_n$ and $\chi_m$ form complete and orthonormal sets by themselves, that is

$$f(\vec{x}) = \sum_n a_n\phi_n(\vec{x}) \text{ for any function} f(\vec{x}) \text{ and } \int_V \phi_n^*(\vec{x})\phi_m(\vec{x}t)\mathrm{d}^d x = \delta_{nm}$$

and

$$g(t) = \sum_n b_n\chi_n(t) \text{ for any function } g(t) \text{ and } \int \chi_n^*(t)\chi_m(t)\mathrm{d}t = \delta_{nm}.$$

Therefore, any solution of the equation

$$\hat{L}\psi(\vec{x}t) = (0)\psi_n(\vec{x}t)$$

can be written as

$$\begin{aligned}&\sum_{n,m}\delta_{\lambda_n\omega_m}a_{nm}\chi_m(0)\phi_n(\vec{x})\\ &\text{and } \psi(\vec{x}0) = \sum_{n,m}\delta_{\lambda_n\omega_m}a_{nm}\chi_m(0)\phi_n(\vec{x}).\end{aligned} \tag{1.8a}$$

Using the orthonormality condition, we obtain

$$\sum_m \delta_{\lambda_n\omega_m}a_{nm}\chi_m(0) = \int \mathrm{d}^d x\phi_n^*(\vec{x})\psi(\vec{x},0). \tag{1.8b}$$

Note that, if $a_{nm}$ can be uniquely determined from the above equation, then $\psi(\vec{x}t)$ will be uniquely determined at all times. This becomes possible if there exists only one eigenstate $\chi_m$ with eigenvalue $\omega_m = \lambda_n$. This is the case for the Schrödinger operator $L_0 = i\frac{\partial}{\partial t}$, which involves only the first time derivative. In this case $\chi_m(t) = e^{-i\omega_m t}$ is uniquely determined once $\omega_m = \lambda_n$ is fixed.

**Homework Q3**

Show that in the case $L_0 = -\frac{\partial^2}{\partial t^2}$ the solution $\psi(\vec{x},t)$ is uniquely determined if we know both $\psi(\vec{x},0)$ and $\left.\frac{\mathrm{d}\psi(\vec{x},t)}{\mathrm{d}t}\right|_{t=0}$.

**Homework Q4**

In the case of the Schrödinger equation, the eigenstates $\chi_m(t) = e^{-i\omega_m t}$ are known and the sum over $\chi_m$ can be performed explicitly when evaluating the Green's function using the expansion method. In this case the Green's function can be written in terms of the eigenvalues $\lambda_n$ and eigenstates $\phi_n$ only. Show that in this case the solution of the initial value problem can be represented as $\psi(\vec{x}t) = \int \mathrm{d}^d x' G(\vec{x}t, \vec{x}'0)\psi(\vec{x}', 0)$.

#### 1.2.1.4 Comment on Non-Linear Equations of Motion

In general, non-linear equations are unsolvable, in the sense that there is no systematic procedure to obtain all the solutions of the equations. Physicists usually approach non-linear equations from a perturbative picture – we assume that the non-linear term in the equation of motion is 'small' and treat it as a correction to the solution of the linear equation. An order by order expansion of the non-linear term can be obtained if the approach works (converges).

**Adiabaticity and Counting Problem** An implicit assumption of this approach is that there is a one-to-one correspondence between the solution of the non-linear equation and the solution of the linear equation. This assumption is called the adiabaticity assumption. Physically, the adiabaticity assumption can be viewed in the following way: assume that we start with a solution of the linear equation $\phi_0$ which is changing into a different function $\phi$ when the non-linear term is turning up gradually. If the change is *always continuous* as a function of the magnitude of the non-linear term $\lambda$ (when $\lambda$ is either increasing or decreasing), that is $\phi = \phi(\lambda)$ and $\phi(\lambda = 0) = \phi_0$, then there exists always a one-to-one correspondence between $\phi_0$ and $\phi$ and adiabaticity is obeyed.

From a counting point of view this seems reasonable – if the total number of solutions of the linear equation (= size of Hilbert space where the differential operators are defined) is the same as the total number of solutions of the non-linear equation. Unfortunately, this is not always true and the problem is more subtle than this naïve picture. Non-perturbative solutions which have no simple connection to solutions in the corresponding linear equation do exist in non-linear equations. Some of these solutions are particularly *stable* (usually because of topological reasons) and are physically important. They are often called *soliton* solutions. For later purposes let us also introduce the term *instanton* – it is a non-perturbative solution of the non-linear equation of motion if you go to imaginary time – meaning that you replace $t \to it$ in your equation of motion. We shall study some of the important examples of solitions and instantons in Chapter 8.

### 1.2.2
### Evaluation of Partition Function for Quadratic Field Theories

In this section we discuss the general method of evaluating the partition function $Z = \int e^{-\frac{H[\phi]}{k_B T}} D\phi$. Before we proceed to the general formulation, we first consider a simple toy model so that we can have a better feeling for the mathematical meaning of the integral.

We consider a field $\phi$ defined on a lattice (like the Ising model) with $N$ independent sites. The energy of the system is

$$H[\phi] = \sum_{i=1,\ldots,N} h[\phi_i] = \sum_{i=1,\ldots,N} a\phi_i + b\phi_i^2 + c\phi_i^4. \tag{1.9a}$$

The partition function is therefore

$$Z = \int d\phi_1 d\phi_2 \cdots d\phi_N e^{-\frac{H[\phi]}{k_B T}} = \prod_i \int d\phi_i e^{-\frac{h[\phi_i]}{k_B T}} = z^N, \tag{1.9b}$$

where

$$z = \int_{-\infty}^{\infty} d\phi e^{-\frac{a\phi + b\phi^2 + c\phi^4}{k_B T}}. \tag{1.9c}$$

Note that in this toy model the partition function is just a product of partition functions of individual sites and the evaluation of the partition function becomes a much simpler problem of evaluating a single integral. Our remaining question is thus: can this integral be performed analytically?

The answer is 'yes' only if $b>0$ and $c=0$ (*quadratic or linear field theories*).

In this case we can write $z = \int_{-\infty}^{\infty} d\phi e^{-\left[\frac{b(\phi + a/2b)^2}{k_B T} - \frac{a^2}{4bk_B T}\right]} = e^{\frac{a^2}{4bk_B T}} \sqrt{\frac{\pi k_B T}{b}}$ and the free energy is given by

$$F = -k_B T \ln(Z) = -N\left[\frac{a^2}{4b} + \frac{k_B T}{2}\ln\left(\frac{k_B T}{b}\right)\right]. \tag{1.10}$$

Note that in this case the problem is like evaluating the partition function of *N decoupled classical harmonic oscillators.*

When $c$ is non-zero, the integral $z = \int_{-\infty}^{\infty} d\phi e^{-\frac{a\phi + b\phi^2 + c\phi^4}{k_B T}}$ cannot be evaluated analytically in general. When $c$ is small, we can expand $z$ in a power series of $c$ and evaluate the series term-by-term. This is the spirit of perturbation theory. In the above example, we obtain

$$z(c) = z(0) + cz' + c^2 z'' + \dots, \tag{1.11}$$

where

$$z(0) = \int_{-\infty}^{\infty} d\phi e^{-\frac{a\phi + b\phi^2}{k_B T}}, \quad z' = \frac{1}{k_B T}\int_{-\infty}^{\infty} \phi^4 d\phi e^{-\frac{a\phi + b\phi^2}{k_B T}}, \quad \text{etc.} \tag{1.12}$$

**Homework Q5**

Treating $c$ as a perturbation, evaluate the first- and second-order corrections in powers of $c$ to the above free energy $F$.

**Question**

What is $F$ if the coefficients $a$ and $b$ are site dependent, that is $a \to a_i$, $b \to b_i$?

**Answer**

$$F = -k_B T \ln(Z) = -\sum_i \left[\frac{a_i^2}{4b_i} + \frac{k_B T}{2}\ln\left(\frac{k_B T}{b_i}\right)\right].$$

Next, we make the problem more realistic by considering a more complicated energy function where different sites are coupled,

$$H[\phi] = \sum_{i,j=1,\ldots,N} b_{ij}\phi_i\phi_j, \quad \text{where } b_{ij} = b(\vec{r}_i - \vec{r}_j). \tag{1.13a}$$

To evaluate the above partition function we try to transform the energy function into the form of independent harmonic oscillators, that is $H[\phi] = \sum_k \lambda_k \phi_k^2$, where we already know how to evaluate the partition function.

The transformation can be readily obtained if we note that the coefficients $b_{ij}$ can be viewed as elements of a matrix $B$ and

$$H[\phi] = \phi^t B\phi, \tag{1.13b}$$

where $\phi = \begin{bmatrix} \phi_1 \\ \cdot \\ \cdot \\ \cdot \\ \phi_N \end{bmatrix}$ is a vector in the Hilbert space spanned by the matrix $B$. The matrix $B$ can in general be diagonalized by a unitary transformation $U$,

$$H[\phi] = \phi^t B\phi = (\phi^t U^+)(UBU^+)(U\phi), \tag{1.14a}$$

where

$$(UBU^+)_{kl} = \lambda_k \delta_{kl} \tag{1.14b}$$

is a diagonal matrix, and $U\phi = \bar{\phi} = \begin{pmatrix} \bar{\phi}_1 \\ \bar{\phi}_2 \\ \cdot \\ \cdot \\ \bar{\phi}_N \end{pmatrix}$.

Therefore, $H[\phi] = \sum_k \lambda_k \bar{\phi}_k^2$, and the partition function is now given by the $N$-dimensional integral $Z = \int d\phi_1 d\phi_2 \cdots d\phi_N e^{-\frac{H[\phi]}{k_B T}}$.

There is one more thing we have to be careful of. Note that the energy function is diagonalized in the $\bar{\phi}$ variables but the integral is performed over the $\phi$ variables. We need a change of integration variables from $\phi$ to $\bar{\phi}$ to evaluate the integral. For linear transformations, we have

$$Z = \int d\phi_1 d\phi_2 \cdots d\phi_N e^{-\frac{H[\phi]}{k_B T}} = \int d\bar{\phi}_1 d\bar{\phi}_2 \cdots d\bar{\phi}_N |\det(U^+)| e^{-\frac{H[\bar{\phi}]}{k_B T}}. \tag{1.15}$$

However, since $U$ is a unitary transformation, $|\det U| = 1$.

$$\text{Therefore, } Z = \int d\bar{\phi}_1 d\bar{\phi}_2 \cdots d\bar{\phi}_N e^{-\frac{H[\bar{\phi}]}{k_B T}} = \prod_k \sqrt{\frac{k_B T}{\lambda_k}} \tag{1.16a}$$

and

$$F = -k_B T \ln(Z) = -\frac{k_B T}{2} \sum_k \ln\left(\frac{k_B T}{\lambda_k}\right). \tag{1.16b}$$

Note that, equivalently, the result can be expressed as

$$F = \frac{k_B T}{2} (\ln(\det B) - \ln(\det(k_B T) I)),$$

which is the form commonly used in the literature.

For the particular form of matrix $b_{ij} = b(|\vec{r}_i - \vec{r}_j|)$, the eigenvectors are $\phi_i(k) = e^{i\vec{k}.\vec{r}_i}$, with eigenvalues given by the Fourier transform $\lambda_k = \sum_i e^{i\vec{k}.\vec{r}_i} b(\vec{r}_i)$. Note that, rigorously speaking, we need $\lambda_k > 0$ for all $k$ for the integrals to converge.

**Exercise**

Show that for $b_{ij} = b(|\vec{r}_i - \vec{r}_j|)$ the eigenvectors are $\phi_i(k) = e^{i\vec{k}.\vec{r}_i}$, with eigenvalues $\lambda_k = \sum_i e^{i\vec{k}.\vec{r}_i} b(\vec{r}_i)$.

#### 1.2.2.1 Non-Linear Energy Functional

Energy functionals with terms of higher order than $\phi^2$ are called non-linear energy functionals and their partition functions cannot be evaluated exactly. A frequently used energy functional is the $\phi^4$ model given by

$$H[\phi] = \sum_{i,j=1,\ldots,N} b_{ij} \phi_i \phi_j + c \sum_i \phi_i^4. \tag{1.17}$$

For small $c$ we can evaluate $Z[\phi]$ approximately by expansion in powers of $c$ as in the simple toy model. The evaluation of perturbation series is more complicated than the toy model because we have to make the unitary transformation $\phi \to \bar{\phi}$. There exist at present very elaborate mathematical methods to do the perturbation expansion systematically. We shall give a brief introduction to these techniques in Chapter 5. However, you are encouraged to evaluate the first- and second-order corrections to $Z[\phi]$ using a straightforward expansion method to get a feeling of the perturbation series.

#### 1.2.2.2 Continuum Limit

Having considered partition functions of a quadratic energy functional on a lattice, we now go to the continuum limit and consider an energy functional of the form $H[\phi] = \int d^d x \phi^+ \hat{H} \phi$.

An explicit example is

$$H[\phi] = \int d^d x ((c\nabla\phi)^2 + m^2\phi^2) = \int d^d x \phi(-c^2\nabla^2 + m^2)\phi. \tag{1.18}$$

Again, we note that we can diagonalize the Hamiltonian to obtain $H[\phi] = \sum_k \lambda_k \bar{\phi}_k^2$, where $\phi_i(k) = e^{i\vec{k}\cdot\vec{r}_i}$ and $\lambda_k = c^2k^2 + m^2$ in this case and

$$F = -k_B T \ln(Z) = -\frac{k_B T}{2} \sum_k \ln\left(\frac{k_B T}{\lambda_k}\right). \tag{1.19a}$$

The difference between lattice models and the continuum case is that, in the lattice models, the dimension of the Hilbert space is $N$ (= number of lattice sites) and we have precisely $N$ eigenvalues and eigenvectors, whereas, in the continuum case, we have an infinite number of eigenstates and the sum over $k$ becomes a continuous integral. In the example we considered, we have

$$F \rightarrow -\frac{k_B T}{2}\frac{V}{(2\pi)^d}\int d^d k \ln\left(\frac{k_B T}{c^2 k^2 + m^2}\right) \tag{1.19b}$$

and we can define a free-energy density as

$$f = \frac{F}{V} = -\frac{k_B T}{2}\frac{1}{(2\pi)^d}\int d^d k \ln\left(\frac{k_B T}{c^2 k^2 + m^2}\right). \tag{1.19c}$$

Note that the integrand diverges at $k \rightarrow \infty$ and the free-energy density is infinity! This result is, of course, unphysical. We shall go back to this problem again later, after we discuss the concept of effective field theory (Chapter 7). Note also that the divergence problem is absent in lattice problems where the integral over $k$ is restricted to the first Brillouin zone.

#### 1.2.2.3 Constraints

Nowadays we often encounter field theory with constraints. A very good example is the (classical) Heisenberg model: $H[\vec{S}_i] = -J\sum_{\langle i\cdot j\rangle}\vec{S}_i\cdot\vec{S}_j$ where the field variables $\vec{S}$ are unit vectors. How do we handle this situation? To have an idea of how we can handle this kind of constraint let us consider a very simple model with just one spin, $H = -S_z B$, where $\vec{B}$ is an external magnetic field. The partition function is

$$Z = \int dS_x dS_y dS_z \delta(S_x^2 + S_y^2 + S_z^2 - 1) e^{\frac{S_z B}{k_B T}}. \tag{1.20}$$

There are two ways to handle this situation; the first way is to solve the constraint $\vec{S}\cdot\vec{S} = 1$ exactly by writing $S_z = \cos\vartheta$, $S_x = \sin\vartheta\cos\phi$, $S_y = \sin\vartheta\sin\phi$, and

$$Z \rightarrow \int d\Omega e^{\frac{\cos\vartheta B}{k_B T}} = \int d\phi \sin\vartheta d\vartheta e^{\frac{\cos\vartheta B}{k_B T}}.$$

The second way is to use the identity $\delta(x) = \frac{1}{2\pi}\int e^{i\lambda x} d\lambda$, and write

$$Z = \int dS_x dS_y dS_z d\lambda e^{i\lambda(S_x^2 + S_y^2 + S_z^2 - 1)} e^{\frac{S_z B}{k_B T}}. \tag{1.21}$$

Note that the partition function now turns into a regular four-dimensional integral and we can employ standard tricks in evaluating integrals to solve the partition function. The first method is readily applicable for one spin but the second method is often used when many spins are present and coupled together and the first method becomes inconvenient. The introduction of $\lambda$ – called the Lagrange multiplier field – is often more convenient in making systematic approximations. You will see a number of examples of handling constraints later in this book.

A mathematical problem: how do we evaluate the integral $\int dx e^{i\lambda x^2 + bx}$?

Physicist's recipe:

Assume that $\int_{-\infty}^{\infty} d\phi e^{-\left[\frac{b(\phi+a/2b)^{24}}{k_B T} - \frac{a^2}{4bk_B T}\right]} = e^{\frac{a^2}{4bk_B T}} \sqrt{\frac{\pi k_B T}{b}}$ is valid even for $b \rightarrow ib$!

To summarize: in this section we have sketched some of the basic mathematical techniques that are used in mathematical analysis of field theories. The techniques we use in more complicated situations are often variants of the above basic techniques. We shall see many examples in later chapters.

# 2
# Basics of Classical Field Theory

## 2.1
## Lagrangian Formulation for Classical Mechanics/Field Theory

One may ask the question of whether for a physical (classical) system there exists always a definite relation between its equation of motion which governs its dynamics and the Hamiltonian function which governs its thermodynamics. For all the existing classical (and quantum) systems we know of, the answer is yes! The reason is that all the classical (and quantum) systems we know of can be described in terms of the so-called Lagrangian formulation. The Lagrangian formulation allows systematic exploration of dynamic and thermodynamic properties of a physical system. We shall explore this approach in the following (see also Ref. [1]).

### 2.1.1
### Basic Ansatz: The Principle of Least Action

The Lagrangian formulation starts with an assumption that all mechanical systems can be characterized by a set of variables $q_i$, $i = 1, \ldots, N$ (positions of $N/3$ particles, for example), and their time derivatives (velocities) $\dot{q}_i$. (Mathematically, that means if all the coordinates $qi$ and velocities $\dot{q}_i$ of the mechanical system are given at some instant, the time evolution of the system is completely determined.) The dynamics of the system is completely specified by a (Lagrangian) function $L(q_1, q_2, \ldots, q_N; \dot{q}_1, \dot{q}_2, \ldots, \dot{q}_N, t)$, or simply $L(q, \dot{q}, t)$, in the following way.

Let the system occupy, at the instants $t_1$ and $t_2$, positions defined by two fixed sets of values of the coordinates $q^{(1)}$ and $q^{(2)}$. The system moves between these two sets of positions in a way such that the integral (called action)

$$S = \int_{t_1}^{t_2} L(q, \dot{q}, t)\mathrm{d}t \tag{2.1}$$

take the least possible value.

*Introduction to Classical and Quantum Field Theory.* Tai-Kai Ng

ISBN: 978-3-527-40726-2

The mathematical tool used to minimize the action $S$ is described in detail in most textbooks on classical mechanics and we shall not repeat it here. The requirement $\delta S = 0$ results in the Lagrange equation of motion

$$\frac{\mathrm{d}}{\mathrm{d}t}\left(\frac{\partial L}{\partial \dot{q}_i}\right) - \frac{\partial L}{\partial q_i} = 0, \quad i = 1, \ldots, N. \tag{2.2}$$

#### 2.1.1.1 Conservation of Energy and Momentum

Conservation of energy and momentum comes out naturally if the Lagrangian of the system satisfies the general properties of translational invariance in time and space (or homogeneity of time and space). First, we consider conservation of energy.

Translational invariance in time is guaranteed if the Lagrangian does not depend explicitly on time, that is $L = L(q, \dot{q})$. In this case, the total time derivative of the Lagrangian is

$$\frac{\mathrm{d}L}{\mathrm{d}t} = \sum_i \frac{\partial L}{\partial q_i}\dot{q}_i + \sum_i \frac{\partial L}{\partial \dot{q}_i}\ddot{q}_i. \tag{2.3}$$

Using the Lagrange equation of motion, we obtain

$$\frac{\mathrm{d}L}{\mathrm{d}t} = \sum_i \frac{\mathrm{d}}{\mathrm{d}t}\left(\frac{\partial L}{\partial \dot{q}_i}\right)\dot{q}_i + \sum_i \frac{\partial L}{\partial \dot{q}_i}\ddot{q}_i = \sum_i \frac{\mathrm{d}}{\mathrm{d}t}\left(\dot{q}_i \frac{\partial L}{\partial \dot{q}_i}\right)$$

$$\text{or} \quad \frac{\mathrm{d}}{\mathrm{d}t}\left(\sum_i \dot{q}_i \frac{\partial L}{\partial \dot{q}_i} - L\right) = 0, \tag{2.4}$$

implying that $H = \sum_i \dot{q}_i \frac{\partial L}{\partial \dot{q}_i} - L$ is a constant of the motion.

$H$ is called the Hamiltonian function or energy of the system. What we have shown is how the energy function of a system can be constructed generally from a given Lagrangian and that it is always conserved as long as the Lagrangian does not depend explicitly on time.

Next, we consider conservation of momentum. The total momentum of a Lagrangian system is conserved as long as the system is translationally invariant in space, that is the mechanical properties of the system are unaffected by a parallel displacement of the whole system in space. To see how it works we choose the coordinates $q_i$ to represent the positions of the particles and the $\dot{q}_i$ the corresponding velocities.

A parallel displacement is a transformation in which every particle in the system is moved by the same amount $\delta\vec{\varepsilon}$ with velocities of the particles remaining unchanged. The Lagrangian remains unchanged in this transformation, that is

$$\delta L = \sum_i \frac{\partial L}{\partial \vec{r}_i} \cdot \delta\vec{\varepsilon} = \delta\vec{\varepsilon} \cdot \sum_i \frac{\partial L}{\partial \vec{r}_i} = 0$$

for arbitrarily small constant displacement $\delta\vec{\varepsilon}$.

This can happen only if $\sum_i \frac{\partial L}{\partial \vec{r}_i} = 0$.

From the Lagrange equation of motion, we obtain correspondingly

$$\frac{d}{dt}\sum_i \frac{\partial L}{\partial \vec{v}_i} = 0. \qquad (2.5a)$$

Thus, we can define the total momentum of the system as

$$\vec{P} = \sum_i \frac{\partial L}{\partial \vec{v}_i}, \qquad (2.5b)$$

which is a constant of the motion as long as the system is translationally invariant in space.

Note that we have only assumed translational invariance in time and space to define energy and momentum and show that they are conserved. We have not even written down an explicit form of the Lagrangian $L$!

#### 2.1.1.2 **Galilean Invariance and the Most General Form of Lagrangian**

Within the Lagrangian formulation we can ask an ambitious question of what is the most general form of Lagrangian we can write down for a system of particles under some given general constraints. The requirement of translational invariance in time and space dictated that a Lagrangian must have the properties $L = L(q, \dot{q})$ and $\sum_i \frac{\partial L}{\partial \vec{r}_i} = 0$. However, we can still write down a lot of Lagrangians satisfying these two properties. The plausible choices of Lagrangian will be strongly reduced if we make one further requirement of Galilean invariance – the Lagrange equation of motion should remain the same in all inertial reference frames, or under the transformation $\vec{r}_i \to \vec{r}_i + \vec{v}t$, $\vec{v}_i \to \vec{v}_i + \vec{v}$.

Before we examine this requirement we first ask the question of under what condition will two different Lagrangians $L_1$, $L_2$ yield the same equation of motion ($\delta S_2 = \delta S_1 = 0$). The answer is simple! It happens when the two Lagrangians differ only by a total time derivative, that is

$$L_2(q, \dot{q}) = L_1(q, \dot{q}) + \frac{d}{dt} f(q), \qquad (2.6)$$

since in this case their actions are related by $S_2 = S_1 + f(q(t_2)) - f(q(t_1))$, and differ by a quantity that is zero upon variation.

Now, let us examine the change in the Lagrangian $L = L(\vec{r}, \vec{v})$ under the transformation $\vec{r}_i \to \vec{r}_i + \delta\vec{v}t$, $\vec{v}_i \to \vec{v}_i + \delta\vec{v}$, where $\delta\vec{v}$ is an infinitesimal constant velocity. To leading order, we obtain

$$\delta L = \sum_i \frac{\partial L}{\partial \vec{r}_i} \cdot \delta\vec{v}t + \sum_i \frac{\partial L}{\partial \vec{v}_i} \cdot \delta\vec{v} = \delta\vec{v} \cdot \sum_i \frac{\partial L}{\partial \vec{v}_i} (= \delta\vec{v} \cdot \vec{P}),$$

if we assume conservation of momentum $\left(\sum_i \frac{\partial L}{\partial \vec{r}_i} = 0\right)$.

If the equation of motion remains the same after the transformation, then $\delta L$ must be a total time derivative, that is

$$\sum_i \frac{\partial L}{\partial \vec{v}_i} = \frac{d}{dt}\vec{f}(\vec{r}_1, \vec{r}_2, \ldots) = \sum_j (\vec{v}_j \cdot \vec{\nabla}_{\vec{r}_j})\vec{f} \text{ for some vector function } \vec{f}.$$

To proceed further, we consider a Taylor expansion in velocities,

$$L = L_0(\vec{r}) + \sum_i \vec{L}_{1i}(\vec{r}).\vec{v}_i + \frac{1}{2}\sum_{ij} \vec{v}_i \overleftrightarrow{L}_{2ij}(\vec{r})\vec{v}_j + \ldots.$$

$\vec{r} = (\vec{r_1}, \vec{r_2} \ldots \vec{r_N})$. Comparing this with the above requirement, we find that only $L_0$ and $\overleftrightarrow{L}_2$ can be non-zero, that is the most general form of Lagrangian that conserves momentum and energy and respects Galilean invariance is

$$L = T - V = \frac{1}{2}\sum_{ij} \vec{v}_i \overleftrightarrow{T}_{2ij}(\vec{r})\vec{v}_j - V(\vec{r}). \tag{2.7a}$$

The velocities can only enter the Lagrangian in *quadratic form*! Note that most generally the mass matrix $\overleftrightarrow{T}_2$ is a function of the coordinates of all other particles in the universe. To respect conservation of momentum, the potential energy $V$ and mass matrix $\overleftrightarrow{T}_2$ must satisfy $\sum_i \frac{\partial V(\vec{r})}{\partial \vec{r}_i} = 0$ and $\sum_i \frac{\partial \overleftrightarrow{T}_2(\vec{r})}{\partial \vec{r}_i} = 0$. $V$ and $\overleftrightarrow{T}_2$ will be further restricted if we impose rotational invariance (conservation of angular momentum) and other constraints. An additional condition that we usually impose is that the kinetic energy term is *local*, and is a sum of contributions from individual particles, that is

$$T = \frac{1}{2}\sum_i \vec{v}_i \overleftrightarrow{T}_{2i}(\vec{r})\vec{v}_i.$$

In this case, rotational invariance further restricts $\overleftrightarrow{T}_2$ to be of the form

$$T_{2ij} = m\delta_{ij} \quad (i, j = \text{direction indices}) \tag{2.7b}$$

and reduces to the usual kinetic energy term for particles.

What you have seen here is a simple example of how one can 'deduce' the allowable forms of physical law from symmetry considerations. The example we have here is a system of classical particles with translational invariance in space and time plus Galilean invariance. This kind of consideration can be extended to different systems (e.g. scalar and vector fields) with different symmetries. We may also replace Galilean invariance by Lorentz invariance if we consider special relativity. The use of symmetries to constrain the allowable forms of physical law is a very important tool in theoretical physics nowadays. You can find a lot of examples in many other textbooks.

**Homework Q1**

Show that the eigenvalues of the mass matrix $\overleftrightarrow{T}_2$ must *all* be positive for the principle of least action to hold.

**Homework Q2**

What is the most general form of action plausible for a single particle if we require the equation of motion to be invariant under Lorentz transformation instead of Galilean transformation? How about two particles if we assume that the action reduces to that of two particles interacting with potential $V(\vec{x}) = \frac{q_1 q_2}{|\vec{x}|}$ in the non-relativistic limit?

(Hint: first determine what are the most general Lorentz-invariant quantities we can construct from the variables $(\vec{r}, t)$, $(\vec{v}, \vec{E})$. You can find this in many textbooks on relativity.)

#### 2.1.1.3 Constraints

We often encounter problems where the forces acting on particles are expressed as constraints. For example, the equation of motion of a particle restricted to move on a curved surface is described by the constraint equation $g(x, y, z) = 0$.

One way to tackle the problem is to eliminate one of the coordinates in terms of the other two, and then express the Lagrangian in terms of the remaining coordinates, for example

$$L(x, y, z; \dot{x}, \dot{y}, \dot{z}) \to L(x, y, z(x, y); \dot{x}, \dot{y}, \dot{z}(x, y)),$$

and then apply the principle of least action to the coordinates $(x, y, \dot{x}, \dot{y})$.

This method is often difficult to implement because very often we cannot obtain a closed-form expression of $z(x, y)$ from $g(x, y, z) = 0$. To overcome this problem, we introduce the method of Lagrange multipliers.

The idea is to write the action as

$$L(q, \dot{q}) \to L_c(q, \dot{q}, \lambda) = L(q, \dot{q}) + \lambda g(q) \tag{2.8a}$$

and then apply the principle of least action $\delta S = 0$ to $L_c$, treating $q, \dot{q}, \lambda$ as independent variables.

The resulting equation of motion becomes

$$\frac{\mathrm{d}}{\mathrm{d}t}\left(\frac{\partial L}{\partial \dot{q}_i}\right) - \frac{\partial L}{\partial q_i} - \lambda \frac{\partial g(q)}{\partial q_i} = 0, \quad i = 1, \ldots, N \tag{2.8b}$$

and

$$g(q) = 0 \text{ (constraint equation)}. \tag{2.8c}$$

These equations have to be solved self consistently to obtain the final equation of motion.

#### 2.1.1.4 Lagrangian Formulation for Classical Field Theory

The Lagrangian formulation can be generalized rather straightforwardly to field theories. First, we consider *lattice field theories*, where the dynamical variables $\phi_i$ are defined on lattice sites $i$.

In this case, the Lagrangian formulation can be applied in a straightforward way,

$$L = L(\phi_1, \phi_2, \ldots, \phi_N; \dot{\phi}_1, \dot{\phi}_2, \ldots, \dot{\phi}_N) = L(\phi, \dot{\phi}). \tag{2.9a}$$

The Lagrange equation of motion is

$$\frac{\mathrm{d}}{\mathrm{d}t}\left(\frac{\partial L}{\partial \dot{\phi}_i}\right) - \frac{\partial L}{\partial \phi_i} = 0, \quad i = 1, \ldots, N = \text{lattice site indices} \tag{2.9b}$$

and

$$\text{the Hamiltonian is } H = \sum_i \dot{\phi}_i \frac{\partial L}{\partial \dot{\phi}_i} - L. \tag{2.9c}$$

Next, we try to generalize the scheme to *continuum field theories*, that is theories where the field $\phi_i \to \phi(\vec{x})$ is defined in continuous space.

We face an immediate problem here that there are actually infinite numbers of variables $\phi(\vec{x})$ because the position vector $\vec{x}$ is a continuous variable. To handle this 'infinity' problem we replace the usual multi-variable calculus by functional calculus.

The basic idea of functional calculus can be understood by considering the following simple example.

Let us consider $L = L[\phi, \dot{\phi}]$, which is a functional (function of a function) of $\phi(\vec{x})$. For example,

$$L = \int d^d x([\dot{\phi}(\vec{x})]^2 - a\phi(\vec{x})^2 - b\phi(\vec{x})^4 - c^2(\nabla\phi(\vec{x}))^2).$$

To derive the Lagrangian equation of motion, we have to define 'derivatives' called functional derivatives,

$$\frac{\partial L}{\partial \phi_i} \rightarrow \frac{\delta L}{\delta \phi(\vec{x})} \quad \text{(partial derivatives} \rightarrow \text{functional derivatives).}$$

The functional derivatives can be defined by introducing one basic rule of 'differentiation' and then assuming that other rules in multi-variable calculus are still obeyed. The basic rule is

$$\underset{\text{(lattice case)}}{\frac{\partial \phi_i}{\partial \phi_j} = \delta_{ij}} \rightarrow \underset{\text{(continuum limit)}}{\frac{\delta \phi(\vec{x}')}{\delta \phi(\vec{x})} = \delta(\vec{x} - \vec{x}')}. \tag{2.10}$$

To see how the above functional derivative rule is used, we consider some examples.

**Example 1**

$$\frac{\delta \int d^d x'(a\phi(\vec{x}')^2 + b\phi(\vec{x}')^4)}{\delta\phi(\vec{x})} = \int d^d x'(2a\phi(\vec{x}') + 4b\phi(\vec{x}')^3)\delta(\vec{x} - \vec{x}')$$
$$= 2a\phi(\vec{x}) + 4b\phi(\vec{x})^3.$$

Note that we have assumed that the usual chain rule of differentiation is valid.

**Example 2**

$$\frac{\delta \int d^d x'([\nabla_{x'}\phi(\vec{x}')]^2)}{\delta\phi(\vec{x})} = \int d^d x' \left( \frac{\delta \nabla_{x'}\phi(\vec{x}')}{\delta\phi(\vec{x})} \nabla_{x'}\phi(\vec{x}') + \nabla_{x'}\phi(\vec{x}') \frac{\delta \nabla_{x'}\phi(\vec{x}')}{\delta\phi(\vec{x})} \right),$$

$$\int d^d x' \left( -\frac{\delta\phi(\vec{x}')}{\delta\phi(\vec{x})} \nabla^2_{x'}\phi(\vec{x}') - \nabla^2_{x'}\phi(\vec{x}') \frac{\delta\phi(\vec{x}')}{\delta\phi(\vec{x})} \right) = -2\nabla^2\phi(\vec{x}).$$

**Homework Q3**

Evaluate $\dfrac{\delta \displaystyle\int d^d x' \tan^2(\phi(\vec{x}'))}{\delta\phi(\vec{x})}$.

With this rule, the Lagrange equation of motion can be written as

$$\frac{d}{dt}\left(\frac{\delta L}{\delta\dot{\phi}(\vec{x})}\right)-\frac{\delta L}{\delta\phi(\vec{x})}=0. \qquad (2.11)$$

**Homework Q4**
Show that the Lagrange equation of motion for

$$L=\int d^dx([\dot{\phi}(\vec{x})]^2-m^2\phi(\vec{x})^2-b\phi(\vec{x})^4-c^2(\nabla\phi(\vec{x}))^2)$$
$$\text{is } \frac{d^2}{dt^2}\phi(\vec{x},t)+m^2\phi(\vec{x},t)+2b\phi(\vec{x},t)^3-c^2\nabla^2\phi(\vec{x},t)=0$$

and the corresponding Hamiltonian of the system is

$$H=\int d^dx\dot{\phi}(\vec{x})\frac{\delta L}{\delta\dot{\phi}(\vec{x})}-L$$
$$=\int d^dx([\dot{\phi}(\vec{x})]^2+m^2\phi(\vec{x})^2+b\phi(\vec{x})^4+c^2(\nabla\phi(\vec{x}))^2).$$

Show that, when $b=0$, the general solution of the Lagrange equation is a superposition of plane waves

$$\phi(\vec{x},t)=Ae^{i(\omega t-\vec{k}.\vec{x})}, \quad \text{with} \quad \omega^2=m^2+c^2\vec{k}^2.$$

It is interesting to note that when compared with mechanics of particles, the kinetic energy of the plane wave is stored in the $\dot{\phi}^2$ term, and the rest of the terms are interpreted as the potential energy stored in the waves.

#### 2.1.1.5 Space–Time Symmetric Lagrangian Formulation
This is often more convenient, in particular for relativistic field theories (see Ref. [2]).

The principle of least action is often formulated in a space–time symmetric way in relativistic theories. We consider

$$L=L[\phi,\partial_\mu\phi], \quad \text{where} \quad \mu=x,y,z,t.$$

In particular, we will be interested in Lagrangians of the form

$$L=\int d^dx\bar{L}(\phi(\vec{x}),\partial_\mu\phi(\vec{x})),$$

where $\bar{L}$ is called the Lagrangian density, and the action is

$$S=\int dt\int d^dx\bar{L}(\phi(\vec{x}),\partial_\mu\phi(\vec{x})), \qquad (2.12)$$

which again places space and time on equal footings. In the space–time symmetric formulation, we work with $S$ directly!

Introducing the 4-vector notation $x=(\vec{x},t)$, we can define the basic rule of a space–time functional derivative as

$$\frac{\delta\phi(x')}{\delta\phi(x)}=\delta(x-x')=\delta(\vec{x}-\vec{x}')\delta(t-t'). \qquad (2.13)$$

It is then easy to see that the requirement $\partial S = 0$ leads to the equation of motion

$$\sum_{\mu} \frac{\partial}{\partial x_{\mu}} \left( \frac{\delta S}{\delta(\partial_{\mu}\phi(\vec{x}))} \right) - \frac{\delta S}{\delta\phi(\vec{x})} = 0. \tag{2.14}$$

Note that $S$ now replaces $L$ in the equation of motion. Although Equation (2.14) looks different from the original Lagrange equation of motion, the final equation of motion is actually the same. You can test it with

$$S = \int \mathrm{d}t\mathrm{d}^d x([\dot{\phi}(\vec{x})]^2 - a\phi(\vec{x})^2 - b\phi(\vec{x})^4 - c^2(\nabla\phi(\vec{x}))^2).$$

In this case, $\frac{\delta S}{\delta\phi(x)} = -2a\phi(x) - 4b\phi(x)^3$, $\frac{\delta S}{\delta(\partial_t\phi(x))} = 2\partial_t\phi(x)$, and $\frac{\delta S}{\delta(\partial_x\phi(x))} = -2c^2\partial_x\phi(x)$.

(Remember that $x = (\vec{x}, t)$.)

Putting this together, we obtain the same equation of motion

$$\frac{\mathrm{d}^2}{\mathrm{d}t^2}\phi(\vec{x}, t) + a\phi(\vec{x}, t) + 2b\phi(\vec{x}, t)^3 - c^2\nabla^2\phi(\vec{x}, t) = 0$$

as obtained in the usual Lagrange formulation.

**Homework Q5**

Show that the Maxwell equations (in the absence of charge and current) can be derived from the Lagrangian density

$$L = -\frac{1}{16\pi}\sum_{\mu\nu} F_{\mu\nu}^2, \quad \text{where } F_{\mu\nu} = \frac{\partial A_{\mu}}{\partial x_{\nu}} - \frac{\partial A_{\nu}}{\partial x_{\mu}} \quad (\mu, \nu = 0, 1, 2, 3),$$

where $(A_0, \vec{A})$ are the scalar and vector potentials. $\vec{B} = \nabla \times \vec{A}$ and $\vec{E} = -\nabla A_0 - \frac{\partial \vec{A}}{\partial t}$. By deriving the Hamiltonian, show that the kinetic energy of the electromagnetic field is stored in the electric field and the potential energy is stored in the magnetic field. *How is gauge invariance reflected in L?*

## 2.2 Conservation Laws in Continuum Field Theory (Noether's Theorem)

As we have seen in the case of particle mechanics, conservation laws are closely tied with symmetries. The same is true in field theory. Here, we shall study how symmetry and conservation laws are related in classical field theory (see also Ref. [2]). We first consider energy–momentum conservation, which is closely tied with translational invariance in space and time.

### 2.2.1 Energy–Momentum Conservation

In the space–time symmetric formulation, translational invariance in space and time is guaranteed as long as the Lagrangian density depends on $x = (\vec{x}, t)$ only implicitly through $\phi$ and $\partial_{\mu}\phi$. For an infinitesimal space–time displacement $x \to x + \delta a$,

$$\bar{L}(\phi(x), \partial_\mu\phi(x)) \to \bar{L}(\phi(x+a), \partial_\mu\phi(x+a)),$$

where $\delta\phi(x) = \sum_\mu \delta a_\mu \partial_\mu \phi(x)$ and $\delta\partial_\mu\phi(x) = \sum_\nu \partial_\mu[\delta a_\nu \partial_\nu \phi(x)]$. Therefore,

$$\delta S = \int dt d^d x \left\{ \left[ \frac{\delta S}{\delta\phi}\delta\phi - \sum_\mu \partial_\mu \left( \frac{\delta S}{\delta\partial_\mu\phi} \right) \delta\phi \right] + \sum_\mu \partial_\mu \left[ \frac{\delta S}{\delta\partial_\mu\phi}\delta\phi \right] \right\}. \tag{2.15a}$$

Note that we may also write

$$\delta S = \int dt d^d x \sum_\mu \frac{\partial\bar{L}}{\partial x_\mu}\delta a_\mu = \int dt d^d x \sum_\nu \sum_\mu \partial_\mu \left[ \frac{\delta S}{\delta\partial_\mu\phi}\partial_\nu\phi\delta a_\nu \right], \tag{2.15b}$$

where we have used the Lagrange equation of motion to eliminate the first term in Equation (2.15a).

For the equation to hold for arbitrary configurations $\phi$, we must have

$$\sum_\mu \frac{\partial\bar{L}}{\partial x_\mu}\delta a_\mu - \sum_\nu \sum_\mu \partial_\mu \left[ \frac{\delta S}{\delta\partial_\mu\phi}\partial_\nu\phi\delta a_\nu \right] = 0 \quad \text{for arbitrary } \delta a_\mu \text{ or}$$

$$\sum_\mu \partial_\mu \left[ \frac{\delta S}{\delta\partial_\mu\phi}\partial_\nu\phi - \delta_{\mu\nu}\bar{L} \right] = 0. \tag{2.15c}$$

The equation represents four continuity equations of the form

$$\frac{\partial}{\partial t}\rho_\nu(\vec{x}, t) + \nabla.\vec{j}_\nu(\vec{x}, t) = 0 \quad (\nu = 0, 1, 2, 3),$$

with $\rho_\nu(\vec{x}, t) = \frac{\delta S}{\delta\phi}\partial_\nu\phi - \delta_{0\nu}\bar{L}$ and $\vec{j}_\nu(\vec{x}, t) = \frac{\delta S}{\delta\nabla\phi}\partial_\nu\phi - \delta_{\vec{n}\nu}\bar{L}$.

The tensor $\Theta_{\mu\nu} = \dfrac{\delta S}{\delta\partial_\mu\phi}\partial_\nu\phi - \delta_{\mu\nu}\bar{L}$ (2.16)

is called the energy–momentum tensor. The continuity equation implies that the integrals $P^\nu = \int d^d x \Theta_{0\nu}(\vec{x}, t)$ are constants of the motion. In particular, $P^0$ is the total energy of the system and $P^i$ ($i = 1, 2, 3$) are the total momenta.

**Example**

$$S = \int dt d^d x ([\dot\phi(x)]^2 - a\phi(x)^2 - b\phi(x)^4 - c^2(\nabla\phi(x))^2).$$

Total energy is $H = \int d^d x [\dot\phi(\vec{x})]^2 + a\phi(\vec{x})^2 + b\phi(\vec{x})^4 + c^2(\nabla\phi(\vec{x}))^2$ and total momentum is $\vec{P} = 2\int d^d x (\dot\phi\nabla\phi)$!

**Homework Q6**

Derive the equation of motion and the expressions of total energy and momentum for the complex scalar field

$$S = \int dt d^d x (|\dot\psi(x)|^2 - a|\psi(x)|^2 - b|\psi(x)|^4 - c^2|\nabla\psi(x)|^2),$$

where $\psi(x)$ is a complex number function and $|\psi|^2 = \psi^*\psi$

(See also, for example, Ref. [3] for the total energy and momenta for the EM field.)

#### 2.2.1.1 Internal Symmetry and Noether's Theorem

The discussion of conservation of momentum and energy can be generalized to discuss conservation laws associated with other symmetries.

Let us imagine transformations $\psi(x) \to \psi(x) + \delta\psi(x)$ and $\partial_\mu\psi(x) \to \partial_\mu\psi(x) + \partial_\mu\delta\psi(x)$ such that the Lagrangian density $\bar{L}$ remains unaffected.

**Example**

In the case of a complex scalar field with action

$$S = \int dt d^d x (|\dot{\psi}(x)|^2 - a|\psi(x)|^2 - b|\psi(x)|^4 - c^2|\nabla\psi(x)|^2),$$

a transformation that leaves $\bar{L}$ invariant is

$$\psi(x) \to \psi(x)e^{i\vartheta} \sim \psi(x) + i\vartheta\psi(x),$$

$$\psi^*(x) \to \psi^*(x)e^{-i\vartheta} \sim \psi^*(x) - i\vartheta\psi^*(x) \quad \text{for small } \vartheta.$$

This symmetry is called (global) U(1) gauge symmetry.

The change in action is in general

$$\begin{aligned} \delta S &= \int dt d^d x \left[\frac{\delta S}{\delta\psi}\delta\psi + \sum_\mu (\partial_\mu \delta\psi)\left(\frac{\delta S}{\delta\partial_\mu\psi}\right)\right] \\ &= \int dt d^d x \left[\frac{\delta S}{\delta\psi} - \sum_\mu \partial_\mu \left(\frac{\delta S}{\delta\partial_\mu\psi}\right)\right]\delta\psi + \int dt d^d x \left[\sum_\mu \partial_\mu \left(\frac{\delta S}{\delta\partial_\mu\psi}\delta\psi\right)\right]. \end{aligned} \tag{2.17}$$

The first term is zero by the equation of motion. Therefore, if the action is invariant under such a transformation, we must have

$$\sum_\mu \partial_\mu \left[\frac{\delta S}{\delta\partial_\mu\psi}\delta\psi\right] = 0, \tag{2.18a}$$

which is a continuity equation with 4-current

$$J_\mu = \frac{\delta S}{\delta\partial_\mu\psi}\delta\psi. \tag{2.18b}$$

We shall be interested in small changes in $\psi$ of the form $\delta\psi = \varepsilon\widehat{A}\,\psi$, where $\varepsilon$ is an infinitesimal number and $\widehat{A}$ is a linear operator. Note that the changes in $\psi$ we discussed above can all be cast in this form. In this case, the Noether current is $J_\mu = \frac{\delta S}{\delta\partial_\mu\psi}\widehat{A}\,\psi$.

What you have seen is the Noether theorem in its simplest form: *The invariance of the action S under a transformation follows from symmetry, and from this symmetry a conservation law can be deduced!*

Note: for the complex fields $\psi = \phi_r + i\phi_i$ we have two independent variables $\phi_r$ and $\phi_i$. It is more often convenient to treat $\psi = \phi_r + i\phi_i$ and $\psi^* = \phi_r - i\phi_i$ as independent variables and write $\frac{\delta S}{\delta\psi} \to \sum_i \left(\frac{\delta S}{\delta\psi_i}\delta\psi_i + \frac{\delta S}{\delta\psi_i^*}\delta\psi_i^*\right)$.

**Example**
For the complex scalar field with action

$$S = \int dtd^dx(|\dot{\psi}(x)|^2 - a|\psi(x)|^2 - b|\psi(x)|^4 - c^2|\nabla\psi(x)|^2),$$

the invariance of $S$ under the transformation

$$\psi(x) \to \psi(x)e^{i\vartheta} \sim \psi(x) + i\vartheta\psi(x),$$

$$\psi^*(x) \to \psi^*(x)e^{-i\vartheta} \sim \psi^*(x) - i\vartheta\psi^*(x)$$

results in the Noether current $J_\mu = \frac{\delta S}{\delta \partial_\mu \psi}\delta\psi + \frac{\delta S}{\delta \partial_\mu \psi^*}\delta\psi^*$, with $J_0 = \rho = i[(\dot{\psi}^*)\psi - \psi^*(\dot{\psi})]$ and $\vec{J} = ic^2[(\nabla\psi^*)\psi - \psi^*(\nabla\psi)]$.

**Question**
What is the physical quantity that is 'conserved' in this case?

**Question**
Is there a conserved current associated with the gauge symmetry of the electromagnetic field? If so, what is this conserved current?

### 2.2.2
### More Complicated Internal Symmetries

We are often interested in symmetries associated with internal structures of the $\psi$ field. The simplest example is perhaps the case when $\psi$ is a complex $N$-component vector; in this case $\psi \to \phi_{jr} + i\phi_{ji}$, $j = 1, \ldots, N$, and

$$\frac{\delta S}{\delta \psi} \to \sum_{j=1,\ldots,N}\left(\frac{\delta S}{\delta \phi_{jr}}\delta\phi_{jr} + \frac{\delta S}{\delta \phi_{ji}}\delta\phi_{ji}\right). \tag{2.19}$$

> P.A.M. Dirac was perhaps the first physicist to realize the importance of field theories with internal structures. His relativistic theory for electrons is a theory with non-trivial internal structure. Nowadays theorists encounter field theories with internal structures routinely.

Let us consider a simple example here:

$$S = \int dtd^dx(\dot{\psi}^+(x)\dot{\psi}(x) - m^2\psi^+(x)\psi(x) - c^2(\nabla\psi^+(x)).(\nabla\psi(x))),$$

$$\text{where } \psi(x) = \begin{pmatrix}\psi_1(x)\\ \psi_2(x)\end{pmatrix}, \ \psi^+(x) = \begin{pmatrix}\psi_1^*(x) & \psi_2^*(x)\end{pmatrix} \tag{2.20}$$

are two-component complex scalar fields. The action is invariant under the transformation

$$\psi \to U\psi,\ \psi^+ \to \psi^+ U^+, \tag{2.21a}$$

where $U$ is a constant $2 \times 2$ unitary matrix. This is the generalization of the phase (U(1)-gauge) transformation for the one-component complex scalar field $\psi(x) \to e^{i\vartheta}\psi(x),\ \psi^*(x) \to \psi^*(x)e^{-i\vartheta}$.

A general $2 \times 2$ unitary matrix can be written as product of a U(1) phase factor and a SU(2) matrix ($\mathrm{U}(2) = \mathrm{U}(1) \times \mathrm{SU}(2)$), or

$$U = e^{i\vartheta} e^{i\vec{L}.\vec{\tau}}, \tag{2.21b}$$

where $\vec{L}$ is a three-component vector and $\vec{\tau}$ = Pauli matrices. For infinitesimal transformations,

$$\psi \to U\psi \sim \psi + \delta\psi = \psi + (i\vartheta + i\vec{L}.\vec{\tau})\psi, \tag{2.21c}$$

and $\delta\psi$ is characterized by a scalar plus a vector field in the internal space described by the $2 \times 2$ unitary matrix.

**Homework Q7**

Show that the associated Noether current has a scalar component plus a vector component given by

$$J_\mu^{(0)} \sim i[(\partial_\mu\psi^+)\psi - \psi^+(\partial_\mu\psi)],\ J_\mu^{(v)} \sim i[(\partial_\mu\psi^+)\tau^v\psi - \psi^+\tau^v(\partial_\mu\psi)].$$

You can probably see that the level of complication increases rapidly when the internal structure of the field becomes more complicated. How many components of conserved current will there be if $\psi$ is a three-component vector?

## References


1 Landau L.D. and Lifshitz, E.M. (1999) *Course of Theoretical Physics*, Vol. 1: *Mechanics*, Butterworth-Heinemann Linacre House, Jordan Hill, Oxford.

2 Weinberg, S. (1995) *The Quantum Theory of Fields*, Vol. 1. Cambridge University Press.

3 Jackson, J.D.(1990) *Classical Electrodynamics*, John Wiley & Sons (Singapore, New York, Chichester, Brisbane, Toronto).

# 3
# Quantization of Classical Field Theories (I)

## 3.1
## Canonical Quantization of Scalar Fields: Bosonic Systems

So far we have talked about classical field theories. A natural question is: is there a quantum analogue of classical field theories, as quantum mechanics is the analogue of Newtonian classical mechanics? This question was raised and settled (at least in principle) at the same time when quantum mechanics was invented. The systematic procedure to derive a 'quantum' theory from the corresponding 'classical' theory is called *canonical quantization* (see e.g. Refs. [1,2]).

Canonical quantization is a general approach to derive a quantum theory from any classical theory that can be described in the Lagrangian framework. Let us consider a classical system described by the Lagrangian $L(q,\dot{q})$. The $q_i$ are the generalized coordinates of the classical system. The generalized momenta of the system are defined by $p_i = \frac{\partial L}{\partial \dot{q}_i}$ and the Hamiltonian is $H = \sum_i \dot{q}_i \frac{\partial L}{\partial \dot{q}_i} - L = \sum_i \dot{q}_i p_i - L$.

The procedure of canonical quantization says that the corresponding quantum system can be obtained from the Hamiltonian of the classical system $H(p, q)$, written in terms of $p_i$, $q_i$ only (the $\dot{q}_i$ are all eliminated in favor of the $p_i$), except that $p_i$, $q_i$ are not classical numbers any more, but are operators obeying the commutation relation $[q_i, p_j] = i\hbar\delta_{ij}$.

The same procedure can be generalized to field theories described by the Lagrangian $L(\phi,\dot{\phi})$. A generalized momentum $\pi(\vec{x})$ is defined by $\pi(\vec{x}) = \frac{\delta L}{\delta\dot{\phi}(\vec{x})}$. The Hamiltonian can be written as $H(\phi,\pi)$, with $\phi(\vec{x}), \pi(\vec{x})$ now operators defined on every point in space, satisfying the commutation rule

$$[\phi(\vec{x}), \pi(\vec{x}')] = i\hbar\delta(\vec{x}-\vec{x}') \tag{3.1a}$$

(or $[\phi(\vec{x}_i), \pi(\vec{x}_j)] = i\hbar\delta_{ij}$ in lattice field theory).

The eigenstates of the operators $\phi(\vec{x})$ (or $\pi(\vec{x})$) span a Hilbert space where the quantum system is defined. The quantum states $|\Psi\rangle$ satisfy the Schrödinger equation

$$i\hbar\frac{\partial}{\partial t}|\Psi(t)\rangle = \widehat{H}|\Psi(t)\rangle \tag{3.1b}$$

as in ordinary quantum mechanics.

*Introduction to Classical and Quantum Field Theory.* Tai-Kai Ng

ISBN: 978-3-527-40726-2

Note that canonical quantization explicitly breaks symmetry between time and space and there is no space–time symmetric formulation of the approach as in the Lagrangian formulation.

To understand the approach and to make a connection with the usual quantum mechanics, we start with a simple example in the following.

**Example**

Field theory on a lattice with

$$L[\phi] = \sum_i \left(\frac{1}{2}\dot{\phi}_i^2 - V(\phi_i)\right). \tag{3.2a}$$

Note that this is a field theory where the fields $\phi_i$ on different sites are decoupled. The momentum conjugates are

$$\pi_i = \frac{\partial L}{\partial \dot{\phi}_i} = \dot{\phi}_i, \tag{3.2b}$$

and the (quantum) Hamiltonian is

$$H = \sum_i \left(\frac{1}{2}\widehat{\pi}_i^2 + V(\widehat{\phi}_i)\right), \tag{3.2c}$$

$$\text{with } [\widehat{\phi}_i, \widehat{\pi}_j] = i\hbar\delta_{ij}. \tag{3.2d}$$

To make the connection with usual quantum mechanics (QM), we introduce the *Schrödinger representation*

$$\widehat{\pi}_i = -i\hbar\frac{\partial}{\partial \phi_i}. \tag{3.2e}$$

The Schrödinger equation for the system is

$$i\hbar\frac{\partial}{\partial t}\psi[\phi_1, \phi_2, \ldots, \phi_N] = \left(\sum_i \widehat{H}_i\right)\psi[\phi_1, \phi_2, \ldots, \phi_N], \tag{3.2f}$$

$$\text{where } \widehat{H}_i = -\frac{\hbar^2}{2}\frac{\partial^2}{\partial \phi_i^2} + V(\phi_i). \tag{3.2g}$$

Since the $\widehat{H}_i$ on different sites are decoupled, we have

$$\psi[\phi] = \prod_i \psi_n(\phi_i), \; E = \sum_i E_{in}, \quad \text{where } \widehat{H}_i\psi_n(\phi) = E_{in}\psi_n(\phi). \tag{3.2h}$$

The wavefunction $\psi_n(\phi_i)$ represents the probability amplitude of the field taking the value $\phi_i$ on site $i$ in quantum state $n$. The total wavefunction $\psi[\phi]$ describes the probability amplitude of the field taking a configuration $\phi = (\phi_1, \phi_2, \ldots, \phi_N)$ on the lattice. Note that a quantum field theory of an $N$-lattice-site system is in this simple case the direct product of $N$ independent systems, each of them described by a Schrödinger equation with a single variable.

More generally, for a Lagrangian of the form

$$L[\phi] = \sum_i \left(\frac{1}{2}\dot{\phi}_i^2\right) - V(\phi_1, \phi_2, \ldots, \phi_N), \tag{3.3a}$$

the Schrödinger equation is, after quantization,

$$i\hbar\frac{\partial}{\partial t}\psi = \left[-\frac{\hbar^2}{2}\left(\sum_i \frac{\partial^2}{\partial\phi_i^2}\right) + V(\phi)\right]\psi, \tag{3.3b}$$

which is a coupled Schrödinger equation with $N$ variables. The diagonalization of this complicated Hamiltonian is a well-defined but highly non-trivial task.

### 3.1.1 General Quadratic Hamiltonian and Bosons

In the case when the 'potential energy' term has the quadratic form $V[\phi] = \frac{1}{2}\sum_{ij}\phi_i V_{ij}\phi_j$ the corresponding Schrödinger equation can be diagonalized by introducing eigenstates of the potential energy matrix (see Chapter 1). Introducing the unitary transformation $U$ which diagonalizes $V$, we obtain

$$V[\phi] = \frac{1}{2}\phi^+ V\phi = \frac{1}{2}\phi^+ U^+(UVU^+)(U\phi) = \frac{1}{2}\sum_k \omega_k^2\bar{\phi}_k\bar{\phi}_k, \tag{3.4a}$$

where $\bar{\phi} = U\phi$. Since this is a linear, unitary transformation, the Schrödinger equation can be written in a diagonalized form in terms of the new variables $\bar{\phi}$,

$$\left(\sum_k \widehat{H}_k\right)\psi[\bar{\phi}_1, \bar{\phi}_2, \ldots, \bar{\phi}_N] = E\psi[\bar{\phi}_1, \bar{\phi}_2, \ldots, \bar{\phi}_N],$$

$$\text{where } \widehat{H}_k = -\frac{\hbar^2}{2}\frac{\partial^2}{\partial\bar{\phi}_k^2} + \frac{\omega_k^2}{2}\bar{\phi}_k^2. \quad \text{(exercise)} \tag{3.4b}$$

The eigenstates and eigenenergies are

$$\begin{aligned} &\psi_n[\bar{\phi}] = \prod_k \psi_n(\bar{\phi}_{ki}), \quad \text{with} \quad E[n] = \sum_k (n_{nk} + \frac{1}{2})\omega_k, \\ &\text{where} \quad \widehat{H}_k\psi_n(\bar{\phi}_k) = (n_{nk} + \frac{1}{2})\omega_k\psi_n(\bar{\phi}_k), \end{aligned} \tag{3.4c}$$

that is $\psi_n[\bar{\phi}] = \Pi_k\,\psi_n(\bar{\phi}_{ki})$ is a product of harmonic oscillator wavefunctions.

Introducing canonical boson operators

$$\widehat{a}_k = \sqrt{\frac{\omega_k}{2\hbar}}\left(\bar{\phi}_k + \frac{i\bar{\pi}_k}{\omega_k}\right), \; \widehat{a}_k^+ = \sqrt{\frac{\omega_k}{2\hbar}}\left(\bar{\phi}_k - \frac{i\bar{\pi}_k}{\omega_k}\right), \tag{3.5a}$$

where $\bar{\pi}_k = -i\hbar\frac{\partial}{\partial\bar{\phi}_k}$,

$$\text{we obtain } H = \sum_k \left(\widehat{a}_k^+\widehat{a}_k + \frac{1}{2}\right)\omega_k, \quad \text{where } \left[\widehat{a}_k, \widehat{a}_{k'}^+\right] = \delta_{kk''} \tag{3.5b}$$

suggesting that the system can be interpreted as a system of free bosons with energy dispersion $\omega_k$.

**Question**

In QM bosons/fermions are terms characterizing 'particles'. However, in our formulation of quantum field theory no 'particles' are present in the beginning. In fact, we have not specified the nature of the $\phi$ fields at all! To what extent can the 'bosons' we derive here correspond to real particles? What is the nature of these particles? We shall discuss these questions here.

Let us start with the simplest case of a local lattice field theory,

$$L[\phi] = \sum_i \left(\frac{1}{2}\dot{\phi}_i^2 - m^2\phi_i^2\right).$$

Following the same analysis, the Hamiltonian of the system can be written as

$$H = \sum_i m\left(\widehat{a}_i^+ \widehat{a}_i + \frac{1}{2}\right),$$

where $[\widehat{a}_i, \widehat{a}_j^+] = \delta_{ij}$, and the eigenstates are characterized by integers $n_i = \langle \widehat{a}_i^+ \widehat{a}_i \rangle$.

The question is: to what extent can we interpret the wavefunction $\psi[n] = \psi[n_1, n_2, \ldots, n_N]$ as representing a system of localized particles, with $n_i = \langle \widehat{a}_i^+ \widehat{a}_i \rangle$ particles located on site $i$, the energy of each particle being $\varepsilon = m$?

In special relativity where $\varepsilon = mc^2$ $(c = 1)$, a particle can be considered as a localized unit of energy, and the mass of the particle measures the amount of this localized energy. In other words, the most fundamental character of a particle is that it represents a quantized energy unit. A quantized quadratic field theory with a Hamiltonian naturally produces energy in quantized units – or produces particles naturally!

Let us now examine how these particles can travel in space. We consider Lagrangians of the form

$$L[\phi] = \sum_i \frac{1}{2}\dot{\phi}_i^2 - \frac{1}{2}\sum_{ij}\phi_i V_{ij}\phi_j, \tag{3.6a}$$

with $V_{ij} = \overline{t}(|\vec{x}_i - \vec{x}_j|)$.

This particular form of interaction can be diagonalized easily by Fourier transformation, where $\phi_{\vec{k}} = \sum_i e^{i\vec{k}.\vec{r}_i}\phi_i$, with eigenvalues $\overline{t}_{\vec{k}} = \sum_i e^{i\vec{k}.(\vec{r}_i - \vec{r}_j)_{V_{ij}}}$. The quantum Hamiltonian is now

$$H = \sum_{\vec{k}} \left(\widehat{a}_{\vec{k}}^+ \widehat{a}_{\vec{k}} + \frac{1}{2}\right)\omega_{\vec{k}}, \quad \text{where } \omega_{\vec{k}}^2 = \overline{t}_{\vec{k}}. \tag{3.6b}$$

The quantum Hamiltonian can be alternatively represented in terms of local boson operators $\widehat{\psi}(\vec{x}_i) = \frac{1}{\sqrt{V}}\sum_{\vec{k}} e^{-i\vec{k}.\vec{r}_i}\widehat{a}_{\vec{k}}$ and $\widehat{\psi}^+(\vec{x}_i) = \frac{1}{\sqrt{V}}\sum_{\vec{k}} e^{i\vec{k}.\vec{r}_i}\widehat{a}^+_{\vec{k}}$. We can then rewrite the Hamiltonian as

$$H = \sum_{ij} \widehat{\psi}^+(\vec{x}_i)T(\vec{x}_i - \vec{x}_j)\widehat{\psi}(\vec{x}_j) + E_G, \tag{3.6c}$$

where $T(\vec{x}_i - \vec{x}_j) = \frac{1}{V}\sum_{\vec{k}} e^{-i\vec{k}.(\vec{x}_i - \vec{x}_j)}\omega_{\vec{k}}$ and $E_G$ is the ground-state energy. Note that $T(\vec{x}_i - \vec{x}_j)$ acts like a hopping term in a lattice model which moves a boson (package of energy) from lattice site $j$ to lattice site $i$. Note that the Hamiltonian of the system can be written totally in terms of the $\psi$ ($\psi^+$) operators, that is *a quantum field theory of the field* $\phi$ *is completely equivalent to a quantum theory of a many-boson system!*

**Homework Q1**

Show that the operators $\widehat{\psi}(\vec{x}_i)$ and $\widehat{\psi}^+(\vec{x}_i)$ defined in this way obey canonical boson commutation rules $[\widehat{\psi}(\vec{x}_i), \widehat{\psi}^+(\vec{x}_j)] = \delta_{ij}$, $[\widehat{\psi}(\vec{x}_i), \widehat{\psi}(\vec{x}_j)] = 0$, and so on.

More generally, we can define general linear transformation between creation and annihilation operators as follows: Let $\phi_n(\vec{x})$ be a complete and orthonormal set of states satisfying $\sum_n \phi_n(\vec{x})\phi^*_n(\vec{x}') = \delta(\vec{x} - \vec{x}')$ and $\hat{a}^+_n$ ($\widehat{a}_n$) be creation (annihilation) operators creating (destroying) a boson in state $\phi_n$. It is easy to show that the operators $\widehat{\psi}(\vec{x}) = \sum_n \phi_n(\vec{x})\widehat{a}_n$ and $\widehat{\psi}^+(\vec{x}) = \sum_n \phi^*_n(\vec{x})\widehat{a}^+_n$ satisfy canonical boson commutation rules $[\widehat{\psi}(\vec{x}), \widehat{\psi}^+(\vec{x}')] = \delta(\vec{x} - \vec{x}')$ and so on, that is the $\psi$ ($\psi^+$) are local boson annihilation (creation) operators.

Alternatively, if $\widehat{\psi}(\vec{x})$ and $\widehat{\psi}^+(\vec{x})$ are boson annihilation and creation operators at position $\vec{x}$, we can define creation (annihilation) operators creating (destroying) a boson in state $\phi_n$ by $\widehat{a}_n = \int \phi_n(\vec{x})\widehat{\psi}(\vec{x})\mathrm{d}^d x$ and $\widehat{a}^+_n = \int \phi^*_n(\vec{x})\widehat{\psi}^+(\vec{x})\mathrm{d}^d x$, which satisfy canonical boson commutation rules $[\widehat{a}_i, \widehat{a}^+_j] = \delta_{ij}$ and so on. (exercise: derive the commutation rules)

The nature of the particles (for example, the charge and spin they carry) depends on the detailed structure of the quantum field theory (i.e. nature of the $\phi$ field) and cannot come out automatically through quantization. We shall see more examples later.

### 3.1.2
### Interaction Between Particles

For a more general quantum field theory with non-quadratic terms

$$L[\phi] = \sum_i \frac{1}{2}\dot{\phi}_i^2 - \frac{1}{2}\sum_{ij} \phi_i t_{ij}\phi_j - V[\phi], \tag{3.7a}$$

the energy spectrum cannot be represented by the simple form

$$H = \sum_{\vec{k}} \left(n_{\vec{k}} + \frac{1}{2}\right)\omega_{\vec{k}}$$

and cannot be interpreted as representing independent (identical) particles or quanta of energy. In the particle representation, this is interpreted as an energy correction coming from interaction between particles. Let us consider

$$V[\phi] = \frac{1}{4}\sum_i \phi_i^4 = \frac{1}{4V^4} \sum_{\vec{k}_1,\vec{k}_2,\vec{k}_3,\vec{k}_4} \phi_{\vec{k}_1}\phi_{\vec{k}_2}\phi_{\vec{k}_3}\phi_{\vec{k}_4}\delta(\vec{k}_1+\vec{k}_2+\vec{k}_3+\vec{k}_4). \tag{3.7b}$$

The $\phi_{\vec{k}}$ fields can be written as $\phi_{\vec{k}} = \sqrt{\frac{\hbar}{2\omega_{\vec{k}}}}(a_{\vec{k}} + a^{+}_{-\vec{k}})$ and the interaction term can be rewritten in terms of the boson operators $a_{\vec{k}}, a^{+}_{\vec{k}}$ and represents scattering events between particles. Note that this particular form of interaction does not conserve boson particle number.

#### Homework Q2

The above formulation can be extended straightforwardly to continuum field theory. Can you work out the quantum Hamiltonian (in boson representation) for the field theory

$$S = \int \mathrm{d}^d x \mathrm{d}t \left[\frac{1}{2}\left(\dot{\phi}^2 - (\nabla\phi)^2 - m^2\phi^2\right) - \frac{1}{4}\phi^4\right] \quad (\phi = \phi(\vec{x}, t))?$$

#### Homework Q3

Use the Noether theorem (assuming that it still applies in the quantum situation) to find what are the conserved quantities in terms of bosons for the above continuum field theory.

### 3.1.3
### Continuum Limit of Lattice Field Theory

An approximation that you will see very often in the study of field theory is the approximation of a lattice field theory by a continuum field theory. This is the simplest example of an effective field theory where we replace an 'exact' lattice theory by an approximate theory in continuous space. This approximation is made because in general mathematical tools for analyzing continuum theory are better developed than lattice theory. The replacement is valid when the fields under consideration are changing slowly enough in space.

To implement the scheme we introduce a 'continuum' field $\widehat{\psi}(\vec{x}) = \frac{1}{V}\sum_{|\vec{k}|<\Lambda} e^{-i\vec{k}.\vec{x}}\widehat{a}_{\vec{k}}$, where the momentum $\vec{k}$ is restricted to be small: $|\vec{k}| < \Lambda \ll a^{-1}$, where $a$ = lattice spacing. In this region, we can expand the energy around $\vec{k} = 0$ to obtain $\omega_{\vec{k}} \sim m + \frac{\hbar^2\vec{k}2}{2m}$ (we assume that $m > 0$). In this limit, the Hamiltonian (3.6b) can be written approximately as

$$H \sim \sum_{|\vec{k}|<\Lambda}\left(\widehat{a}_{\vec{k}}^{\,+}\widehat{a}_{\vec{k}} + \frac{1}{2}\right)\left(m_0 + \frac{\hbar^2\vec{k}_2}{2m}\right) \sim m\int \mathrm{d}^d x \widehat{\psi}^{+}(\vec{x})\widehat{\psi}(\vec{x})$$
$$+ \frac{\hbar^2}{2m}\int \mathrm{d}^d x \nabla\widehat{\psi}^{+}(\vec{x}).\nabla\widehat{\psi}(\vec{x}) + E_G, \tag{3.8}$$

which looks like the quantum Hamiltonian for a free, non-relativistic particle except that $\widehat{\psi}(\widehat{\psi}^+)$ are operators but not complex numbers. The $\widehat{\psi}(\widehat{\psi}^+)$ operators satisfy boson commutation rules $[\widehat{\psi}(\vec{x}), \widehat{\psi}^+(\vec{x}')] = \delta(\vec{x}-\vec{x}')$. Note that because of the cutoff in momentum, the Hamiltonian and operators are, strictly speaking, well defined only in momentum space and position is 'smeared out' with uncertainty $\Delta x \sim \Lambda^{-1}$.

**Homework Q4**

Work out the approximate continuum Hamiltonian for the lattice model with $L[\phi] = \sum_i \frac{1}{2}\dot{\phi}_i^2 - \frac{t}{2}\sum_{i,\delta} \phi_i \phi_{i+\delta} - E_0 \sum_i \phi_i^2$, where the lattice is (i) a square lattice and $\delta = \pm\widehat{x}, \pm\widehat{y}$ represents nearest-neighbor sites and (ii) a triangular lattice where $\delta = \pm\widehat{x}, \pm(\widehat{x} + \sqrt{3}\widehat{y})/2, \mp(\widehat{x} - \sqrt{3}\widehat{y})/2$.

**Homework Q5**

Going backward, suppose that we have a quantum many-boson Hamiltonian that conserves boson number:

$$\frac{\hbar^2}{2m}\int d^d x \nabla\widehat{\psi}^+(\vec{x}).\nabla\widehat{\psi}(\vec{x}) + \iint d^d x d^d x' u(\vec{x}-\vec{x}')\widehat{\psi}^+(\vec{x})\widehat{\psi}^+(\vec{x}')\widehat{\psi}(\vec{x}')\widehat{\psi}(\vec{x}).$$

Can you write down an equivalent classical field theory, which, after quantization, gives rise to this Hamiltonian?

## 3.2 Introduction to Quantum Statistics

You have seen an example of how a quantum field theory is equivalent to a system of bosons, and we shall study later quantum field theories for fermions. It will be helpful if we first introduce the fundamental concept of exchange of quantum particles and quantum statistics here.

The basic idea is that, in quantum mechanics, identical particles exist which cannot be distinguished from each other in any physical measurements. An obvious example is the bosons we discussed in the previous section coming from the Lagrangian $L[\phi] = \sum_i \left(\frac{1}{2}\dot{\phi}_i^2 - m^2\phi_i^2\right)$. Let us consider an eigenstate of the system with one boson (b1) located at site $i$ and the other (b2) located at site $j$. The wavefunction $\psi[n]$ that describes this state is the product of harmonic oscillator wavefunctions $\psi_n(\phi_i)$ on all sites, with $n=1$ on sites $i$ and $j$, and $n=0$ on the rest of the sites. There is no way to distinguish which particle (1 or 2) occupies which site ($i$ or $j$) in the sense that the wavefunctions $\psi_{12}(\vec{x}_i, \vec{x}_j)$ and $\psi_{21}(\vec{x}_i, \vec{x}_j)$ are indistinguishable because they represent the same quantum state in quantum mechanics!

However, exchange of particles has a physical meaning because we can physically move the positions of the particles from $\vec{x}$ to $\vec{x} + \delta\vec{x}$,

$$\psi(\vec{x}_i, \vec{x}_j) \rightarrow \psi(\vec{x}_i + \delta\vec{x}_i, \vec{x}_j + \delta\vec{x}_j).$$

In particular, we may imagine moving successively the positions of the two particles such that $\vec{x}_1 \to \vec{x}_2$ and $\vec{x}_2 \to \vec{x}_1$ at the end of the operation. The movement can be chosen to be slow enough (adiabatic motion) such that no dynamical change to the wavefunction occurs. The inability to distinguish the wavefunctions after exchanging the particles implies that the final wavefunction must be the same as the initial wavefunction except for an overall phase which is allowed by quantum mechanics, that is

$$\psi_{12}(\vec{x}_i, \vec{x}_j) = \psi_{21}(\vec{x}_i, \vec{x}_j)e^{i\theta}.$$

The phase reflects the topological/geometrical properties of the system (see Chapter 6 on Berry phase for a more detailed discussion. See also Ref. [3]). For exchanging the positions of (identical) particles, it can be shown that the exchange phase $\theta$ can be any real number in one and two dimensions, but can only be $n\pi$ ($n =$ integer) in three (or above) dimensions, that is

$$\psi_{12}(\vec{x}_i, \vec{x}_j) = \pm\psi_{21}(\vec{x}_i, \vec{x}_j), \tag{3.9}$$

where the + sign stands for bosons and the – sign stands for fermions. In 3D, the result $\theta = n\pi$ is often obtained from the argument that the wavefunction must go back to itself if we exchange the particles twice, that is

$$\psi_{12}(\vec{x}_i, \vec{x}_j) = e^{i\vartheta}\psi_{21}(\vec{x}_i, \vec{x}_j) = e^{2i\vartheta}\psi_{12}(\vec{x}_i, \vec{x}_j) = \psi_{12}(\vec{x}_i, \vec{x}_j).$$

In dimensions less than three, the argument that the wavefunction must go back to itself after exchanging twice is not valid because of the topological reason that the path of deformation cannot be reduced to a 'nil' path without crossing a singular point and particles with arbitrary statistics (anyons) are allowed. This point will be explained further in Chapter 6. You can also look at Ref. [3] for details.

The exchange of particles can be generalized to systems with an arbitrary number of identical particles, with essentially the same result, that is

$$\psi_{1,2,\ldots,i,\ldots,j,\ldots,N}(\vec{x}_1, \ldots, \vec{x}_N) = e^{i\theta}\psi_{1,2,\ldots,j,\ldots,i,\ldots,N}(\vec{x}_1, \ldots, \vec{x}_N)$$

after exchanging two particles. The exchange phase is not affected by the presence of other (identical) particles in 3D. The much more complicated situation of 2D anyons will not be discussed in this book.

### 3.2.1
### Fock Space for Bosons and Fermions

Next we ask the question of how we can construct (and label) the Hilbert space for a many-particle system, given the restrictions on the exchange symmetry? The only systematic procedure physicists have developed so far is to construct the many-particle Hilbert space starting from the Hilbert space of a one-particle system. The procedure works only for bosons and fermions. Physicists are still trying to find general approaches that work also for anyons in 2D.

To see how the approach (for fermions and bosons) works, we first consider a two-particle system. If the two particles are distinguishable, the Hilbert space of the two-particle system is simply the product of the Hilbert space of the one-particle systems, that is $S_2 = S_1 \otimes S_1$. If the $|k\rangle$ represent a complete set of states spanning $S_1$, then $|k1\rangle|k'2\rangle$ (1, 2 are particle indices) span the Hilbert space $S_2$ of two distinguishable particles.

The Hilbert space of fermions and bosons is constructed by taking only states in $S_2$ that satisfy the proper exchange symmetry. For example, the linear combinations $|k1\rangle|k'2\rangle + (-)|k2\rangle|k'1\rangle$ are symmetric $(+)$ and anti-symmetric $(-)$ under particle exchange $1 \leftrightarrow 2$ and are chosen as permitted states in the Hilbert space of bosons and fermions, respectively. Similar constructions can be extended to systems with arbitrary numbers of particles. The Hilbert space constructed in this way is called a Fock space. The particle indices are removed from the wavefunction after (anti)-symmetrization. For example, in the case of two particles, the states are denoted by $|k, k'\rangle = \frac{1}{\sqrt{2}}(|k1\rangle|k'2\rangle + (-)|k2\rangle|k'1\rangle)$.

**Homework Q6a**

If there are $N$ states in the one-particle Hilbert space, what are the sizes of the Hilbert space of $M$ particles which are (i) distinguishable, (ii) indistinguishable bosons, (iii) indistinguishable fermions?

Next, we introduce *creation* and *annihilation* operators $a_k^+$ and $a_k$, which increase (decrease) the particle number in the system by one. For example,

$$a_k^+ |0\rangle = |k\rangle, a_{k'}^+ |k\rangle = |k', k\rangle = a_{k'}^+ a_k^+ |0\rangle.$$

The procedure

$$|k\rangle = |k1\rangle \to |k', k\rangle = \frac{1}{\sqrt{2}}(|k1\rangle|k'2\rangle \pm |k2\rangle|k'1\rangle)$$

has to be defined carefully. We may define it as

$$\begin{aligned} &|k1\rangle \to (|k1\rangle|k'2\rangle \pm P_{12}|k1\rangle|k'2\rangle), \\ &\quad \text{where } P_{12}|k1\rangle|k'2\rangle = |k2\rangle|k'1\rangle. \end{aligned} \tag{3.10a}$$

With this definition, we can compare this with

$$a_k^+ |k'\rangle = |k, k'\rangle = a_k^+ a_{k'+} |0\rangle = |k'1\rangle|k2\rangle \pm P_{12}|k'1\rangle|k2\rangle. \tag{3.10b}$$

We see that the two ways of producing the (same) wavefunction differ by a phase factor

$$a_k^+ a_{k'}^+ |0\rangle = \pm a_{k'}^+ a_k^+ |0\rangle \text{ ('+' for boson, '−' for fermion)}. \tag{3.10c}$$

It can be shown that the same difference in phase factor comes out in other procedures of going from one-particle to two-particle states also, as long as they are performed systematically. This construction can be extended to states with arbitrary numbers of particles, where it can be shown that

$$a_k^+ a_{k'}^+ |\chi\rangle = \pm a_{k'}^+ a_k^+ |\chi\rangle \text{ ('+' for boson, '−' for fermion)}$$

for an arbitrary wavefunction $|\chi\rangle$.

Therefore, we have an operator identity

$$a_k^+ a_{k'}^+ \mp a_{k'}^+ a_k^+ = 0 \text{ ("−" for boson, "+" for fermion).} \quad (3.11a)$$

Similarly, one can show that $a_k a_{k'} \mp a_{k'} a_k = 0$ and

$$a_k a_{k'}^+ \mp a_{k'}^+ a_k = \delta_{kk'} \quad (3.11b)$$

**Homework Q6b**

Demonstrate Equation (3.11a) and (3.11b) for states with two particles and three particles.

Note that, from the commutation relation, it is easy to show that fermions satisfy the Pauli exclusion principle – we can only put one fermion in any given quantum one-particle state.

It is often more convenient to work in the 'particle representation' in quantum field theories, that is a representation in terms of boson or fermion creation and annihilation operators but not in terms of number-field variables. However, exceptions do exist. An example is the electromagnetic (EM) field where it is often more convenient to work with electric and magnetic fields rather than photons. One has to understand that the two approaches are equivalent and know how to switch back and forth between the two representations.

## 3.2.2
## Introduction to Grassmann Field and Quantum Field Theory for Fermions

We have shown an example of a classical field theory, which, after quantization, is equivalent to a many-particle theory of bosons. It is interesting to ask whether a similar procedure can be found for fermions, that is can we find an example of a classical field theory, which, after quantization, is equivalent to a many-particle theory of fermions.

The straightforward answer to the question is 'no', if we adhere rigidly to the canonical quantization scheme. The reason can be seen by asking the reverse question: can one recover a classical theory if we start from a quantum theory of fermions by taking the limit $\hbar \to 0$?

In the canonical quantization scheme, quantum mechanics enters through the commutation rule $[q, p] = i\hbar$ or $[a, a^+] = 1$.

$$\left(\text{Recall that } \hat{a} = \sqrt{\frac{\omega_k}{2\hbar}}\left(q + \frac{ip}{\omega_k}\right),\ \hat{a}^+ = \sqrt{\frac{\omega_k}{2\hbar}}\left(q - \frac{ip}{\omega_k}\right).\right)$$

In the limit $\hbar \to 0$, we recover the classical result $[q, p] = 0$, or $[(\sqrt{\hbar}a), (\sqrt{\hbar}a^+)] = 0$, which is a mathematical relation satisfied by classical numbers. This is the minimal mathematical consistency needed to recover a classical theory from quantum theory.

The problem with fermions is that the creation and annihilation operators satisfy an anti-commutation rule $\{c, c^+\} = 1$. In the limit $\hbar \to 0$, the 'classical' fields satisfy $\{(\sqrt{\hbar}c), (\sqrt{\hbar}c^+)\} = (\hbar cc^+ + \hbar c^+ c) = 0$, which is not a relation satisfied by classical numbers. This fundamental mathematical difficulty makes it impossible to relate fermions to any standard classical field theory.

We can obtain a physical idea of why it is impossible to obtain a theory of fermions from a standard classical theory from another angle. Let us consider a one-site field theory (site $i$) described by a field variable $\phi_i$. After (canonical) quantization we obtain a complete set of states described by wavefunctions $\psi_n(\phi_i)$. If $\phi_i$ is a continuous variable, the size of the Hilbert space is infinite and we obtain an infinite number of eigenstates ($n = 1, \ldots, \infty$). In the language of quantum field theory where each eigenstate corresponds to a particular number of particles on site $i$, the existence of an infinite number of states means that we can put an infinite number of particles on site $i$. Therefore, these particles must be bosons. To obtain a quantum theory of fermions we must find a 'classical' system where there are only two possible eigenstates after quantization, or we have to find a 'variable' $\phi_i$ which admits only two eigenstates with *any* operator.

It turns out that the above problems can be solved and a 'classical' field theory for fermions can be constructed by introducing the so-called Grassmann variables – algebraic quantities satisfying strange algebraic rules called Grassmann algebra.

We introduce independent Grassmann variables $x_i$, $x_j$ satisfying the anti-commutation rule

$$x_i x_j + x_j x_i = 0. \tag{3.12}$$

Therefore, $x_i^2 = 0$ by definition.

As a result, any function $f(x)$ of the Grassmann variable $x$ is characterized by the first two terms in a Taylor expansion, $f(x) = f(0) + f'(0)x$. All higher-order terms are zero and there are only two independent functions we can construct using Grassmann variables. We can also introduce the following rules for differentiation and integration of Grassmann variables,

$$\partial_x f(x) = f'(0), \quad \partial_x^2 f(x) = 0, \quad \text{and} \quad \int \mathrm{d}x f(x) = f'(0), \tag{3.13a}$$

with similar 'Grassmann' exchange rules for differentials,

$$x_i(\mathrm{d}x_j) + (\mathrm{d}x_j)x_i = 0, \quad x_i \frac{\partial}{\partial x_j} + \frac{\partial}{\partial x_j} x_i = 0, \quad \text{etc.} \tag{3.13b}$$

A 'classical' theory constructed by Grassmann variables will solve the problems we discuss – it satisfies the anti-commutation rule and has only two independent eigenstates for any operator! The theory can be quantized by imposing the anti-commutation rule

$$x_i p_j + p_j x_i = i\hbar\delta_{ij}$$

for appropriately chosen pairs of Grassmann variables $(x, p)$.

To see how the approach works, let us consider a one-site quantum field theory with

$$H = e_0 c^+ c, \quad \text{with} \quad \{c, c^+\} = 1.$$

To construct the classical Grassmann field theory from this Hamiltonian, we introduce canonical 'position' and 'momentum' operators

$$x = \sqrt{\frac{\hbar}{2\varepsilon_0}}(c+c^+),\ p = -i\sqrt{\frac{\hbar\varepsilon_0}{2}}(c-c^+), \tag{3.14a}$$

where $\{x, x\} = \{p, p\} = 0$ and $\{x, p\} = i\hbar$. These two fermionic operators are examples of Majorana fermions in the literature (see e.g. Ref. [4]). In terms of $x$ and $p$,

$$H = \frac{\varepsilon_0}{2\hbar}\left(\varepsilon_0 xx - ixp + ipx + \frac{pp}{\varepsilon_0}\right) = \varepsilon_0\left(\frac{1}{2} + \frac{i}{\hbar}px\right). \tag{3.14b}$$

The corresponding 'classical' Hamiltonian is obtained when $x$, $p$ are replaced by Grassmann numbers with $\{x, p\} = 0$, and so on. Note that $\hbar$ appears in the 'classical' Hamiltonian, reflecting the intrinsic 'quantum' nature of fermion systems.

It is important to know that Grassmann variables cannot represent physical 'observables'. In quantum mechanics observables are always represented by operators with eigenvalues that are ordinary classical numbers. In this sense, the introduction of Grassmann numbers cannot help us in actual computation. However, it offers a very useful tool in formal mathematical manipulations when fermionic quantum field theory is formulated in path integral language. We shall see many examples in later chapters.

### Many-Site System: Linear Transformation of Grassmann Fields

Next, we consider an $N$-site system where each site $i$ is characterized by Grassmann fields $c_i, c_i^*$. The Hamiltonian is

$$H = \sum_{\langle i,j\rangle}(t_{ij}c_i^* c_j + t_{ji}^* c_j^* c_i) = C^* T C, \tag{3.15a}$$

where the $t_{ij}$ are ordinary numbers. Can we 'diagonalize' this Hamiltonian?

If $c_i, c_i^*$ are ordinary numbers we can diagonalize the Hamiltonian by a unitary transformation $U$ such that $UTU^+$ is diagonalized. Let us follow the same procedure here, that is we write

$$C^* T C = (C^* U^+)(UTU^+)(UC) = \sum_k \varepsilon_k c_k^* c_k. \tag{3.15b}$$

The question is: what are the $c_k = \sum_i U_{ki} c_i$? Are they independent Grassmann numbers satisfying

$$c_k c_{k'}^* + c_{k'}^* c_k = 0 \quad \text{for all } k, k'? \tag{3.15c}$$

**Homework Q7a**

Show that Equation (3.15c) is correct.

The correctness of Equation (3.15c) shows that Grassmann fields can undergo linear transformations as for ordinary classical number fields. It is also straightforward to show that the corresponding quantum operators $c_k = \sum_i U_{ki} c_i$, $c_k^+ = \sum_i U_{ik}^* c_i^+$ satisfy

$$c_k c_{k'}^+ + c_{k'}^+ c_k = \delta_{kk'} \quad \text{if} \quad c_i c_j^+ + c_j^+ c_i = \delta_{ij}. \tag{3.15d}$$

**Homework Q7b**

Show that Equation (3.15d) is correct.

A more realistic example is the Hamiltonian

$$H = \frac{\hbar^2}{2m} \int d^3x \nabla \widehat{\psi}^+(\vec{x}).\nabla \widehat{\psi}(\vec{x}) + \int d^3x V(\vec{x}) \widehat{\psi}^+(\vec{x}) \widehat{\psi}(\vec{x}),$$

where $\hat{\psi}(\hat{\psi}^+)$ are operators satisfying anti-commutation relations $\{\widehat{\psi}(\vec{x}), \widehat{\psi}(\vec{x}')\} = 0$, $\{\widehat{\psi}(\vec{x}), \widehat{\psi}^+(\vec{x}')\} = \delta(\vec{x}-\vec{x}')$, and so on. The Hamiltonian represents a system of (spinless) fermions with kinetic energy $p^2/2m$ moving under a potential $V(\vec{x})$. The Hamiltonian can be diagonalized using an ordinary unitary transformation as in the above example. A Lagrangian formulation of a fermion quantum field theory is also available and will be discussed in the next chapter.

**Homework Q7c**

Consider a quantum field theory described by the Hamiltonian $H = \frac{\hbar^2}{2m} \int d^3x \nabla \widehat{\psi}^+(\vec{x}).\nabla \widehat{\psi}(\vec{x})$.

What are the ground-state wavefunction and energy of the system when $\psi$ represents (i) bosons, (ii) fermions, if the density of particles in the system is $n$? What if

$$H = \frac{\hbar^2}{2m} \int d^3x \nabla \widehat{\psi}^+(\vec{x}).\nabla \widehat{\psi}(\vec{x}) + \int d^3x \left(\frac{1}{2} k \vec{x}^2\right) \widehat{\psi}^+(\vec{x}) \widehat{\psi}(\vec{x})?$$

What are the corresponding ground-state wavefunctions and energies if there are 10 particles in the system?

## 3.3 Path Integral Quantization of Mechanics and Field Theory

The path integral was invented by Richard Feynman (see e.g. Ref. [4]) as an alternative way of formulating quantum mechanics. In modern terms, we consider the solution of the Schrödinger equation

$$i\hbar\frac{\partial}{\partial t}|\psi(t)\rangle = H|\psi(t)\rangle \tag{3.16a}$$

with given initial state $|\psi(0)\rangle$. The solution can be written as

$$|\psi(t)\rangle = \hat{U}(t,0)|\psi(0)\rangle, \tag{3.16b}$$

where $\hat{U}(t,t') = e^{-i\frac{\hat{H}}{\hbar}(t-t')}$ is the time-evolution operator. The path integral is essentially a technique to compute $\hat{U}(t,t')$.

To compute $\hat{U}(t,t')$ as a path integral, we apply the following steps:

Step I: Choose a representation for $\hat{U}$, that is choose a complete set of eigenstates $|m\rangle$, $\sum_m |m\rangle\langle m| = 1$, and then determine $\langle n|\hat{U}(t,t')|m\rangle$ for arbitrary $|n\rangle$ and $|m\rangle$.

For example, in one-particle quantum mechanics, we usually choose $|m\rangle = |x\rangle$, $1 = \int \mathrm{d}x|x\rangle\langle x|$, and we need to evaluate $\langle x|\hat{U}(t,t')|x'\rangle$.

Step II: To evaluate $\langle x|\hat{U}(t,t')|x'\rangle$, we divide the time $t-t'$ into $N$ steps, $t-t' = N\cdot\Delta t$. We shall be interested in the limit $N \to \infty$, $\Delta t \to 0$.
Let $t_0 = t'$, $t_k = t' + k\cdot\Delta t$, $t_N = t$,

$$\hat{U}(t,t') = \hat{U}(t_N,t_0) = \hat{U}(t_N,t_{N-1})\hat{U}(t_{N-1},t_{N-2})\cdots\hat{U}(t_2,t_1)\hat{U}(t_1,t_0). \tag{3.17a}$$

We then insert $1 = \sum_m |m\rangle\langle m|$ between each time interval, that is

$$\begin{aligned}\langle n|\hat{U}(t,t')|m\rangle = &\sum_{m_1,m_2,\ldots,m_{n-2},m_{n-1}} \langle n|\hat{U}(t_N,t_{N-1})|m_{N-1}\rangle\langle m_{N-1}|\hat{U}(t_{N-1},t_{N-2})|m_{N-2}\rangle \\ &\times\cdots\times\langle m_2|\hat{U}(t_2,t_1)|m_1\rangle\langle m_1|\hat{U}(t_1,t_0)|m\rangle.\end{aligned} \tag{3.17b}$$

For example, in one-particle quantum mechanics we have

$$\begin{aligned}\langle x|\hat{U}(t,t')|x'\rangle = &\int\cdots\int \mathrm{d}x_1\mathrm{d}x_2\cdots\mathrm{d}x_{N-2}\mathrm{d}x_{N-1}\langle x|\hat{U}(t_N,t_{N-1})|x_{N-1}\rangle \\ &\langle x_{N-1}|\hat{U}(t_{N-1},t_{N-2})|x_{N-2}\rangle \\ &\times\cdots\times\langle x_2|\hat{U}(t_2,t_1)|x_1\rangle\langle x_1|\hat{U}(t_1,t_0)|x'\rangle.\end{aligned}$$

Step III: We evaluate $\langle m_{k+1}|\hat{U}(t_{k+1},t_k)|m_k\rangle$ in the limit $t_{k+1} - t_k = \Delta t \to 0$.

$$\begin{aligned}&\langle m_{k+1}|\hat{U}(t_{k+1},t_k)|m_k\rangle = \langle m_{k+1}|e^{-i\hat{H}\cdot\Delta t/\hbar}|m_k\rangle \\ &\sim \langle m_{k+1}|(1-\frac{i}{\hbar}\hat{H}\cdot\Delta t + O(\Delta t^2))|m_k\rangle \\ &= \langle m_{k+1}|m_k\rangle - \frac{i}{\hbar}\langle m_{k+1}|\hat{H}|m_k\rangle\cdot\Delta t + O(\Delta t^2).\end{aligned} \tag{3.18a}$$

Next, we try to write it as $e^{-\frac{i}{\hbar}S\cdot\Delta t}$ (correct in the limit $t_{k+1} - t_k = \Delta t \to 0$) and see whether one can identify $S$ with something meaningful at the end.

We consider one-particle quantum mechanics with $\hat{H} = \frac{p^2}{2m} + V(x)$ as example. In this case,

$$\langle x_{k+1}|(1-\frac{i}{\hbar}\hat{H}\cdot\Delta t)|x_k\rangle = \int \frac{\mathrm{d}p_x}{2\pi\hbar}\langle x_{k+1}|p_k\rangle\langle p_k|(1-\frac{i}{\hbar}\hat{H}(\hat{p},\hat{x})\cdot\Delta t)|x_k\rangle. \tag{3.18b}$$

where $|p\rangle$'s are momentum eigenstates and $\langle x_{k+1}|p_k\rangle = e^{\frac{i}{\hbar}p_x\cdot x_{k+1}}$. Therefore,

$$\begin{aligned}\langle x_{k+1}|\left(1-\frac{i}{\hbar}\hat{H}\cdot\Delta t\right)|x_k\rangle &\sim \int\frac{\mathrm{d}p_x}{2\pi\hbar}e^{\frac{i}{\hbar}p_k\cdot(x_{k+1}-x_k)-\frac{i}{\hbar}H(p_k,x_k)\cdot\Delta t}\\ \sim \int\frac{\mathrm{d}p_x}{2\pi\hbar}e^{\frac{i}{\hbar}[p_k.\dot{x}_k-H(p_k,x_k)]\cdot\Delta t} &= \int\frac{\mathrm{d}p_x}{2\pi\hbar}e^{\frac{i}{\hbar}[p_k\cdot\dot{x}_k-\frac{p_k^2}{2m}-V(x_k)]\cdot\Delta t}.\end{aligned} \tag{3.18c}$$

Note that the integral over $p_x$ can be performed 'exactly':

$$\begin{aligned}&\int_{-\infty}^{\infty}\mathrm{d}x e^{\frac{i}{\hbar}(\alpha\cdot x-\beta\cdot x^2)} = \int_{-\infty}^{\infty}\mathrm{d}x e^{-\frac{i}{\hbar}\left\{\beta\cdot(x-\frac{\alpha}{2\beta})^2-\frac{\alpha^2}{4\beta}\right\}} = C\cdot e^{\frac{i}{\hbar}\cdot\frac{\alpha^2}{4\cdot\beta}}\\ &\text{and } \int\frac{\mathrm{d}p_x}{2\pi\hbar}e^{\frac{i}{\hbar}\left[p_k\cdot\dot{x}_k-\frac{p_k{}^2}{2m}-V(x_k)\right]\cdot\Delta t} = C\cdot e^{\frac{i}{\hbar}\left[\frac{m}{2}\cdot x_k^2-V(x_k)\right]\cdot\Delta t},\end{aligned} \tag{3.18d}$$

where $C = \int_{-\infty}^{\infty}\mathrm{d}x\cdot e^{-\frac{i}{\hbar}\beta(x-\frac{\alpha}{2\beta})^2}$

(note that $C$ is in fact not a mathematically well-defined integral).

Putting everything back into $\langle x|\hat{U}(t,t')|x'\rangle$, we obtain

$$\langle x|\hat{U}(t,t')|x'\rangle \sim \int Dx(t'')e^{\frac{i}{\hbar}\int_{t''}^{t'}L(t'')\mathrm{d}t''}, \tag{3.19a}$$

where

$$L(t'') = \frac{m}{2}\dot{x}(t'')^2-V(x(t'')) \tag{3.19b}$$

is the classical Lagrangian, $\int Dx(t'') = \int\cdots\int \mathrm{d}x_1\mathrm{d}x_2\cdots\mathrm{d}x_{N-2}\mathrm{d}x_{N-1}$. The expression is correct to some undetermined constant $(C)^\infty$.

### 3.3.1 Imaginary-Time Path Integral and Partition Function

The path integral becomes much better defined if we go to 'imaginary time', $t\to -i\tau$. In this case,

$$e^{-\frac{i}{\hbar}\hat{H}\cdot t}\to e^{-\frac{\hat{H}\cdot\tau}{\hbar}} = e^{-\beta\hat{H}},$$

where we have identified $\tau = \hbar\beta$. Note that the time-evolution operator becomes the partition function when time is imaginary. For example, the partition function for the one-particle system is

$$Z = \mathrm{Tr}(e^{-\beta H}) = \int \mathrm{d}x \langle x|e^{-\beta H}|x\rangle = \int \mathrm{d}x \langle x|\hat{U}(-i\beta, 0)|x\rangle. \tag{3.20}$$

Following the same procedure as in real time, we obtain

$$\begin{aligned}\langle x|\hat{U}(-i\beta,0)|x\rangle &= \int\cdots\int \mathrm{d}x_1\mathrm{d}x_2\cdots\mathrm{d}x_{N-2}\mathrm{d}x_{N-1}\langle x|\hat{U}(-i\tau_N,-i\tau_{N-1})|x_{N-1}\rangle\\ &\quad\times\langle x_{N-1}|\hat{U}(-i\tau_{N-1},-i\tau_{N-2})|x_{N-2}\rangle\cdots\langle x_1|\hat{U}(-i\tau_1,0)|x\rangle.\end{aligned} \tag{3.21}$$

As an example, we consider again $\hat{H} = \frac{p^2}{2m} + V(x)$. In this case,

$$\langle x_{k+1}|e^{-\frac{i}{\hbar}H\cdot\Delta t}|x_k\rangle \sim \int \frac{\mathrm{d}p_k}{2\pi\hbar} e^{\frac{i}{\hbar}\left[p_k\cdot\dot{x}_k - \frac{p_k{}^2}{2m} - V(x_k)\right]\cdot\Delta t}. \tag{3.22a}$$

With $t \to -i\tau$, we obtain

$$\langle x_{k+1}|e^{-\frac{i}{\hbar}H\cdot\Delta t}|x_k\rangle \Rightarrow \int \frac{\mathrm{d}p_k}{2\pi\hbar} e^{\frac{1}{\hbar}\left[ip_k\cdot\dot{x}_k - (\frac{p_k{}^2}{2m} + V(x_k))\right]\cdot\Delta\tau}. \tag{3.22b}$$

Note that the integral is of the form $\int_{\infty}^{-\infty} \mathrm{d}x e^{-\beta\cdot x^2 + i\alpha\cdot x}$ with $\beta > 0$ and is now well-defined mathematically:

$$\begin{aligned}\int_{-\infty}^{\infty} \mathrm{d}x e^{-\beta\cdot x^2 + i\alpha\cdot x} &= \int_{-\infty}^{\infty} \mathrm{d}x e^{-\beta\cdot(x - \frac{i\alpha}{2\beta})^2}\cdot e^{-\frac{\alpha^2}{4\beta}}\\ &= \mathrm{C}\cdot e^{-\frac{\alpha^2}{4\cdot\beta}} = C\cdot e^{-\frac{1}{\hbar}[\frac{m}{2}\cdot\dot{x}\cdot_k{}^2 + V(x_k)]\cdot\Delta\tau},\end{aligned} \tag{3.22c}$$

where $C = \int_{-\infty}^{\infty} \mathrm{d}x\cdot e^{-\beta(x - \frac{i\alpha}{2\beta})^2} = \sqrt{\frac{\pi}{\beta}}$.

Putting together as before, we obtain

$$Z \sim \int Dx(\tau')e^{-\int_0^\beta H(\tau')\mathrm{d}\tau'}, \tag{3.23a}$$

$$\text{where}\, H(\tau') = -L(-i\tau') = \frac{m}{2}\dot{x}(\tau')^2 + V(x(\tau')) \tag{3.23b}$$

is the Hamiltonian.

Note that the path integral can be used to evaluate the time evolution of the system (dynamics) and the partition function (thermodynamics) by simply switching between real and imaginary time. This property is very useful in computations. We shall see many examples in later chapters.

### 3.3.2
### Application to Quantum Field Theory

Let us consider a field theory with

$$
\begin{aligned}
L[\phi] &= \sum_i \frac{1}{2}\dot{\phi}_i^2 - V[\phi] \text{ (lattice field theory)} \quad \text{or} \\
L[\phi] &= \int \mathrm{d}^d x \left(\frac{1}{2}\dot{\phi}(\vec{x})^2 - V[\phi(\vec{x})]\right) \text{ (continuum field theory).}
\end{aligned}
\tag{3.24}
$$

Correspondingly,

$$
H[\phi,\pi] = \sum_i \frac{1}{2}\pi_i^2 + V[\phi] \quad \text{or}
$$

$$
H[\phi,\pi] = \int \mathrm{d}^d x \left(\frac{1}{2}\pi(\vec{x})^2 + V[\phi(\vec{x})]\right).
$$

Recall that $[\phi_i, \pi_i] = i\hbar\delta_{ij}$, $[\phi(\vec{x}), \pi(\vec{x}')] = i\hbar\delta(\vec{x}-\vec{x}')$. (3.25)

As in the previous section, we consider evaluation of the time-evolution operator $U(t, t')$.

Step I: Choose a complete set of states $|m\rangle$ $\left(\sum_m |m\rangle\langle m| = 1\right)$.
We again choose the Schrödinger representation

$$
\begin{aligned}
1 &= \int \mathrm{d}\phi_1 \mathrm{d}\phi_2 \cdots \mathrm{d}\phi_N |\phi_1, \phi_2, \ldots, \phi_N\rangle\langle\phi_1, \phi_2, \ldots, \phi_N| \\
&\quad (\text{or } \widehat{I} = \int \Pi \mathrm{d}\phi(\vec{x}) |\phi(\vec{x}), \ldots\rangle\langle\phi(\vec{x}), \ldots|).
\end{aligned}
\tag{3.26a}
$$

To simplify notation, we write it as $\widehat{I} = \int D\phi|\phi\rangle\langle\phi|$.

Step II: To evaluate $\langle\phi|\hat{U}(t,t')|\phi'\rangle$, we divide the time $t-t'$ into $N$ steps, $t-t'$ $N{\cdot}\Delta t$ and consider the limit $N \to \infty$, $\Delta t \to 0$. We then insert $\widehat{I} = \int D\phi|\phi \times \phi|$ between each time interval:

$$
\begin{aligned}
\langle\phi|\hat{U}(t,t')|\phi'\rangle &= \int \cdots \int D\phi_1 D\phi_2 \cdots D\phi_{N-2} D\phi_{N-1} \langle\phi|\hat{U}(t_N, t_{N-1})|\phi_{N-1}\rangle \\
&\quad \times \langle\phi_{N-1}|\hat{U}(t_{N-1}, t_{N-2})|\phi_{N-2}\rangle \\
&\quad \times \cdots \times \langle\phi_2|\hat{U}(t_2,t_1)|\phi_1\rangle\langle\phi_1|\hat{U}(t_1,t_0)|\phi'\rangle.
\end{aligned}
\tag{3.26b}
$$

Step III: We evaluate $\langle\phi_{k+1}|\hat{U}(t_{k+1}, t_k)|\phi_k\rangle$ in the limit $t_{k+1} - t_k = \Delta t \to 0$:

$$
\begin{aligned}
&\langle\phi_{k+1}|\hat{U}(t_{k+1}, t_k)|\phi_k\rangle = \langle\phi_{k+1}|e^{-i\hat{H}\cdot\Delta t/\hbar}|\phi_k\rangle \\
&\sim \langle\phi_{k+1}|(1 - \frac{i}{\hbar}\hat{H}\cdot\Delta t + O(\Delta t^2))|\phi_k\rangle \\
&= \langle\phi_{k+1}|\phi_k\rangle - \frac{i}{\hbar}\langle\phi_{k+1}|\hat{H}\rangle \cdot \Delta t + O(\Delta t^2).
\end{aligned}
\tag{3.26c}
$$

To evaluate these matrix elements, let us be careful and go back to our clumsy notation

$$
\begin{aligned}
&\langle\phi_{k+1}|\left(1-\frac{i}{\hbar}\hat{H}\cdot\Delta t\right)|\phi_k\rangle\\
&=\langle\phi_{1(k+1)},\phi_{2(k+1)},\ldots,\phi_{N(k+1)}|\left(1-\frac{i}{\hbar}\hat{H}\cdot\Delta t\right)|\phi_{1k},\phi_{2k},\ldots,\phi_{Nk}\rangle\\
&=\Pi_l\int\frac{d\pi(x_l)}{(2\pi\hbar)}\langle\phi_{1(k+1)},\phi_{2(k+1)},\ldots,\phi_{N(k+1)}|\pi_{1k},\pi_{2k},\ldots,\pi_{Nk}\rangle\langle\pi_{1k},\pi_{2k},\ldots,\pi_{Nk}|\\
&\quad\times\left(\widehat{I}-\frac{i}{\hbar}\hat{H}(\pi,\phi)\cdot\Delta t\right)|\phi_k\rangle.
\end{aligned}
\tag{3.26d}
$$

Note that $\langle\phi_{k+1}|\pi_k\rangle=\langle\phi_{1(k+1)},\phi_{2(k+1)},\ldots,\phi_{N(k+1)}|\pi_{1k},\pi_{2k},\ldots,\pi_{Nk}\rangle e^{\frac{i}{\hbar}\sum_j\pi_k(x_j)\cdot\phi_{k+1}(x_j)}$.

**Exercise**

Verify the above equality!

Therefore,

$$
\begin{aligned}
&\langle\phi_{k+1}(x_i)|\left(1-\frac{i}{\hbar}\hat{H}\cdot\Delta t\right)|\phi_k(x_j)\rangle\\
&\sim\prod_l\int\frac{d\pi(x_l)}{2\pi\hbar}e^{\sum_j\frac{i}{\hbar}\pi_k(x_j)\cdot\phi_{k+1}(x_j)}e^{\sum_j-\frac{i}{\hbar}\pi_k(x_j)\cdot\phi_k(x_j)}e^{-\frac{i}{\hbar}H(\pi_k,\phi_k)\Delta t}\\
&\sim\prod_l\int\frac{d\pi(x_l)}{2\pi\hbar}e^{\frac{i}{\hbar}\left[\sum_l\pi_k(x_l)\cdot\dot{\phi}_k(x_l)-H(\pi_k,\phi_k)\right]\Delta t}=C^N e^{\frac{i}{\hbar}\left[\sum_l\cdot\dot{\phi}_k(x_l)^2-V[\phi_k]\right]\Delta t},
\end{aligned}
$$

for Hamiltonians of form given by (3.24). (3.26e)

Putting everything back into $\langle\phi|\hat{U}(t,t')|\phi'\rangle$, we obtain

$$
\langle\phi|\hat{U}(t,t')|\phi'\rangle\sim\int D\phi(t'')e^{\frac{i}{\hbar}\int_{t''}^{t'}L(t'')dt''},
\tag{3.27}
$$

where $L(t'')=\frac{1}{2}\sum_i\dot{\phi}_i(t'')^2-V[\phi(t'')]$

$$
\left(\text{or } L(t'')=\frac{1}{2}\int d^d x\dot{\phi}(\vec{x},t'')^2-V[\phi(t'')]\right)
$$

is the classical Lagrangian for the field theory. It is easy to show that the corresponding partition function is given by

$$
\begin{aligned}
&Z\sim\int D\phi(t'')e^{-\int_0^\beta H(t'')dt''},\text{where}\\
&H(t'')=\frac{1}{2}\sum_i\dot{\phi}_i(t'')^2+V[\phi(t'')]\text{ is the Hamiltonian.}
\end{aligned}
\tag{3.28}
$$

**Homework Q8**

System with vanishing canonical momenta.

Consider a field theory with Lagrangian

$L[\phi] = -\int d^d x V[\phi(x)]$ such that $\pi(x) = \frac{\delta L}{\delta \dot{\phi}(x)} = 0$; the Hamiltonian becomes $H = \int d^d x V[\phi(x)]$.

Work out the corresponding Schrödinger equation for this system and the path integral representation of the system. (Hint: to recover the correct Hamiltonian with no kinetic energy, you may consider $L[\phi] = \int d^d x \left[\frac{\dot{\phi}^2}{2\alpha} - V(\phi(x))\right]$, and let $\alpha \to 0$ at the end of the calculation.)

## References


1 Itzykson, C. and Zuber, J.-B. (1980) *Quantum Field Theory*, McGrow-Hill, USA.

2 Weinberg, S. (1995) *The Quantum Theory of Fields*, Vol. 1., Cambridge University Press.

3 Wilczek, F. (1990) *Fractional Statistics and Anyon Superconductivity*, World Scientific (Singapore. New Jersey. London. Hong Kong).

4 Feynman, R.F. (1982) *Statistical Mechanics – A set of lectures*, The Benjamin/Cummings Publishing Company, Inc. Advanced Book Program (London. Amsterdam. Don Mills. Ontario. Sydney. Tokyo).

# 4
# Quantization of Classical Field Theorie (II)

## 4.1
## Path Integral Quantization in Coherent State Representations of Bosons and Fermions

A path integral for field theories can also be formulated by introducing classical fields corresponding directly to the creation and annihilation operators $\psi(\vec{x}), \psi^{+}(\vec{x})$ (or $a_k$, $a_k^{+}$) which is essentially a change of basis. These are called coherent state representations. We shall first introduce coherent states in the following. We start again with a single-site Lagrangian

$$L[\phi] = \frac{1}{2}\left[\dot{\phi}^2 - m^2\phi^2\right]. \tag{4.1a}$$

The Hamiltonian of the system is, in terms of creation and annihilation operators,

$$H = m\left(\widehat{a}^{+}\widehat{a} + \frac{1}{2}\right), \quad \text{where } \left[\widehat{a}, \widehat{a}^{+}\right] = 1. \tag{4.1b}$$

To formulate the path integral in classical fields corresponding directly to $\widehat{a}, \widehat{a}^{+}$ operators, a natural way is to construct eigenstates for these operators, that is states satisfying $\widehat{a}|a\rangle = a|a\rangle$ and $\langle a|a* = \langle a|\widehat{a}^{+}$, and ask whether these eigenstates form a complete set, that is whether

$$1 = \sum_a |a\rangle\langle a| \quad \text{or} \quad 1 = \int \frac{dada^*}{2\pi}|a\rangle\langle a|,$$

where $a = x + iy$ is a complex number and $\displaystyle\int \frac{dada^*}{2\pi} = \iint \frac{dxdy}{\pi}$.

In the study of harmonic oscillators, the states satisfying $\widehat{a}|a\rangle = a|a\rangle$ are called coherent states. The state $\widehat{a}|k\rangle = k|k\rangle$ is the ground state of the Hamiltonian

$$H = m\left(\widehat{a}^{+}\widehat{a} + \frac{1}{2}\right) - m(k^*\widehat{a} + k\widehat{a}^{+}) = m\left(\widehat{A}^{+}\widehat{A} + \frac{1}{2} - |k|^2\right),$$

where $\widehat{A} = \widehat{a} - k$ and $\widehat{A}^{+} = \widehat{a}^{+} - k^*$; $k$ is a complex number. Note that $[\widehat{A}, \widehat{A}^{+}] = 1$ and the ground state is given by $\widehat{A}|0_k\rangle = 0|0_k\rangle$ or $\widehat{a}|0_k\rangle = k|0_k\rangle$, which confirms our assertion.

*Introduction to Classical and Quantum Field Theory.* Tai-Kai Ng

ISBN: 978-3-527-40726-2

The coherent states have rather particular mathematical properties. We can expand the coherent state in the Fock representation to obtain

$$|a\rangle = e^{-\frac{|a|^2}{2}} \sum_{n=0}^{\infty} \frac{a^n}{\sqrt{n!}} |n\rangle, \tag{4.2}$$

where $\widehat{a}^+ \widehat{a}|n\rangle = n|n\rangle$.

**Homework Q1a**
Show that the above state satisfies $\widehat{a}|a\rangle = a|a\rangle$.

It is easy to see that the states $|a\rangle, |a'\rangle$ with $a \neq a'$ are, in general, non-orthogonal to each other and the set of states $|a\rangle$ forms an over-complete set. However, it turns out that the completeness relation, $1 = \int \frac{dada^*}{2\pi} |a\rangle\langle a|$, stills hold despite the over-completeness.

**Homework Q1b**
Show that $\langle b|a\rangle = e^{b^* a - \frac{1}{2}(|b|^2 + |a|^2)}$ and $|n\rangle = \int \frac{dada^*}{2\pi} |a\rangle\langle a|n\rangle$ for any state $|n\rangle$, that is, completeness.

With this groundwork we can proceed to construct the path integral in the coherent state representation for the single-site model. Following previous analysis, we study $\langle a|U(t, t')|a'\rangle$ in the basis set $\{|a\rangle\}$.

Dividing into time steps as before, we obtain

$$\begin{aligned}\langle a|\hat{U}(t, t')|a'\rangle = \int \cdots \int & \frac{da_1 da_1^*}{2\pi} \frac{da_2 da_2^*}{2\pi} \cdots \frac{da_{N-1} da_{N-1}^*}{2\pi} \\ & \times \langle a|\hat{U}(t_N, t_{N-1})|a_{N-1}\rangle\langle a_{N-1}|\hat{U}(t_{N-1}, t_{N-2})|a_{N-2}\rangle \\ & \times \cdots \times \langle a_2|\hat{U}(t_2, t_1)|a_1\rangle\langle a_1|\hat{U}(t_1, t_0)|a'\rangle,\end{aligned} \tag{4.3a}$$

where $\langle a_{k+1}|\hat{U}(t_{k+1}, t_k)|a_k\rangle \sim \langle a_{k+1}|(1 - \frac{i}{\hbar}\hat{H}(\widehat{a}^+, \widehat{a}) \cdot \Delta t + O(\Delta t^2))|a_k\rangle$.

To evaluate the matrix element, we use the property of coherent states,

$$\langle a_{k+1}|\left(1 - \frac{i}{\hbar}\hat{H}(\widehat{a}^+, \widehat{a}) \cdot \Delta t\right)|a_k\rangle = \langle a_{k+1}|a_k\rangle\left(1 - \frac{i}{\hbar}H(a_{k+1}^*, a_k)\Delta t\right), \tag{4.3b}$$

where we have assumed that the Hamiltonian is written in a 'normal-ordered' form with all the $\widehat{a}^+$ operators written on the left of the $\widehat{a}$ operators and used $\widehat{a}|a\rangle = a|a\rangle$ and $\langle b|\widehat{a}^+ = \langle b|b^*$.

Using the result from Homework Q1b, we obtain

$$\begin{aligned}& \left(1 - \tfrac{i}{\hbar}H(a_k^*, a_k)\Delta t\right)\langle a_{k+1}|a_k\rangle \\ & \sim e^{-\frac{i}{\hbar}H(a_k^*, a_k)\Delta t} e^{a_{k+1}^* a_k - \frac{1}{2}|a_{k+1}|^2 - \frac{1}{2}|a_k|^2}\end{aligned} \tag{4.3c}$$

to leading order in $\Delta t$.

Next, we write

$$a^*_{k+1}a_k - \frac{1}{2}\left(|a_k|^2 + |a_{k+1}|^2\right) = -\frac{1}{2}a^*_{k+1}(a_{k+1}-a_k) + \frac{1}{2}(a^*_{k+1}-a^*_k)a_k$$
$$\sim -\left(\frac{1}{2}a^*_{k+1}\dot{a}_k - \dot{a}^*_k a_k\right)\Delta t. \tag{4.3d}$$

Therefore, we obtain

$$\langle a_{k+1}|\hat{U}(t_{k+1},t_k)|a_k\rangle \sim e^{-\left[\frac{1}{2}(a^*_k\dot{a}_k - \dot{a}^*_k a_k) + \frac{i}{\hbar}H(a^*_k,a_k)\right]\Delta t}.$$

Putting together all the results as before, we obtain

$$\langle a|\hat{U}(t,t')|a'\rangle \sim \int Da^+(t'')Da(t'')e^{\frac{i}{\hbar}\int_{t'}^{t} L(t'')\mathrm{d}t''}, \tag{4.4a}$$

$$\text{where } L = i\hbar a^+\dot{a} - H(a^+,a), \tag{4.4b}$$

with boundary condition $a(t') = a'$, $a(t) = a$. We have integrated by parts:

$$-\frac{1}{2}\int \mathrm{d}t''(a^*(t'')\dot{a}(t'') - \dot{a}^*(t'')a(t'')) = -\int \mathrm{d}t'' a^*(t'')\dot{a}(t''),$$

and assumed that $a = a'$ to obtain the last result.

### 4.1.1 Imaginary Time and Partition Function

Putting $t \to -i\tau$, we obtain for the partition function

$$Z = \mathrm{Tr}(\langle a|\hat{U}(-i\hbar\beta,0)|a\rangle) \sim \int Da^+(\tau'')Da(\tau'')e^{-\int_0^\beta L(\tau'')\mathrm{d}\tau''}, \tag{4.5a}$$

$$\text{where } L(\tau) = a^+\partial_\tau a + H(a^+,a). \tag{4.5b}$$

**Exercise**

Derive the expression for the partition function $Z$ following previous analysis.

#### 4.1.1.1 Quantum Field Theory for Bosons

The above approach can be generalized to multiple site by letting $\widehat{a}(\widehat{a}^+) \to \widehat{\psi}(\vec{x}_i)(\widehat{\psi}^+(\vec{x}_i))$, where $i$ runs over all points in space.

Introducing coherent states $\hat{\psi}(\vec{r})|\psi(\vec{r})\rangle = \psi(\vec{r})|\psi(\vec{r})\rangle$ at each spatial point $\vec{r}$, we have the completeness relation

$$\int \frac{\mathrm{d}\psi(\vec{r}_1)\mathrm{d}\psi^*(\vec{r}_1)}{2\pi}\frac{\mathrm{d}\psi(\vec{r}_2)\mathrm{d}\psi^*(\vec{r}_2)}{2\pi}\cdots|\psi(\vec{r}_2)\psi(\vec{r}_1)\cdots\rangle\langle\psi(\vec{r}_1)\psi(\vec{r}_2)\cdots|\cdots = 1, \tag{4.6a}$$

with

$$\langle\psi(\vec{r})|\phi(\vec{r})\rangle = e^{\psi(\vec{r})^*\phi(\vec{r})-\frac{1}{2}(|\psi|^2+|\phi|^2)}. \tag{4.6b}$$

Following procedures as before, we obtain the path integral formalism for many-boson systems described by the Hamiltonian

$$H = \int d^d r h[\bar{\psi}(\vec{r}), \psi(\vec{r})] :$$

$$\langle\psi|\hat{U}(t,t')|\psi'\rangle \sim \int D\bar{\psi}(\vec{r},t'')D\psi(\vec{r},t'')e^{\frac{i}{\hbar}\int_{t'}^{t} L(t'')dt''}, \tag{4.7a}$$

$$\text{where } L = \int d^d r[i\hbar\bar{\psi}^+(\vec{r})\partial_t\psi(\vec{r}) - h(\bar{\psi}(\vec{r}), \psi(\vec{r}))]. \tag{4.7b}$$

At imaginary time, we obtain

$$Z = \mathrm{Tr}(\langle\psi|\hat{U}(-i\hbar\beta, 0)|\psi\rangle) \sim \int D\bar{\psi}(\vec{r},\tau'')D\psi(\vec{r},\tau'')e^{-\int_0^\beta L(\tau'')d\tau''}, \tag{4.8a}$$

$$\text{where } L(\tau) = \int d^d r[\bar{\psi}^+(\vec{r})\partial_\tau\psi(\vec{r}) + h(\bar{\psi}(\vec{r}), \psi(\vec{r}))], \tag{4.8b}$$

with boundary conditions

$$\psi(\vec{r},0) = \psi(\vec{r},\beta) \quad \text{and} \quad \bar{\psi}(\vec{r},0) = \bar{\psi}(\vec{r},\beta). \tag{4.8c}$$

#### 4.1.1.2 Quantum Field Theory for Fermions

As in the case of bosons, we first introduce a coherent state representation for fermions in a lattice model. The coherent state is defined by

$$|\{\psi_i\}\rangle = \prod_i(|0_i\rangle + |1_i\rangle\psi_i) \quad \text{and} \quad \langle\{\bar{\psi}\}_i| = \prod_i(\langle 0_i| + \bar{\psi}_i\langle 1_i|), \tag{4.9a}$$

where $|0_i\rangle$ and $|1_i\rangle$ are the states with no fermion on site $i$ and one fermion on site $i$, respectively. $\psi_i, \bar{\psi}_i$ are Grassmann numbers. Using the Grassmann number rules we discussed in the last chapter, we find that

$$\widehat{\psi_i}|\psi_i\rangle = 0|0_i\rangle + |0_i\rangle\psi_i = |1_i\rangle\psi_i\psi_i + |0_i\rangle\psi_i = \psi_i|\psi_i\rangle, \quad \text{etc.}, \tag{4.9b}$$

illustrating that the constructed states are coherent states.

We can also evaluate the integral

$$\int(|0_i\rangle + |1_i\rangle\psi_i)(\langle 0_i| + \bar{\psi}_i\langle 1_i|)e^{-\bar{\psi}_i\psi_i}d\bar{\psi}_i d\psi_i$$
$$= \int(|0_i\rangle\langle 0_i|(1+\psi_i\bar{\psi}_i) + |0_i\rangle\langle 1_i|\bar{\psi} + |1_i\rangle\langle 0_i|\psi_i + |1_i\rangle\langle 1_i|\psi_i\bar{\psi}_i)d\bar{\psi}_i d\psi_i$$
$$= |0_i\rangle\langle 0_i| + |1_i\rangle\langle 1_i| = \widehat{I}.$$

Therefore, we have a completeness relation

$$\int|\psi\rangle\langle\bar{\psi}|e^{-\bar{\psi}\psi}d\bar{\psi}d\psi = \int|\{\psi_i\}\rangle\langle\{\bar{\psi}_i\}|e^{-\sum_i\bar{\psi}_i\psi_i}\prod_i d\bar{\psi}_i d\psi_i = \widehat{I} \tag{4.9c}$$

for the quantum theory for fermions.

Another important result is

$$\int \langle -\bar{\psi}|\,\widehat{A}\,|\psi\rangle e^{-\bar{\psi}\psi} \mathrm{d}\bar{\psi}\mathrm{d}\psi = \int \langle \{-\bar{\psi}_i\}|A|\{\psi_i\}\rangle e^{-\sum_i \bar{\psi}_i \psi_i} \prod_i \mathrm{d}\bar{\psi}_i \mathrm{d}\psi_i = \mathrm{Tr}\,\widehat{A}, \tag{4.10}$$

where $\widehat{A}$ is any linear operator.

**Homework Q2**

Derive Equation (4.10).

Using these results, and repeating the construction of path integrals as before, we obtain for the quantum Hamiltonian (normal ordered)

$$H = \int \mathrm{d}^d r h[\bar{\psi}(\vec{r}), \psi(\vec{r})]$$

$$\langle \psi|\,\hat{U}(t,t')|\psi'\rangle \sim \int D\bar{\psi}(\vec{r},t'')D\psi(\vec{r},t'')e^{\frac{i}{\hbar}\int_{t'}^{t} L(t'')\mathrm{d}t''}, \tag{4.11a}$$

$$\text{where } L = \int \mathrm{d}^d r[i\hbar\bar{\psi}^+(\vec{r})\partial_t\psi(\vec{r}) - h[\bar{\psi}(\vec{r}), \psi(\vec{r})]]. \tag{4.11b}$$

The approach can be generalized to imaginary time as for bosons with

$$Z = \mathrm{Tr}(\langle \psi|\,\hat{U}(-i\hbar\beta, 0)|\psi\rangle) \sim \int D\bar{\psi}(\vec{r},\tau'')D\psi(\vec{r},\tau'')e^{-\int_0^\beta L(\tau'')\mathrm{d}\tau''},$$

$$\text{where } L(\tau) = \int \mathrm{d}^d r[\bar{\psi}^+(\vec{r})\partial_\tau\psi(\vec{r}) + h(\bar{\psi}(\vec{r}), \psi(\vec{r}))]$$

with boundary condition

$$\psi(\vec{r},0) = \psi' \text{ and } \bar{\psi}(\vec{r},\beta) = -\psi \tag{4.11c}$$

(see Ref. [2]). The result is formally the same as bosons, except for the minus sign in the boundary condition and that the $\psi, \bar{\psi}$ are Grassmann numbers here.

Let us pause and summarize our findings here so far. We learnt that quantum field theories (QFTs) are equivalent to many-particle quantum mechanical systems – a natural outcome of particle–wave duality. A quantum field theory of classical variables is a theory of bosons, and a quantum field theory of Grassmann variables is a theory of fermions. As a result, we have freedom in choosing 'particle' or 'field' representations when we perform calculations – at least in the case of bosons. There are also two different formulations which we can use when performing calculations. We can use the traditional Hamiltonian formulation, and we can also use the path integral formulation. In the case of bosons, the path integral can be formulated in terms of field variables, or a coherent state representation of many-particle states. Path integral formulations for fermions are usually formulated in terms of a coherent state representation.

The different representations and formulations we outlined here all have their own strengths and weaknesses. There is no single superior approach. We shall illustrate these different approaches using two examples in the following; the first example is phonons. The second example is Dirac fermions. We shall consider for simplicity these examples in 1D. The concept of an effective field theory will be introduced when we discuss these examples.

## 4.2
## Two Simple Examples of QFT

### 4.2.1
### Phonons

We start with a model of identical atoms located on a regular 1D array, representing atoms in a 1D solid [3]. The atoms mutually interact and the resulting vibration modes are collective motions involving many atoms. A simplified model for this system is a model of a one-dimensional harmonic chain

$$L = \sum_i \frac{1}{2} m\dot{x}_i^2 - \frac{K}{2} \sum_i (x_i - x_{i+1})^2, \tag{4.12}$$

where $x_i$ and $\dot{x}_i$ are the displacement away from the $i$th atom's equilibrium position and the velocity, respectively. We assume for simplicity that the atom can move only in the longitudinal direction. The classical solution is obtained by solving the equation of motion

$$-m\ddot{x}_l = m\omega^2 x_l = K(2x_l - x_{l-1} - x_{l+1}) \tag{4.13a}$$

with solutions of the form $x_l(t) = x_0 \cos(kal - \omega_k t)$, where the normal mode frequencies are given by

$$\omega_k^2 = \frac{K}{m} 2(1 - \cos(ka)) = \frac{4K}{m} \sin^2\left(\frac{ka}{2}\right). \tag{4.13b}$$

We may begin the quantum mechanical solution by looking for a transformation which 'diagonalizes' the Lagrangian, that is a transformation $x_k = \sum_j u_{kj} x_j$ such that the Lagrangian becomes of the form

$$L = \sum_k \frac{1}{2} \dot{x}_k^2 - \frac{1}{2} \sum_k \omega_k^2 x_k{}^2. \tag{4.14}$$

A Lagrangian of this form describes decoupled harmonic oscillators where the generalization to quantum mechanics is easy. Translation invariance (on the lattice) suggests that we look for a Fourier transformation

$$x_k = \frac{1}{N^{1/2}} \sum_l e^{-ikal} x_l \quad \left(\text{and } x_l = \frac{1}{N^{1/2}} \sum_k e^{ikal} x_k\right). \tag{4.15a}$$

It is easy to show that

$$\sum_l x_l x_{l+m} = \frac{1}{N} \sum_{k,k'} x_k x_{k'} \sum_l e^{ila(k+k')} e^{imak'} = \sum_k x_k x_{-k} e^{iamk}, \tag{4.15b}$$

and the potential energy term is

$$\frac{K}{2}\sum_i (x_i - x_{i+1})^2 = \frac{K}{2}\sum_k x_k x_{-k}(2 - e^{ika} - e^{-ika}) = \frac{m}{2}\sum_k \omega_k^2 x_k x_{-k}. \quad (4.15c)$$

Therefore, $L = \sum_k \frac{1}{2} m\dot{x}_k^2 - \omega_k^2 x_k{}^2$, where the $\omega_k$ are the normal frequencies we solved earlier. To quantize the system, we write

$$H = \sum_k \frac{1}{2m} p_k^2 + \frac{1}{2}\sum_k \omega_k^2 x_k{}^2, \quad \text{where } [x_k, p_{k'}] = i\hbar\delta_{k,k'}. \quad (4.16)$$

Note that the commutation relation can also be obtained directly from the identification

$$p_k = \frac{1}{N^{1/2}}\sum_l e^{ikal} p_l, \quad (4.17a)$$

with the fundamental commutation rule $[x_i, p_j] = i\hbar\delta_{ij}$. In this case,

$$[x_k, p_{k'}] = \frac{1}{N}\sum_{l,m} e^{-ikal} e^{ik'am}[x_l, p_m] = \frac{i}{N}\sum_l e^{ial(k'-k)} = i\hbar\delta_{k,k'}. \quad (4.17b)$$

We next define the boson creation and destruction operators as

$$a_k = \left(\frac{m\omega_k}{2\hbar}\right)^{1/2}\left(x_k + \frac{i}{m\omega_k} p_{-k}\right), \quad a_k^+ = \left(\frac{m\omega_k}{2\hbar}\right)^{1/2}\left(x_{-k} - \frac{i}{m\omega_k} p_k\right), \quad (4.18)$$

where $[a_k, a_{k'}^+] = \delta_{k,k'}$, $[a_k^+, a_{k'}^+] = 0$, $[a_k, a_{k'}] = 0$, and the Hamiltonian in terms of $a$, $a^+$ is

$$H = \sum_k \omega_k \left(a_k^+ a_k + \frac{1}{2}\right). \quad (4.19)$$

These 'bosons' are called phonons. They are quantized vibration modes (sound waves) in the solid. Note that the model is a particular case of the lattice field theory we have studied ($L[\phi] = \sum_i \frac{1}{2} m\dot{\phi}_i^2 - V[\phi]$) if we identify $x_i \to \phi_i$ with $V[\phi] = \frac{K}{2}\sum_i \times (\phi_i - \phi_{i+1})^2$. The nature of the 'bosons' which comes out from quantization can be identified once the nature of the $\phi$ field is known.

#### 4.2.1.1 Continuum Limit

Let us go to the small-wave-vector limit and ask what the form of the corresponding continuum field theory is. An easy way to see the result is to note that $x_l - x_{l-1} \sim \frac{dx}{d(la)} a$ for slowly varying $x_l = \frac{1}{N^{1/2}}\sum_{|k|<\Lambda} e^{ikal} x_k$.

Therefore, approximating the sum by an integral, $\sum_i() \to \frac{1}{a}\int dx(\,)$, we obtain

$$L \sim \frac{1}{2a}\int dx\left(m\dot{\phi}(x)^2 - Ka^2(\nabla\phi(x))^2\right), \quad (4.20)$$

which is similar to the Klein–Gordon model with zero mass and velocity $c^2 = Ka^2/m$. The main difference of the two models is that $\phi$ is real here, but is a complex scalar field in the Klein–Gordon model.

The Klein–Gordon model describes relativistic bosons with energy spectrum $E = \sqrt{m^2c^4 + c^2p^2}$, whereas $\omega_k^2 = \frac{4K}{m}\sin^2(\frac{ka}{2}) \sim \frac{Ka^2}{m}k^2$ at small $k$ for phonons.

Is this a coincidence? Is there any reason why a relativistic theory for bosons should come out from the low-energy limit of a lattice model of atoms? You will see more similar examples later in this book.

### 4.2.2
### Dirac Fermions in 1D

Historically, physicists have tried to generalize the non-relativistic Schrödinger equation $\hat{H}|\psi\rangle = i\hbar\frac{\partial}{\partial t}|\psi\rangle$ to obtain an equation consistent with special relativity.

$H = \frac{p^2}{2m}$ for non-relativistic free particles, whereas, in relativity, we have instead $H = \sqrt{p^2c^2 + m^2c^4}$.

Therefore, a natural way of generalizing the Schrödinger equation is to replace

$$\hat{H}|\psi\rangle = E|\psi\rangle \text{ by } \hat{H}^2|\psi\rangle = -\hbar^2\frac{\partial^2}{\partial t^2}|\psi\rangle$$

$$\text{or } c^2\nabla^2\psi(x) + m^2c^4\psi(x) = -\hbar^2\frac{\partial^2}{\partial t^2}\psi(x). \tag{4.21}$$

The resulting equation is called the Klein–Gordon equation. However, the Klein–Gordon equation has an unsatisfactory feature that the magnitude of the wavefunction $|\psi|^2$ does not satisfy a continuity equation of the form

$$\frac{\partial}{\partial t}|\psi|^2 + \nabla\cdot\vec{j} = 0.$$

Indeed, one can use Noether's theorem (Chapter 2) to show that it satisfies a continuity equation with $\rho = \psi^*\dot{\psi} - \dot{\psi}^*\psi$. Note that $\rho$ defined in this way is *not* positive definite and could not be interpreted as a probability density.

We now understand that $\psi(x)$ should not be considered as a single-particle wavefunction, but should be considered as a classical field which is to be quantized to become a many-body boson system. $\rho$ measures the total 'charge' carried by the bosons, which can be either positive or negative.

Dirac approached the problem in a different way. He looked for a relativistic generalization of the Schrödinger equation with three criteria in mind.

**Criteria**

(1) The equation should be a first-order differential equation of the form

$$\hat{H}|\psi\rangle = i\hbar\frac{\partial}{\partial t}|\psi\rangle.$$

(2) The equation must be 'symmetric' with respect to space and time derivatives to respect relativity.

(3) $\hat{H}^2 = \hat{p}^2c^2 + m^2c^4$ for free particles.

Dirac started with an equation of the form

$$i\hbar\frac{\partial}{\partial t}|\psi\rangle = (\vec{\alpha}\cdot\vec{p}+\beta m)|\psi\rangle \tag{4.22}$$

$$\Rightarrow -\hbar^2\frac{\partial^2}{\partial t^2}|\psi\rangle = (\vec{\alpha}\cdot\vec{p}+\beta m)(\vec{\alpha}\cdot\vec{p}+\beta m)|\psi\rangle = (p^2c^2+m^2c^4)|\psi\rangle.$$

Therefore, we must have

$$(\vec{\alpha}\cdot\vec{p})(\vec{\alpha}\cdot\vec{p})+m^2\beta^2+(\vec{\alpha}\cdot\vec{p})m\beta+m\beta(\vec{\alpha}\cdot\vec{p}) = p^2c^2+m^2c^4,$$

which gives

$$\alpha_i\beta+\beta\alpha_i = 0, \quad i=1,2,3 \tag{4.23a}$$

and

$$\{\alpha_i,\alpha_j\} = 2\delta_{i,j}c^2, \quad \beta^2 = c^4, \tag{4.23b}$$

that is $\alpha$ and $\beta$ are operators (matrices) themselves.

In 3D, the minimum size of the matrices $\alpha$ and $\beta$ that satisfy the above algebra is $4\times 4$.

(From now on, we set $\hbar = c = 1$.)

The standard set of matrices physicists use are (Dirac matrices)

$$\beta = \begin{pmatrix} I & 0 \\ 0 & -I \end{pmatrix} = \begin{pmatrix} 1 & 0 & 0 & 0 \\ 0 & 1 & 0 & 0 \\ 0 & 0 & -1 & 0 \\ 0 & 0 & 0 & -1 \end{pmatrix}, \quad \vec{\alpha} = \begin{pmatrix} 0\vec{\sigma} \\ \vec{\sigma}0 \end{pmatrix}, \tag{4.24a}$$

where $\vec{\sigma}$ are Pauli matrices.

For simplicity, we shall consider the Dirac equation in 1D here. (This is a very important topic in condensed matter physics because many 1D quantum field theories are exactly solvable and realizable experimentally.)

In this case, we need only two matrices:

$$\alpha_x = \begin{pmatrix} 0 & 1 \\ 1 & 0 \end{pmatrix}, \quad \beta = \begin{pmatrix} 1 & 0 \\ 0 & -1 \end{pmatrix}, \tag{4.24b}$$

that is the minimum size of the matrices we need is $2\times 2$.

There is a deep reason why the minimum dimension of the Dirac matrices differs in 3D and 1D. The reason is that, in 3D, there is a spin-statistics theorem which tells us that fermions must have half-integer spins and the minimum spin for fermions in 3D is $^1/_2$. The wavefunction has four components corresponding to $\uparrow$ and $\downarrow$ states of spin-$^1/_2$ particles and anti-particles. The spin-statistics theorem is a consequence of rotational symmetry in space and no such theorem exists in 1D, where we cannot even define rotation. Therefore, we can have spinless fermions in 1D and the two-component wavefunction corresponds to particles and anti-particles. We shall learn more about the spin-statistics theorem in Chapter 6.

Therefore, in 1D, the Dirac equation for free particles becomes

$$i\frac{\partial}{\partial t}\begin{pmatrix}\psi_+\\ \psi_-\end{pmatrix}=\begin{pmatrix}m & \hat{p}\\ \hat{p} & -m\end{pmatrix}\begin{pmatrix}\psi_+\\ \psi_-\end{pmatrix},\quad \text{where } \hat{p}=-i\frac{\partial}{\partial x}. \tag{4.25}$$

To solve the equation, we try solutions of the form

$$\psi_+ = \psi_{0+}e^{ikx}e^{iEt},$$

$$\psi_- = \psi_{0-}e^{ikx}e^{iEt}. \tag{4.26a}$$

Putting it into the equation, we obtain

$$i\frac{\partial}{\partial t}\begin{pmatrix}\psi_{0+}\\ \psi_{0-}\end{pmatrix}=\begin{pmatrix}m & k\\ k & -m\end{pmatrix}\begin{pmatrix}\psi_{0+}\\ \psi_{0-}\end{pmatrix}=E\begin{pmatrix}\psi_{0+}\\ \psi_{0-}\end{pmatrix} \tag{4.26b}$$

$$\Rightarrow \det\begin{vmatrix}E-m & k\\ k & E+m\end{vmatrix}=0$$

$$\text{or } E^2 = m^2+k^2. \tag{4.26c}$$

Note that $E=\pm\sqrt{m^2+k^2}$ represents two independent solutions of the Dirac equation (see figure below).

For example, when $m=0$,

$$E>0 \text{ solution (particle)} \begin{pmatrix}\psi_+\\ \psi_-\end{pmatrix}=\frac{1}{\sqrt{2}}\begin{pmatrix}1\\ 1\end{pmatrix}=\chi_+, \tag{4.27a}$$

$$E<0 \text{ solution (anti-particle)} \begin{pmatrix}\psi_+\\ \psi_-\end{pmatrix}=\frac{1}{\sqrt{2}}\begin{pmatrix}1\\ -1\end{pmatrix}=\chi_-. \tag{4.27b}$$

Energy spectrum:

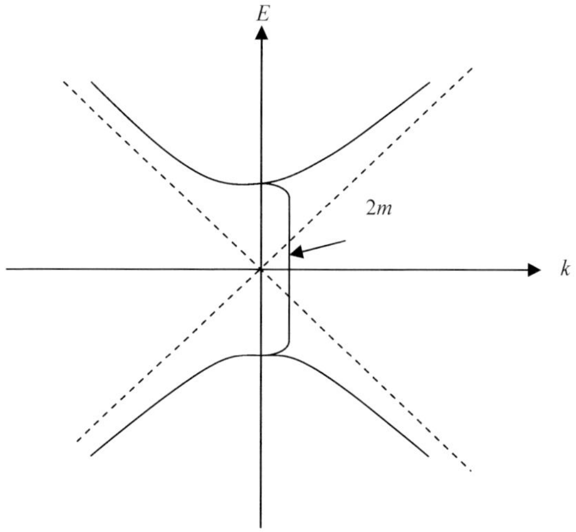

The appearance of negative energy solutions in the Dirac equation posed a problem to physicists at the time. Note that matters always carry positive energy in physics (recall that $E = mc^2$) and the meaning of a negative energy solution was not clear. In classical physics it can be assumed that although negative energy solutions to the Einstein equation $E^2 = p^2c^2 + m^2c^4$ do exist, they never appear in our real, physical world. This argument is not applicable in quantum mechanics because a negative energy solution is required to construct a complete set of the solutions and would unavoidably enter the 'real' world (see Homework Q3b).

Dirac proposed that to solve the negative energy problem one must consider the equation as describing a many-particle system *of fermions*. The ground state (=vacuum) is formed by filling all negative energy states, while the positive energy state remains empty, very much like electrons filling up valence states in a semiconductor.

A continuity equation can be derived from the Dirac equation by combining the Dirac equation with its conjugate equation

$$-i\frac{\partial}{\partial t}(\psi_+^+,\psi_-^+) = (\psi_+^+,\psi_-^+)\begin{pmatrix} m & -\hat{p} \\ -\hat{p} & -m \end{pmatrix};$$

we obtain

$$\begin{aligned} i\frac{\partial}{\partial t}(\bar{\psi}\psi) &= i\frac{\partial}{\partial t}\left[(\psi_+^+,\psi_-^+)\begin{pmatrix} \psi_+ \\ \psi_- \end{pmatrix}\right] \\ &= -(\psi_+^+,\psi_-^+)\begin{pmatrix} m & -\overleftarrow{\hat{p}} \\ -\overleftarrow{\hat{p}} & -m \end{pmatrix}\begin{pmatrix} \psi_+ \\ \psi_- \end{pmatrix} + (\psi_+^+,\psi_-^+)\begin{pmatrix} m & \overrightarrow{\hat{p}} \\ \overrightarrow{\hat{p}} & -m \end{pmatrix}\begin{pmatrix} \psi_+ \\ \psi_- \end{pmatrix} \\ &= -i\left(\frac{\partial}{\partial x}\bar{\psi}\right)\alpha_x\psi - i\bar{\psi}\alpha_x\left(\frac{\partial}{\partial x}\psi\right) \\ &= -i\frac{\partial}{\partial x}(\bar{\psi}\alpha_x\psi). \end{aligned} \tag{4.28}$$

Thus, a continuity equation is obtained if we identify $\rho = \bar{\psi}\psi$ and $j = \bar{\psi}\alpha_x\psi$.

Note that $\rho = \bar{\psi}\psi$ is now positive definite, that is it can be interpreted as a probability density. We shall see later that this result is modified in the many-body interpretation of the Dirac equation.

#### 4.2.2.1 **Covariant form of Dirac equation**

We start with the original Dirac equation (1D):

$$i\frac{\partial}{\partial t}\psi = -i\alpha_x\frac{\partial}{\partial x}\psi + m\beta\psi.$$

$$\text{Let } \beta^{-1} = \beta = \gamma_0, \quad \beta\alpha_x = \begin{pmatrix} 0 & 1 \\ -1 & 0 \end{pmatrix} = \gamma_1;$$

we obtain

$$i\gamma_0\frac{\partial}{\partial t}\psi + i\gamma_1\frac{\partial}{\partial x}\psi = m\psi \tag{4.29a}$$

$$\text{or} \quad i\gamma^\mu\partial_\mu\psi = m\psi \quad (x_0 = t, x_1 = x). \tag{4.29b}$$

This is called the covariant form of the Dirac equation. A similar form can also be derived in 3D.

### 4.2.3
### Quantization of Dirac Equation

To 'quantize' the Dirac equation we regard the $\bar{\psi}, \psi$ fields in the Dirac equation as actually 'classical' Grassmann number fields and quantization means that the $\bar{\psi}, \psi$ fields should be treated as operators satisfying anti-communication relations

$$\{\psi(\vec{x}), \psi(\vec{x}')\} = 0, \quad \{\psi(\vec{x}), \bar{\psi}(\vec{x}')\} = \delta(\vec{x}-\vec{x}'), \quad \text{etc.} \tag{4.30}$$

in the quantum theory. We should be careful with the commutation relations because $\psi, \bar{\psi}$ are two-component vectors. It is more convenient to work in a basis where the Dirac equation is 'diagonalized'.

As in usual classical number fields, we introduce a basis $\{\chi_+, \chi_-\}$ where the Dirac equation is diagonalized, that is

$$H_{\mathrm{Dirac}}\chi_\pm(k;x) = \pm E_k\chi_\pm(k;x), \quad \chi_\pm(k;x) = e^{ikx}\begin{pmatrix} \psi_{0+}^{\pm} \\ \psi_{0-}^{\pm} \end{pmatrix}. \tag{4.31a}$$

Furthermore, we recall that the completeness relation can also be applied to Grassmann number functions and we can write

$$\psi(x) = \int \frac{\mathrm{d}k}{2\pi}[a(k)x_+(k;x) + b'(k)x_-(k;x)], \tag{4.31b}$$

where $a(k)$, $b'(k)$ are Grassmann numbers representing the 'amplitude' of the eigen-wavefunctions. Note that the Hamiltonian can be written in terms of $a(k)$, $b'(k)$ as

$$H_{\mathrm{Dirac}} = \int \mathrm{d}kE_k[a^+(k)a(k) - b'^+(k)b'(k)], \tag{4.31c}$$

where $E_k = +\sqrt{k^2+m^2}$.

The classical theory can be 'quantized' by imposing the anti-commutation rules

$$\{\hat{a}(k), \hat{a}^+(k')\} = \delta(k-k'), \quad \{\hat{b}'(k), \hat{b}'^+(k')\} = \delta(k-k'), \tag{4.32a}$$

$$\{\hat{a}(k), \hat{b}'^+(k')\} = \{\hat{a}(k), \hat{a}(k')\} = 0, \quad \text{etc.} \tag{4.32b}$$

The ground state of the many-body Dirac Hamiltonian $H_{\mathrm{Dirac}}$ is formed by filling up all the negative energy states.

Note that energy can always be lowered by putting more particles in the negative energy state if the particles are bosons. In this case the ground state is unstable since one can put an arbitrary number of bosons in any state. A well-defined ground state can be obtained for fermions because of the Pauli exclusion principle.

> Note: this is a general stability criterion. For bosons, the energy spectrum must be positive definite; for fermions, the energy spectrum can be both positive and negative.

Therefore, the ground state is specified by $\langle \hat{b}_k'^{+} \hat{b}'_k \rangle = 1$, $\langle \hat{a}_k^{+} \hat{a}_k \rangle = 0$ for all $k$.

To describe excited states it is convenient to introduce the 'hole' or anti-particle operator for the $b$. We introduce

$$\hat{b}_k^{+} = \hat{b}'_{-k}, \quad \hat{b}_k = \hat{b}'^{+}_{-k}, \tag{4.33a}$$

that is, destroying a fermion particle = creating a hole (anti-particle).

In terms of $\hat{a}_k$, $\hat{b}_k$ operators, the Dirac Hamiltonian is

$$\hat{H}_{\text{Dirac}} = \int \mathrm{d}k E_k[\hat{a}^{+}(k)a(k) - \hat{b}'(k)\hat{b}'^{+}(k)] = \int \mathrm{d}k E_k[\hat{a}^{+}(k)a(k) + \hat{b}^{+}(k)\hat{b}(k)] + E_{\mathrm{G}}, \tag{4.33b}$$

where $E_{\mathrm{G}}$ ($\to -\infty$) is the ground-state energy.

Now let us examine the meaning of the charge $Q = \bar{\psi}\psi$. Note that

$$\begin{aligned} Q &= \int \mathrm{d}x \bar{\psi}\psi \\ &= \int \mathrm{d}k[\hat{a}^{+}(k)a(k) + \hat{b}'^{+}(k)b'(k)] \\ &= \int \mathrm{d}k[\hat{a}^{+}(k)a(k) + \hat{b}(k)\hat{b}^{+}(k)] \\ &= \int \mathrm{d}k[\hat{a}^{+}(k)a(k) - \hat{b}^{+}(k)\hat{b}(k) + 1]. \end{aligned} \tag{4.34a}$$

Therefore, we find that

(1) The ground state has infinite charge $[\hat{a}^{+}a = \hat{b}^{+}b = 0,\ Q_{\mathrm{G}} = \int \mathrm{d}k]$.
(2) Holes carry opposite 'charge' to particles.
(3) The charge of 'excitation' is

$$Q_{\text{ex}} = Q - Q_{\mathrm{G}} = \int \mathrm{d}k[\hat{a}^{+}(k)a(k) - \hat{b}^{+}(k)\hat{b}(k)], \tag{4.34b}$$

which is *not* positive definite. In modern applications of the Dirac equation we assume that $Q_{\mathrm{G}}$ is not measurable and the physical observable is $Q_{\text{ex}}$.

**Homework Q3a**

Show that charge is conserved in the Dirac equation but not the number of fermions using the continuity equation (or Noether theorem).

#### 4.2.3.1 Lattice Dirac Fermions

**Question**

Can we find a lattice model where the low-energy effective theory has the same form as the Dirac Hamiltonian?

**Answer**

Yes.

The Dirac equation and the Klein–Gordon equation were initially constructed from symmetry requirements of special relativity. They are allowed by special relativity but they are not the only equations that are allowed. We may ask for example why our universe admits electrons which are described by the Dirac equation with only four possible components but not with a more elaborate structure. To answer these questions we need to know beyond special relativity. We need to know more about the microscopic structure of the medium (or vacuum) which forms the background of our universe. This is the reason why physicists are interested in studying microscopic models whose low-energy effective theory has the same form as the Dirac equation, the Klein–Gordon equation, or the Maxwell equations. We want to understand the plausible microscopic structures (beyond special relativity) that can create the above equations. We shall raise questions along this line again in later chapters.

The simplest model with a low-energy effective theory of the same form as the Dirac Hamiltonian in 1D is perhaps *a lattice model of 1D electrons on a periodic lattice.* The lattice Hamiltonian is

$$H = -t\sum_i (c_{i+1}^+ c_i + h.c.) + u\sum_i (-1)^i c_i^+ c_i, \tag{4.35}$$

where the $i$ are site indices; the $c_i(c_i^+)$ are fermion annihilation (creation) operators on lattice site $i$. To diagonalize the Hamiltonian, we first introduce sublattices A and B where lattice sites $i = 2n + 1$ belong to the A sublattice and lattice sites $i = 2n$ belong to the B sublattice. We then introduce the sublattice Fourier transforms

$$c_k^{\mathrm{A}} = \sum_{i=2n+1} e^{ikx} c_i \quad \text{and} \quad c_k^{\mathrm{B}} = \sum_{i=2n} e^{ikx} c_i, \quad \text{etc.}$$

With the sublattice fields it is easy to show that the Hamiltonian can be written as

$$H = \sum_k \begin{pmatrix} c_k^{\mathrm{A}+} & c_k^{\mathrm{B}+} \end{pmatrix} \begin{pmatrix} u & -t\cos k \\ -t\cos k & -u \end{pmatrix} \begin{pmatrix} c_k^{\mathrm{A}} \\ c_k^{\mathrm{B}} \end{pmatrix}, \tag{4.36a}$$

where the eigenenergies are $E_k = \pm\sqrt{t^2\cos^2 k + u^2}$.

The energy spectrum of this Hamiltonian has smallest gap at $k \sim \pi/2$, where cos $k = 0$. For a momentum around this point $k = \pi/2 + p$, where $p$ is small, the Hamiltonian is approximately

$$H \sim \sum_p \begin{pmatrix} c_p^{\mathrm{A}+} & c_p^{\mathrm{B}+} \end{pmatrix} \begin{pmatrix} u & \sin p \\ \sin p & -u \end{pmatrix} \begin{pmatrix} c_k^{\mathrm{A}} \\ c_k^{\mathrm{B}} \end{pmatrix} \sim \sum_p \bar{\psi}_p(\beta u + t\alpha_x p)\psi_p \tag{4.36b}$$

and has the same form as Dirac's Hamiltonian.

It turns out that it is much more difficult to perform an analogous construction in three dimensions because of the appearance of spins. You can look at Ref. [4] for more details.

Summarizing, we have shown in this section two examples of field theories which can be 'quantized' quite straightforwardly. An example of a quantum field theory where a more careful quantization procedure is needed is the electromagnetic field.

This is discussed in Chapter 6. The subtle mathematical structure of gauge field theories will be discussed in more detail in Chapter 12.

**Homework Q3b**

Solve the scattering problem in the (1D) Dirac model with $H = (\alpha_x p + \beta m) + V_0\theta(x + a/2)\theta(x - a/2)$. Show that, for an incoming state with positive energy, the scattered state involves negative energy states. What happens in the non-relativistic limit $p \to 0$? Repeat the calculation with the lattice Hamiltonian

$$H = -t\sum_i (c_{i+1}^+ c_i + h.c.) + u\sum_i (-1)^i c_i^+ c_i + V_0\delta(x_i).$$

Do you detect any qualitative difference between the two solutions?

## 4.3 Simple Applications of Path Integral Formulation

In the above examples, we apply either the canonical quantization scheme (phonons) or work directly with a particle representation with annihilation and creation operators (Dirac fermions). In both cases we work with the Hamiltonian formulation. To illustrate how to use the path integrals (or Lagrangian formulation), we shall evaluate the free energy of the above two models in the path integral formulation in the 'particle', or coherent state, representation. The techniques employed in the calculation will be used also in later chapters. For comparison, we first consider the phonon and fermion problems in the *Hamiltonian* formulation.

For phonons the partition function is given by

$$Z = \mathrm{Tr}(e^{-\beta H}) = \prod_k e^{-\beta\frac{\omega_k}{2}} \sum_n e^{-\beta\omega_k n} = \prod_k e^{-\beta\frac{\omega_k}{2}} \left(\frac{1}{1 - e^{-\beta\omega_k}}\right)$$

$$= \exp\left(-\sum_k \left(\frac{\beta\omega_k}{2} + \ln(1 - e^{-\beta\omega_k})\right)\right).$$

Therefore, the free energy per unit volume is given by

$$F = \frac{1}{V}\sum_k \left(\frac{\omega_k}{2} + \frac{1}{\beta}\ln(1 - e^{-\beta\omega_k})\right). \tag{4.37a}$$

**Homework Q4**

Using a similar approach, show that the free energy per unit volume for Dirac fermions is

$$F = -\frac{2}{V}\sum_k \left(\frac{E_k}{2} + \frac{1}{\beta}\ln(1 + e^{-\beta E_k})\right). \tag{4.37b}$$

We now evaluate the free energies again using the path integral formalism at imaginary time in the coherent state representation. First, we consider phonons.

The action for phonons at imaginary time is

$$S_b = \int_0^\beta d\tau \sum_k \bar{\psi}(k,\tau)\frac{\partial}{\partial\tau}\psi(k,\tau) + \omega_k\bar{\psi}(k,\tau)\psi(k,\tau) + \sum_k \frac{\beta\omega_k}{2}, \tag{4.38a}$$

where $Z(\beta) = \int D\bar{\psi}D\psi e^{-S_0}$; $\bar{\psi}$ and $\psi$ are complex numbers.

The action for Dirac fermions is

$$S_f = \int_0^\beta d\tau \sum_{k,\alpha=a,b} \bar{\psi}_\alpha(k,\tau)\frac{\partial}{\partial\tau}\psi_\alpha(k,\tau) + E_k\bar{\psi}_\alpha(k,\tau)\psi_\alpha(k,\tau) - \beta\sum_k E_k, \tag{4.38b}$$

where we work with 'diagonalized' representations for simplicity.

To evaluate the path integral we introduce Fourier-transformed fields

$$\psi(k,i\omega) = \int_0^\beta d\tau e^{i\omega\cdot\tau}\psi(k,\tau),\ \bar{\psi}(k,i\omega) = \int_0^\beta d\tau e^{-i\omega\cdot\tau}\bar{\psi}(k,\vec{r}). \tag{4.39a}$$

The requirement that the initial state and the final state are the same in the path integral leads to the following boundary conditions (see previous section):

*for bosons:*

$$\psi(\vec{r},0) = \psi(\vec{r},\beta);\ \text{therefore, } \omega_n = \frac{2n\pi}{\beta}, \tag{4.39b}$$

*for fermions:*

$$\psi(\vec{r},0) = -\psi(\vec{r},\beta);\ \text{therefore, } \omega_n = \frac{(2n+1)\pi}{\beta}. \tag{4.39c}$$

In terms of the Fourier-transformed field, the actions become

$$S_b = \sum_{k,i\omega_n}[-i\omega_n + \omega_k]\bar{\psi}(k,i\omega_n)\psi(k,i\omega_n) + \sum_k \frac{\beta\omega_k}{2} \tag{4.40a}$$

and

$$S_f = \sum_{k,\alpha,i\omega_n}[-i\omega_n + E_k]\bar{\psi}_\alpha(k,i\omega_n)\psi_\alpha(k,i\omega_n) - \beta\sum_k E_k. \tag{4.40b}$$

Correspondingly,

$$Z_b(\beta) \to \int D\psi(k,i\omega_n)D\bar{\psi}(k,i\omega_n)e^{-S_b} \tag{4.41a}$$

and

$$Z_f(\beta) \to \prod_\alpha \int D\psi_\alpha(k,i\omega_n)D\bar{\psi}_\alpha(k,i\omega_n)e^{-S_f}. \tag{4.41b}$$

Note that $Z_b(\beta)$ can be written as

$$Z_b(\beta) = \prod_{k,i\omega_n} z(k, i\omega_n) e^{-\frac{\beta\omega_k}{2}}, \tag{4.42a}$$

$$\text{where } z(k, i\omega_n) = \int d\psi(k, i\omega_n) d\bar{\psi}(k, i\omega_n) e^{-[-i\omega_n + \omega_k]\bar{\psi}(k,i\omega_n)\psi(k,i\omega_n)}, \tag{4.42b}$$

which is an integral of the form

$$I = \int dz dz^* e^{-\alpha z^* z} = \frac{2\pi}{\alpha}.$$

**Homework Q5a**

Show that $I = 2\pi/\alpha$.

Therefore, $Z_b(\beta) = \Pi_{k,i\omega_n} \left(\frac{-\pi}{i\omega - \omega_k}\right) e^{-\frac{\beta\omega_k}{2}} = Ce^{-\sum_{k,i\omega_n}\left(\ln(i\omega_n - \omega_k) + \frac{\beta\omega_k}{2}\right)}$ and

$$F(\beta) = \frac{1}{\beta V} \sum_{k,i\omega_n} \ln(i\omega_n - \omega_k) + \frac{1}{V} \sum_k \frac{\omega_k}{2}. \tag{4.43}$$

Similarly, $Z_f(\beta) = \Pi_{k,\alpha,i\omega_n} \tilde{z}_\alpha(k, i\omega_n) e^{\frac{\beta E_k}{2}}$, where

$$\tilde{z}_\alpha(k, i\omega_n) = \int d\psi_\alpha(k, i\omega_n) d\bar{\psi}_\alpha(k, i\omega_n) e^{-[-i\omega_n + E_k]\bar{\psi}_\varepsilon(k,i\omega_n)\psi_\varepsilon(k,i\omega_n)}.$$

The integral can be evaluated using the Grassmann variable integration rules $\int x dx = 1, \int dx = 0$.

We obtain

$$\begin{aligned}
\tilde{z}_\alpha(k, i\omega_n) &= \int d\psi_\alpha(k, i\omega_n) d\bar{\psi}_\alpha(k, i\omega_n)(1 - [-i\omega_n + E_k]\bar{\psi}_\varepsilon(k, i\omega_n)\psi_\varepsilon(k, i\omega_n) \\
&= -\int d\psi_\alpha(k, i\omega_n) d\bar{\psi}_\alpha(k, i\omega_n)[-i\omega_n + E_k]\bar{\psi}_\varepsilon(k, i\omega_n)\psi_\varepsilon(k, i\omega_n) \\
&= i\omega_n - E_k.
\end{aligned} \tag{4.44}$$

Therefore, $Z_f(\beta) = \Pi_{k,i\omega_n} (i\omega_n - E_k)^2 e^{\beta E_k} = Ce^{2\sum_{k,i\omega_n}\left(\ln(i\omega_n - \omega_k) + \frac{\beta\omega_k}{2}\right)}$

$$\text{and} \quad F(\beta) = \frac{-2}{\beta V} \sum_{k,i\omega_n} \ln(i\omega_n - E_k) - \frac{1}{V} \sum_k E_k. \tag{4.45}$$

Lastly, we have to evaluate the sum over frequencies. There are well-established rules for this purpose (see for example Ref. [3] for derivation of the rules).

To evaluate a boson series ($\omega_n = \frac{2n\pi}{\beta}$) such as

$$S = -\frac{1}{\beta} \sum_{i\omega_n} f(i\omega_n), \tag{4.46a}$$

we just find all the simple poles of $f(z)$ and at these poles $z_j$ find the residues $r_j$ of $f(z \to z_j)$, and

$$S = \sum_j r_j n_B(z_j), \quad \text{where } n_B(\varepsilon) = \frac{1}{e^{\beta\varepsilon} - 1}. \tag{4.46b}$$

For the fermion series $\left(\omega_n = \frac{(2n+1)\pi}{\beta}\right)$,

$$S = -\frac{1}{\beta}\sum_{i\omega_n} f(i\omega_n),$$

the result is similar but with two important differences. We obtain

$$S = -\sum_j r_j n_F(z_j), \quad \text{where } n_F(\varepsilon) = \frac{1}{e^{\beta\varepsilon} + 1}. \tag{4.46c}$$

Note the sign difference and the appearance of fermion instead of boson occupation number.

**Homework Q5b**

$$\text{Show that } \frac{1}{\beta}\sum_{i\omega_n} \ln(i\omega_n - \varepsilon) = \begin{cases} \frac{1}{\beta}\ln(1 - e^{-\beta\varepsilon}) & \text{(bosons)}, \\ \frac{1}{\beta}\ln(1 + e^{-\beta\varepsilon}) & \text{(fermions)}. \end{cases}$$

(Note that a trick is needed to perform this sum since the logarithmic function does not have simple poles.)

With this, we arrive at the same expression for the free energy that we obtained using the Hamiltonian approach,

$$F_b = \frac{1}{V}\sum_k \left(\frac{\omega_k}{2} + \frac{1}{\beta}\ln(1 - e^{-\beta\omega_k})\right) \quad \text{and}$$

$$F_f = -\frac{2}{V}\sum_k \left(\frac{E_k}{2} + \frac{1}{\beta}\ln(1 + e^{-\beta E_k})\right).$$

## Dark Energy and Idea of Supersymmetry

Note that the vacuum energies of free bosons and Dirac fermions are large (~infinity) and of opposite signs, and would exactly cancel each other if $\frac{1}{V}\sum_k \frac{\omega_k}{2} = \frac{1}{V'}\sum_{k'} E_{k'}$. This happens in the lattice model if $\omega_k = E_k$ (note that, in the lattice model for fermions, the particle and anti-particle are coming from different sublattices, which results in a factor of two difference in volume ($V' = 2V$)).

This cancellation forms the basis of the *supersymmetric* theories where bosons and fermions always appear in pairs with spectra which are the same at high energy but differ by a small amount at low energy (small broken supersymmetry). In this case the large ground-state energy (dark energy) is avoided and provides the basis of a plausible scenario of why the observed cosmological constant which measures this energy is so small.

## 4.4
## Symmetry and Conservation Laws in Quantum Field Theory

The Lagrangian formulation plus Noether's theorem provide a natural framework to study how symmetries lead to conservation laws in classical field theories. However, it is not obvious whether the conservation laws derived in a classical field theory remain valid in the corresponding quantum field theory. The proof of Noether's theorem relies on the classical equation of motion, which implies that the classical system follows a path which optimizes the action ($\delta S = 0$). However, the corresponding quantum system does not restrict itself to the optimal path in the path integral formulation.

It turns out that, within the canonical quantization scheme, Noether's theorem still works at the quantum level. The only change is that the classical conserved currents become operators. The main reason behind this result is that the classical equation of motion in the canonical quantization scheme stays unchanged after quantization, except that the classical variables become operators. (You may view this as a consequence of the correspondence principle.) To see how this occurs, we consider a classical Lagrangian system with a Hamiltonian function $H(q, p)$, where $q = \{q_i\}$ and $p = \{p_i\}$ are canonical coordinates and momenta of the system. The equations of motion for $q$ and $p$ are Hamilton's equations of motion

$$\dot{q}_i = \frac{\delta H(q,p)}{\delta p_i} \quad \text{and} \quad \dot{p}_i = -\frac{\delta H(q,p)}{\delta q_i}. \tag{4.47}$$

Therefore, the equation of motion for any classical quantity $f(q, p)$ which is a function of $q$ and $p$ is

$$\dot{f} = \left[\sum_i \frac{\delta f}{\delta q_i}\frac{\delta H(q,p)}{\delta p_i} - \frac{\delta f}{\delta p_i}\frac{\delta H(q,p)}{\delta q_i}\right]. \tag{4.48}$$

Next, we consider the quantum (Heisenberg) equations of motion for $q$ and $p$,

$$\dot{q}_i = \frac{1}{i\hbar}[q_i, H(q,p)] \quad \text{and} \quad \dot{p}_i = \frac{1}{i\hbar}[p_i, H(q,p)]. \tag{4.49}$$

To evaluate these commutators, it is most convenient to use the representations $\left(p_i = p_i,\ q_i = i\hbar\frac{\delta}{\delta p_i}\right)$ and $\left(q_i = q_i,\ p_i = -i\hbar\frac{\delta}{\delta q_i}\right)$ consecutively. We obtain

$$\dot{q}_i = \frac{1}{i\hbar}[q_i, H(q,p)] = \frac{\delta}{\delta p_i} H\left(q = i\hbar\frac{\delta}{\delta p}, p\right) \tag{4.50a}$$

and

$$\dot{p}_i = \frac{1}{i\hbar}[p_i, H(p,q)] = -\frac{\delta}{\delta q_i} H\left(q, p = -i\hbar\frac{\delta}{\delta q}\right), \tag{4.50b}$$

which have the same form as the classical equations of motion except that $q$ and $p$ are operators. Using again $\dot{f} = \left[\sum_i \frac{\delta f}{\delta q_i}\dot{q}_i + \frac{\delta f}{\delta p_i}\dot{p}_i\right]$ for any operator $f(\widehat{q}, \widehat{p})$, we find that the equation of motion for an arbitrary operator $f(\widehat{q}, \widehat{p})$ has the same form as the equation of motion for the classical function $f(q, p)$.

The analysis can be extended straightforwardly to the case of continuous fields $\phi(\vec{x}, t)$ and $\pi(\vec{x}, t)$. We shall leave it as an exercise for readers.

The above result can be extended to fermionic systems via the coherent state representation. In this case the Lagrangian is of the form

$$L = \int d^d r[i\hbar\bar{\psi}(\vec{r})\partial_t\psi(\vec{r}) - h[\bar{\psi}(\vec{r}), \psi(\vec{r})]] \tag{4.51}$$

and the canonical coordinate and momenta are $\psi(\vec{r})$ and $\pi(\vec{r}) = i\hbar\bar{\psi}(\vec{r})$, respectively. The classical equations of motion are

$$\dot{\psi}(\vec{r}) = \frac{-i}{\hbar}\frac{\delta}{\delta\bar{\psi}(\vec{r})}H(\bar{\psi}, \psi) \quad \text{and} \quad \dot{\bar{\psi}}(\vec{r}) = \frac{-i}{\hbar}\frac{\delta}{\delta\psi(\vec{r})}H(\bar{\psi}, \psi), \tag{4.52a}$$

where

$$H(\bar{\psi}, \psi) = \int d^d r h[\bar{\psi}(\vec{r}), \psi(\vec{r})]. \tag{4.52b}$$

The derivatives in Equation (4.52a) are performed according to the Grassmann number rule. It can be shown that the resulting classical equation of motion is the same as the quantum equation of motion

$$\dot{\psi}(\vec{r}) = \frac{1}{i\hbar}[\psi(\vec{r}), H(\bar{\psi}, \psi)], \quad \dot{\bar{\psi}}(\vec{r}) = \frac{1}{i\hbar}[\bar{\psi}(\vec{r}), H(\bar{\psi}, \psi)]. \tag{4.53}$$

**Homework Q6**

Show that the quantum and classical equations of motion are the same for a fermionic system with Hamiltonian

$$H = \frac{\hbar^2}{2m}\int d^d r(\nabla\bar{\psi}).(\nabla\psi) + \iint d^d r d^d r' u(\vec{r} - \vec{r}')\bar{\psi}(\vec{r})\psi(\vec{r})\bar{\psi}(\vec{r}')\psi(\vec{r}').$$

Because of this result, a classical conservation law represented by the equation of motion $\frac{\partial\rho}{\partial t} = -\nabla\cdot\vec{j}$ must have a quantum counterpart, that is $\frac{\partial\widehat{\rho}}{\partial t} = -\nabla\cdot\vec{j}$, where $\widehat{\rho}$ and $\widehat{j}$ are the corresponding density and current operators. The classical equation of motion becomes an operator equation of motion in QM.

## References


1 Weinberg, S. (1995) *The Quantum Theory of Fields*, Vol. 1, Cambridge University Press.

2 Nagaosa, N. (1995) *Quantum Field Theory in Condensed Matter Physics*, Springer (Berlin, Heidelberg, New York, Barcelonc, Hong Kong, London, Milan, Paris, Singapore, Tokyo).

3 Mahan, G.D. (1990) *Many Particle Physics*, Plenum Press (N.Y. and London).

4 Wen, X.G. (2004) *Theory of Quantum Many-body Systems*, Oxford University Press.

# Part Two

*Introduction to Classical and Quantum Field Theory.* Tai-Kai Ng

ISBN: 978-3-527-40726-2

# 5
# Perturbation Theory, Variational Approach and Correlation Functions

## 5.1
## Introduction to Perturbation Theory

In its simplest form, the problem of perturbation theory can be formulated as follows: suppose that we have a Hamiltonian $H = H_0 + \lambda V$ which we cannot diagonalize exactly. However, we know how to diagonalize $H_0$. The question is, can we compute the eigenstates $|\psi_{n\lambda}\rangle$ and eigenenergies $E_{n\lambda}$ of the Hamiltonian $H$ starting from eigenstates of $H_0$, at least when $\lambda$ is small?

The basic assumption of perturbation theory is that with an appropriate (clever) choice of $H_0$, the eigenstates $|\psi^{(0)}_{n\lambda}\rangle$ of $H_0$ and the eigenstates $|\psi_{n\lambda}\rangle$ of $H$ will be 'close enough' to each other such that we may start our calculation from $H_0$ and $|\psi^{(0)}_{n\lambda}\rangle$ treating the effects of $\lambda\widehat{V}$ as small corrections. The question of how to include the effects of $\lambda\widehat{V}$ systematically in computing various quantities is the subject of perturbation theory. Mathematically, such a procedure works in principle if there exists a one-to-one correspondence between the eigenstates $|\psi^{(0)}_{n\lambda}\rangle$ and $|\psi_{n\lambda}\rangle$. The states will evolve continuously from $|\psi^{(0)}_{n\lambda}\rangle$ to $|\psi_{n\lambda}\rangle$ as $\lambda$ increases continuously from zero to its final value (*adiabatic assumption*).

The problem becomes more complicated if we think about how to compare theoretical calculations with experimental results. For a (time-independent) measure denoted by the operator $\widehat{A}$, we need to compute $\langle\psi_{n\lambda}|\,\widehat{A}\,|\psi_{n\lambda}\rangle$ or, for macroscopic systems, the thermodynamic average $\mathrm{Tr}(\widehat{A}\,e^{-\beta\widehat{H}})/Z$, and we need to know how the eigenenergies and eigenfunctions of *all* eigenstates are affected by $V$ in perturbation theory. If we consider time-dependent measurements, the situation is even more complicated. In these experiments, a small perturbation denoted by $\widehat{B}$ is often introduced at time $t'$, and the response of the system is probed by a measurement $\widehat{A}$ at a later time $t$. We need to compute time-dependent quantities of the form

$$\langle\psi_{n\lambda}|\,\widehat{A}(t)\,\widehat{B}(t')|\psi_{n\lambda}\rangle \sim \langle\psi_{n\lambda}|\,\widehat{A}\,\widehat{U}(t,t')\,\widehat{B}\,|\psi_{n\lambda}\rangle$$

*Introduction to Classical and Quantum Field Theory.* Tai-Kai Ng

ISBN: 978-3-527-40726-2

or the corresponding thermodynamic averages. $\widehat{U}$ is the time-evolution operator we discussed in previous two chapters. The problem of perturbation theory is how can we compute systematically the corrections to these dynamic correlation functions from $V$?

In this chapter we shall assume that perturbation theory works and shall introduce the basic techniques of perturbation theory in path integral and Dyson's approaches. Elaborate techniques have been developed to reorganize and sum over the perturbation series. The topic is well covered in many textbooks (see e.g. Refs. [1–4]). We shall not go into these details here because of the introductory nature of the book. You will learn more in later chapters when we encounter real examples.

### 5.1.1
### Path Integral Approach

We start with the path integral approach in the coherent state representation (for either bosons or fermions). First, we introduce Gaussian integrals for correlation functions.

Definition: an $n$-point correlation function is defined as

$$\underbrace{\langle\bar{\psi}\bar{\psi}\cdots\psi\psi\rangle}_{n\ \bar{\psi}\text{or}\ \psi\ \text{together}}=\frac{\int D\bar{\psi}D\psi(\bar{\psi}\bar{\psi}\cdots\psi\psi)e^{-S_0}}{\int D\bar{\psi}D\psi e^{-S_0}},\tag{5.1}$$

where $S_0=-\sum_{k,i\omega_n}[i\omega_n-\varepsilon_k]\bar{\psi}(k,i\omega_n)\psi(k,i\omega_n)$.

For example, we have the two-point functions

$$G(\vec{k},\vec{k}',i\omega_n,i\omega'_n)=\langle\bar{\psi}(\vec{k},i\omega_n)\psi(\vec{k}',i\omega'_n)\rangle,$$

$$F(\vec{k},\vec{k}',i\omega_n,i\omega'_n)=\langle\psi(\vec{k},i\omega_n)\psi(\vec{k}',i\omega'_n)\rangle$$

and a four-point function

$$\Gamma(\vec{k}_1,\vec{k}_2,\vec{k}_3,\vec{k}_4,i\omega_{n_1},i\omega_{n_2}i\omega_{n_3},i\omega_{n_4})=\langle\bar{\psi}(\vec{k}_1,i\omega_{n_1})\bar{\psi}(\vec{k}_2,i\omega_{n_2})\psi(\vec{k}_3,i\omega_{n_3})\psi(\vec{k}_4,i\omega_{n_4})\rangle.$$

First, we consider evaluating a two-point function (one-particle Green's function)

$$\begin{aligned}G(\vec{k},\vec{k}',i\omega_n,i\omega'_n)&=\langle\bar{\psi}(\vec{k},i\omega_n)\psi(\vec{k}',i\omega'_n)\rangle\\&=\frac{1}{Z}\int D\bar{\psi}D\psi[\bar{\psi}(\vec{k},i\omega_n)\psi(\vec{k}',i\omega'_n)]e^{-\sum_{k,i\omega_n}[-i\omega_n+\varepsilon_k]\bar{\psi}(\vec{k},i\omega_n)\psi(\vec{k},i\omega_n)}.\end{aligned}\tag{5.2}$$

To evaluate $G$, we note that integrals over different $(\vec{k},i\omega_n)$ fields are all independent; thus,

$$\begin{aligned}G(\vec{k},\vec{k}',i\omega_n,i\omega'_n)=&\frac{1}{Z_2}\int d\bar{\psi}(\vec{k},i\omega_n)d\psi(\vec{k},i\omega_n)d\bar{\psi}(\vec{k}',i\omega'_n)d\psi(\vec{k}',i\omega'_n)\\&\times\left[\bar{\psi}(\vec{k},i\omega_n)\psi(\vec{k}',i\omega'_n)\right]e^{-\left[-i\omega_n+\frac{k^2}{2m}-\mu\right]\bar{\psi}(\vec{k},i\omega_n)\psi(\vec{k},i\omega_n)}\\&\times e^{-\left[-i\omega'_n+\frac{k'^2}{2m}-\mu\right]\bar{\psi}(\vec{k}',i\omega'_n)\psi(\vec{k}',i\omega'_n)},\end{aligned}\tag{5.3a}$$

where

$$Z_2 = \int d\bar{\psi}(\vec{k}, i\omega_n) d\psi(\vec{k}, i\omega_n) d\bar{\psi}(\vec{k}', i\omega'_n) d\psi(\vec{k}', i\omega'_n)$$
$$\times e^{-[-i\omega_n + \varepsilon_k]\bar{\psi}(\vec{k}, i\omega_n)\psi(\vec{k}, i\omega_n)} e^{-[-i\omega'_n + \varepsilon_{k'}]\bar{\psi}(\vec{k}', i\omega'_n)\psi(\vec{k}', i\omega'_n)}. \tag{5.3b}$$

We first consider boson fields. To simplify notation, we let

$$z_1 = \psi(\vec{k}, i\omega_n), \quad z_1^* = \bar{\psi}(\vec{k}, i\omega_n),$$
$$z_2 = \psi(\vec{k}', i\omega'_n), \quad z_2^* = \bar{\psi}(\vec{k}', i\omega'_n).$$

Therefore,

$$G(\vec{k}, \vec{k}', i\omega_n, i\omega'_n) = \langle \bar{\psi}(\vec{k}, i\omega_n)\psi(\vec{k}', i\omega'_n)\rangle$$
$$= \frac{\int dz_1^* dz_1 dz_2^* dz_2 [z_1^* z_2] e^{-[-i\omega_n + \varepsilon_k] z_1^* z_1} e^{-[-i\omega'_n + \varepsilon_{k'}] z_2^* z_2}}{\int dz_1^* dz_1 dz_2^* dz_2 e^{-[-i\omega_n + \varepsilon_k] z_1^* z_1} e^{-[-i\omega'_n + \varepsilon_{k'}] z_2^* z_2}}. \tag{5.4}$$

Recall that $\int dz^* \int dz = 2\int r dr d\theta$ and $\iint re^{i\theta} f(r) dr d\theta = 0$ for any $\theta$-independent function $f$.

Therefore, the integral equals zero unless $z_1 = z_2$. In this case,

$$G(\vec{k}, \vec{k}', i\omega_n, i\omega'_n) = \langle \bar{\psi}(\vec{k}, i\omega_n)\psi(\vec{k}', i\omega'_n)\rangle$$
$$= \frac{\int dz_1^* dz_1 [z_1^* z_1] e^{-[-i\omega_n + \varepsilon_k] z_1^* z_1}}{\int dz_1^* dz_1 e^{-[-i\omega_n + \varepsilon_k] z_1^* z_1}} \tag{5.5a}$$

or

$$G(\vec{k}, \vec{k}', i\omega_n, i\omega'_n) = \frac{-1}{i\omega_n - \varepsilon_k} \delta_{\vec{k},\vec{k}'} \delta_{i\omega_n, i\omega'_n}. \tag{5.5b}$$

**Homework Q1a**

Derive the above result.

Note that the Green's function is by convention defined with a minus sign as

$$G(\vec{k}, \vec{k}', i\omega_n, i\omega'_n) = -\langle \psi(\vec{k}', i\omega'_n)\bar{\psi}(\vec{k}, i\omega_n)\rangle = -\langle \bar{\psi}(\vec{k}, i\omega_n)\psi(\vec{k}', i\omega'_n)\rangle.$$

With this convention, $G(\vec{k}, \vec{k}', i\omega_n, i\omega'_n) = \dfrac{1}{i\omega_n - \varepsilon_k} \delta_{\vec{k},\vec{k}'} \delta_{i\omega_n, i\omega'_n}$. (5.5c)

For fermions similar considerations apply. It can be shown that $G(\vec{k}, \vec{k}', i\omega_n, i\omega'_n) \neq 0$ only when $\vec{k} = \vec{k}'$, $\omega_n = \omega'_n$, and

$$G(\vec{k}, \vec{k}', i\omega_n, i\omega'_n) = -\langle \psi(\vec{k}', i\omega'_n)\bar{\psi}(\vec{k}, i\omega_n)\rangle = \langle \bar{\psi}(\vec{k}, i\omega_n)\psi(\vec{k}', i\omega'_n)\rangle$$
$$= \frac{\int dz_1^* dz_1 [z_1^* z_1] e^{-[-i\omega_n + \varepsilon_k] z_1^* z_1}}{\int dz_1^* dz_1 e^{-[-i\omega_n + \varepsilon_k] z_1^* z_1}}, \tag{5.6a}$$

where $z_1, z_i^*$ are now Grassmann numbers. Evaluating the integral, we obtain

$$G(\vec{k}, \vec{k}', i\omega_n, i\omega'_n) = \frac{1}{i\omega_n - \varepsilon_k} \delta_{\vec{k},\vec{k}'} \delta_{i\omega_n, i\omega'_n}, \tag{5.6b}$$

which is the same expression as for bosons.

**Homework Q1b**
Derive the above result.

The calculation can be extended straightforwardly to high-order correlation functions.

**Homework Q2**
Show that

$$\begin{aligned}
&\Gamma(\vec{k}_1, \vec{k}_2, \vec{k}_3, \vec{k}_4, i\omega_{n_1}, i\omega_{n_2} i\omega_{n_3}, i\omega_{n_4}) \\
&= \langle \bar{\psi}(\vec{k}_1, i\omega_{n_1}) \bar{\psi}(\vec{k}_2, i\omega_{n_2}) \psi(\vec{k}_3, i\omega_{n_3}) \psi(\vec{k}_4, i\omega_{n_4}) \rangle \\
&= \mp \langle \bar{\psi}(\vec{k}_1, i\omega_{n_1}) \psi(\vec{k}_3, i\omega_{n_3}) \rangle \langle \bar{\psi}(\vec{k}_2, i\omega_{n_2}) \psi(\vec{k}_4, i\omega_{n_4}) \rangle \delta_{13}\delta_{24} \\
&+ \langle \bar{\psi}(\vec{k}_1, i\omega_{n_1}) \psi(\vec{k}_4, i\omega_{n_4}) \rangle \langle \bar{\psi}(\vec{k}_2, i\omega_{n_2}) \psi(\vec{k}_3, i\omega_{n_3}) \rangle \delta_{14}\delta_{23}
\end{aligned}$$

(− for fermion, + for boson. This is an example of the so-called Wick's theorem (see below)).

#### 5.1.1.1 Perturbation Theory for Interacting Systems

Next, we consider the evaluation of a partition function in the presence of interaction:

$$Z(\beta) = \int D\psi(\vec{k}, i\omega_n) D\bar{\psi}(\vec{k}, i\omega_n) e^{-(S_0 + S_{\text{int}})}. \tag{5.7a}$$

We shall consider $S_{\text{int}} = \frac{1}{2} g \int d\tau d\vec{r} (\bar{\psi}(\vec{r}, \tau)\psi(\vec{r}, \tau))^2$ and the $\psi$ are boson fields as an example.

The basic idea of perturbation theory is to expand $S_{\text{int}}$ in a power series and evaluate the series term-by-term, that is we write

$$e^{-(S_0 + S_{\text{int}})} = e^{-S_0} \cdot \sum_{n=0}^{\infty} \frac{(-1)^n}{n!} (S_{\text{int}})^n \tag{5.7b}$$

and

$$Z(\beta) \to \int D\psi D\bar{\psi} \sum_{n=0}^{\infty} \frac{(-1)^n}{n!} (S_{\text{int}})^n \cdot e^{-S_0}, \tag{5.7c}$$

where $\psi(\bar{\psi})_n = \psi(x_n)(\bar{\psi}(x_n))$, and so on.

For the above form of interaction, $(S_{\text{int}})^n \to \left(\frac{-g}{2}\right)^n \left[\int_0^\beta d\tau \int d^d r(\bar{\psi}\psi \cdots \psi\psi)\right]^{4n}$ and

$$Z(\beta) \to \sum_{n=0}^{\infty} \frac{(-g)^n}{2^n n!} \int D\psi D\bar{\psi} \left[\int_0^\beta d\tau \int d^d r(\bar{\psi}\psi \cdots \psi\psi)\right]^{4n} \cdot e^{-S_0}. \tag{5.7d}$$

**Homework Q3**

With the result of Homework Q2, evaluate the first-order correction to the free energy of a fermion system with fermion density $n$ in terms of integrals involving $u(\vec{r})$ for

$$S_{\text{int}} = \frac{1}{2}\int d\tau d\vec{r} d\vec{r}' \bar{\psi}(\vec{r},\tau)\psi(\vec{r},\tau)u(\vec{r}-\vec{r}')\bar{\psi}(\vec{r}',\tau)\psi(\vec{r}',\tau).$$

Can you repeat the above analysis for a boson system?

#### 5.1.1.2 Wick's Theorem

You can see that the evaluation of perturbation series is a very tedious process, especially when we go to higher order. The mathematics can be simplified by a trick, which results in the so-called Wick's theorem.

The trick is to introduce the generating functional

$$Z(j,\bar{j}) = \frac{\int D\psi D\bar{\psi} e^{-S_0 - \int dx (j(x)\bar{\psi}(x) + \bar{j}(x)\psi(x))}}{\int D\psi D\bar{\psi} e^{-S_0}} = e^{\int\int dx dx' \bar{j}(x) G(x,x') j(x')}, \tag{5.8}$$

where $x = (\vec{x}, t)$ and $G(x, x')$ is the inverse of the single-particle operator $K(x, x')$ defined by $S_0 = \int\int dx dx' \bar{\psi}(x) K(x, x')\psi(x')$ and $\int dx''(x) K(x, x'') G(x'', x') = \delta(x-x')$. Note that there is no need to specify the exact form of $K(x, x')$ here.

**Homework Q4**

Derive Equation (5.8).

With Equation (5.8), we obtain

$$\frac{\int D\bar{\psi} D\psi (\bar{\psi}_{x1}\bar{\psi}_{x2} \cdots \psi_{yn-1}\psi_{yn}) e^{-S_0}}{\int D\bar{\psi} D\psi e^{-S_0}} = \frac{\delta^{2n}}{\delta j_1 \delta j_2 \cdots \delta \bar{j}_{n-1} \delta \bar{j}_n} Z(j,\bar{j})\Big|_{j\bar{j}=0}$$

$$= G(x_1, y_1)G(x_2, y_2) \cdots G(x_n, y_n) + P[G(x_1, y_1)G(x_2, y_2) \cdots G(x_n, y_n)], \tag{5.9a}$$

where

$$P[G(x_1, y_1)G(x_2, y_2) \cdots G(x_n, y_n)]$$
$$= \sum \text{all possible permutations of } x_n, y_m \text{ for bosons,}$$
$$P[G(x_1, y_1)G(x_2, y_2) \cdots G(x_n, y_n)]$$
$$= \sum (-1)^P \text{ all possible permutations of } x_n, y_m \text{ for fermions.} \tag{5.9b}$$

It is easy to see that the result for the four-point function (Homework Q2) can be produced readily from Wick's theorem.

In the above example with

$$Z(\beta) = \sum_{n=0}^{\infty} \frac{(-g)^n}{2^n n!} \int D\psi D\bar{\psi} \left[ \int_0^{\beta} d\tau \int d^d r (\bar{\psi}\bar{\psi} \cdots \psi\psi) \right]^{4n} \cdot e^{-S_0},$$

we find that to second order in g

$$\begin{aligned} Z(\beta) = Z_0(\beta) \Bigg\{ 1 - \left(\frac{g}{2}\right) \int d1 G(1,1)^2 \\ + \frac{1}{2}\left(\frac{g}{2}\right)^2 \iint d1d2[4G(1,1)^2 G(2,2)^2 + 16G(1,1)G(1,2)G(2,1)G(2,2) \\ + 4G(1,2)^2 G(2,1)^2] \end{aligned} \tag{5.10}$$

(recall that the $\psi$ are boson fields),

where $G(i, j) = G(x_i, x_j)$, $di = d^d x_i$ and so on. The multiplicative factors come from summing over different permutations. These different terms are conveniently represented by diagrams (Feynman diagrams) as follows:

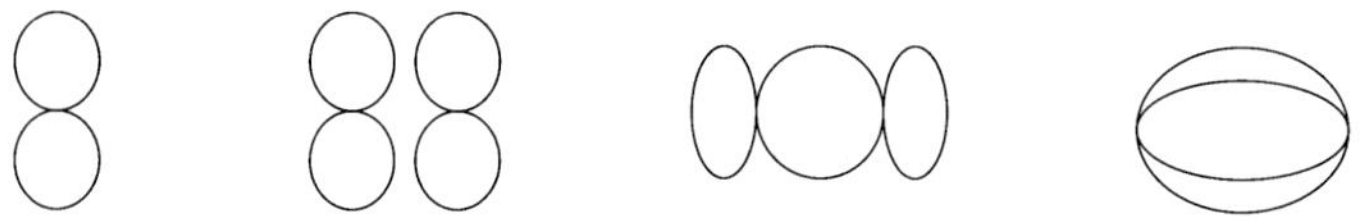

$G(1,1)^2$ $G(1,1)^2 G(2,2)^2$ $G(1,1)G(1,2)G(2,1)G(2,2)$ $G(1,2)^2 G(2,1)^2$

**Question**

Are the following diagrams allowed in perturbation theory? Why? Write down the corresponding integrals for the (allowed) diagrams.

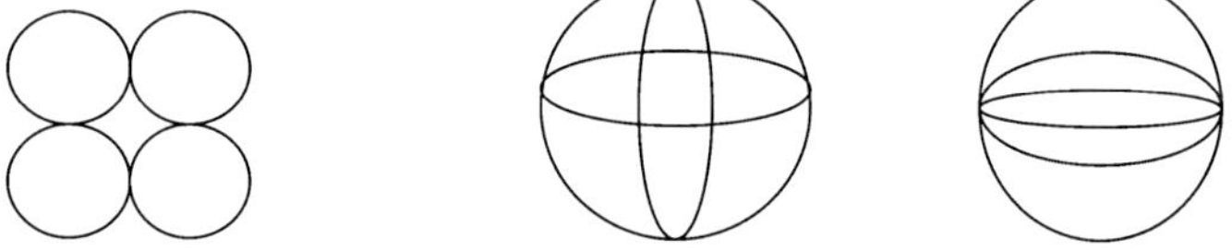

The corresponding free energy is given by

$$F = -\frac{1}{\beta} \ln Z \sim F_0 + \frac{1}{\beta}\left[ g \int d1 G(1,1)^2 - 2g^2 \iint d1d2 G(1,1)G(1,2)G(2,1)G(2,2) \right.$$

$$\left. - \frac{g^2}{2} \iint d1d2 G(1,2)^2 G(2,1)^2 + \ldots \right]. \tag{5.11}$$

Note that the second-order diagram

is cancelled by (first-order diagram)$^2$ in $F$. This is not accidental. There is a theorem which states that all diagrams which can be separated into more than one piece become cancelled in evaluating the free energy and only diagrams which are connected together as one piece remain (linked-cluster theorem). You can find a simple proof of the theorem in Ref. [3] using the idea of 'replicas'.

A similar technique is developed to evaluate the correlation functions. For example, in the case of the one-particle Green's function, we have

$$G(\vec{k},\vec{k}', i\omega_n, i\omega'_n) = \langle \bar{\psi}(\vec{k}, i\omega_n)\psi(\vec{k}', i\omega'_n)\rangle = \frac{\int D\bar{\psi} D\psi[\bar{\psi}(\vec{k}, i\omega_n)\psi(\vec{k}', i\omega'_n)]e^{-(S_0+S_{\text{int}})}}{\int d\bar{\psi} d\psi e^{-(S_0+S_{\text{int}})}}, \tag{5.12}$$

and perturbation series have to be performed at both the numerator and the denominator to evaluate $G$. The perturbation theory can be much simplified with the help of Wick's theorem and the linked-cluster theorem. The calculation of the one-particle Green's function will be discussed in more detail in Dyson's approach to perturbation theory. It will be discussed also in Chapter 7 and Appendix A. You can also look at, for example, Refs. [1, 2] for details.

### 5.1.2 Dyson's Approach

Freeman Dyson invented a different presentation of perturbation theory based on the Hamiltonian formulation. Although not obvious at first sight, the approach is mathematically equivalent to Feynman's path integral approach. Alternative ways to construct perturbation theory were discovered later (see e.g. Ref. [5]). The idea behind the different methods is more or less similar although the technical details are different. It is very often a matter of convenience which particular approach should be used.

Similar to Feynman's path integral approach, Dyson studied the time-evolution operator $U(t)$. He developed a perturbation theory to compute the time-evolution operator. Wavefunctions of the system appear in his approach explicitly – this is an important difference between the Dyson and path integral treatments. Instead of treating the perturbation $\lambda\widehat{V}$ as time independent, Dyson assumed a time-dependent perturbation $\lambda\widehat{V}(t) \sim \lambda\widehat{V}e^{-\eta|t|}$, where $\eta \to 0$ is a small positive number. Note that $\lambda\widehat{V}(t)$ vanishes at times $t \to \pm\infty$. The corresponding time-evolution operator is called the $S$ matrix, $S(t, t') = U_\eta(t, t')$.

With $\lambda\widehat{V}(t)$ we can envision that, at time $t \to -\infty$, $H \to H_0$ and the eigenstates of the system are $|\psi^{(0)}_{n\lambda}\rangle$. As time increases to $t \sim 0$, $H \to H_0 + \lambda V$ and $|\psi^{(0)}_{n\lambda}\rangle \to |\psi_{n\lambda}\rangle$ if the assumption of adiabaticity is robust. In this case, $|\psi_{n\lambda}(t)\rangle \sim S(t,-\infty)|\psi_{n\lambda}^{(0)}\rangle$, if $t$ is 'far enough' from $-\infty$. To make this correspondence exact, Dyson took the limit $\eta \to 0$ *at the end of the calculation.*

Therefore, expectation values can be computed as

$$\begin{aligned}\langle\psi_{n\lambda}|\,\widehat{A}\,|\psi_{n\lambda}\rangle &\sim \langle\psi_{n\lambda}^{(0)}|S(-\infty,t)\,\widehat{A}\,S(t,-\infty)|\psi_{n\lambda}^{(0)}\rangle \\ &= \langle\psi_{n\lambda}^{(0)}|S(-\infty,\infty)S(\infty,t)\,\widehat{A}\,S(t,-\infty)|\psi_{n\lambda}^{(0)}\rangle\end{aligned} \tag{5.13a}$$

and

$$\begin{aligned}\langle\psi_{n\lambda}|\,\widehat{A}(t)\,\widehat{B}(t')|\psi_{n\lambda}\rangle &\sim \langle\psi_{n\lambda}^{(0)}S(-\infty,t)\,\widehat{A}\,S(t,t')\,\widehat{B}\,S(t',-\infty)|\psi_{n\lambda}^{(0)}\rangle \\ &= \langle\psi_{n\lambda}^{(0)}|S(-\infty,\infty)S(\infty,t)\,\widehat{A}\,S(t,t')\,\widehat{B}\,S(t',-\infty)|\psi_{n\lambda}^{(0)}\rangle.\end{aligned} \tag{5.13b}$$

Note that we have to compute the time evolution of the system through a closed time loop ($t \to -\infty$ to $t \to \infty$ and back to $t \to -\infty$) in this approach. This is called the closed-time-loop Green's function technique or the Keldysh Green's function technique [1, 6]:

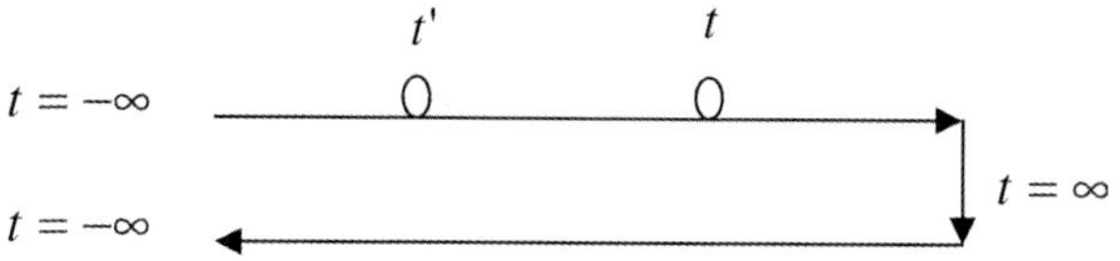


The quantity $S(\infty,-\infty)|\psi_{n\lambda}^{(0)}\rangle$ is of particular interest in distinguishing between 'equilibrium' and 'non-equilibrium' systems. Let us imagine that $n=0$ and $|\psi_{0\lambda}^{(0)}\rangle$ is the ground state of $H_0$. The interaction term $\lambda\widehat{V}(t) \sim \lambda\widehat{V}e^{-\eta|t|}$ is turned on gradually starting from time $t \to -\infty$. At $t \sim 0$, it is assumed that $|\psi_{0\lambda}^{(0)}\rangle \to |\psi_{0\lambda}\rangle$ becomes the ground state of $H$. As time increases again, the interaction term is turned off slowly and $H \to H_0$ as $t \to \infty$. The question is what happens to the wavefunction? Does it return to $|\psi_{0\lambda}^{(0)}\rangle$?

The answer is 'yes' if the ground state stays unique and changes adiabatically as a function of the strength of interaction $\lambda\widehat{V}$ (non-degenerate). In this case $|\psi_{0\lambda}\rangle \to e^{i\phi}|\psi_{0\lambda}^{(0)}\rangle$ as time $t \to \infty$ ($e^{i\phi}$ can be considered as a Berry phase term; see the next chapter) and $S(\infty,-\infty)|\psi_{n\lambda}^{(0)}\rangle = e^{i\phi}|\psi_{n\lambda}^{(0)}\rangle$. When this happens, our calculation of expectation values can be simplified to

$$\begin{aligned}\frac{\langle\psi_{0\lambda}|\,\widehat{A}\,|\psi_{0\lambda}\rangle}{\langle\psi_{0\lambda}|\psi_{0\lambda}\rangle} &= \frac{\langle\psi_{0\lambda}^{(0)}|S(\infty,-\infty)S(\infty,t)\,\widehat{A}\,S(t,-\infty)|\psi_{0\lambda}^{(0)}\rangle}{\langle\psi_{0\lambda}^{(0)}|S(\infty,-\infty)|\psi_{0\lambda}^{(0)}\rangle} \\ &= \langle\psi_{0\lambda}^{(0)}|S(\infty,t)\,\widehat{A}\,S(t,-\infty)|\psi_{0\lambda}^{(0)}\rangle\end{aligned} \tag{5.14a}$$

and

$$\frac{\langle\psi_{0\lambda}|\,\widehat{A}(t)\,\widehat{B}(t')|\psi_{0\lambda}\rangle}{\langle\psi_{0\lambda}|\psi_{0\lambda}\rangle} = \langle\psi_{0\lambda}^{(0)}|S(\infty,t)\,\widehat{A}\,S(t,t')\,\widehat{B}\,S(t',-\infty)|\psi_{0\lambda}^{(0)}\rangle. \tag{5.14b}$$

This is what Dyson originally proposed. The goal of perturbation theory is to expand the $S$ matrix in power series of $\lambda\widehat{V}$ and evaluate the above expectation values.

For $n \neq 0$ (excited states) the assumption $S(\infty, -\infty)|\psi_{n\lambda}^{(0)}\rangle = e^{i\phi}|\psi_{n\lambda}^{(0)}\rangle$ does not hold in general, and the perturbation has to start with the more complicated closed-time-loop Green's function. An example where the adiabatic approximation fails is the scattering problem where a particle in an initial plane-wave state $e^{i\vec{k}.\vec{x}}$ is scattered to other plane-wave states with the same energy. In the rest frame of the particle, the scattering potential is zero at $t \to -\infty$ (the particle is infinitely far away). The scattering potential is gradually turned on as time $t \to 0$ and turned off again as $t \to \infty$, as the particle leaves the scatterer. The final state of the particle becomes different from the initial state in the process, although $H \to H_0$ at both $t \to \pm\infty$. A similar consideration applies when the system evolves under a time-dependent Hamiltonian. In these cases a closed-time-loop Green's function has to be used for computation. There is also a corresponding path integral approach to non-equilibrium problems (Keldysh path integral), which we shall not discuss here.

#### 5.1.2.1 Time-Evolution Operator at Imaginary Time

In situations of physical interest the system under investigation is often not an eigenstate of a particular Hamiltonian but should be represented by a density matrix $\rho = \sum_{m,n} |m\rangle \rho_{mn} \langle n|$. The physical observables are represented by

$$\langle A \rangle = \mathrm{Tr}(\rho A) = \sum_{m,n} \rho_{mn} \langle n|A|m\rangle,$$

where $\rho_{mn} = \delta_{nm} e^{-\beta(E_n - \mu N)}$ at thermal equilibrium. What is interesting is that at equilibrium a (imaginary) time-evolution operator can also be defined to compute the density matrix and can be treated in perturbation theory as the usual $S$ matrix in real time. The thermal time-evolution operator is defined as

$$U(\tau) = e^{\tau(H_0 - \mu N)} e^{-\tau(H - \mu N)} \tag{5.15}$$

and relates the density matrix $\rho = e^{-\beta(H-\mu N)}$ of the full Hamiltonian $H$ to the 'unperturbed' density matrix $\rho_0 = e^{-\beta(H-\mu N)}$ by $\rho = \rho_0 U(\beta)$ and $\langle A \rangle = \mathrm{Tr}(\rho A) = \mathrm{Tr}(\rho_0 U(\beta) A)$. We shall return to the thermal time-evolution operator later.

#### 5.1.2.2 Perturbation Expansion for S Matrix

Perturbation expansion for the $S$ matrix is usually formulated in the interaction picture. The reason can be understood from the following consideration. We divide our Hamiltonian into two parts: the unperturbed Hamiltonian $H_0$ and the interaction $\lambda\widehat{V}(t)$. We are not very interested in how the system evolves under $H_0$, but want to know how the system changes under the perturbation $\lambda\widehat{V}(t)$.

To separate the time evolution from the two terms, let us start with the Schrödinger wavefunction

$$|\psi_S(t)\rangle = e^{-i\frac{Ht}{\hbar}}|\psi(0)\rangle$$

and define the wavefunction in the interaction representation as

$$|\psi_I(t)\rangle = e^{i\frac{H_0 t}{\hbar}} e^{-i\frac{Ht}{\hbar}}|\psi(0)\rangle. \tag{5.16}$$

Using the Schrödinger equation, it is easy to show that

$$\begin{aligned} i\hbar\frac{\partial}{\partial t}|\psi_{\mathrm{I}}(t)\rangle &= -e^{i\frac{H_0 t}{\hbar}}(H_0-H)e^{-i\frac{Ht}{\hbar}}|\psi(0)\rangle = \left(e^{i\frac{H_0 t}{\hbar}}\lambda V e^{-i\frac{H_0 t}{\hbar}}\right)|\psi_{\mathrm{I}}(t)\rangle \\ &= V_{\mathrm{I}}(t)|\psi_{\mathrm{I}}(t)\rangle, \end{aligned} \tag{5.17}$$

that is the time evolution of $|\psi_{\mathrm{I}}(t)\rangle$ is determined by the interaction $V$ only. Note that operators corresponding to observables have to be redefined also because of the change in representation. To make sure that

$$\langle\psi_{\mathrm{I}}(t)|A_{\mathrm{I}}(t)|\psi_{\mathrm{I}}(t)\rangle = \langle\psi_{\mathrm{S}}(t)|A_{\mathrm{S}}(t)|\psi_{\mathrm{S}}(t)\rangle,$$

$$\text{we need } A_{\mathrm{I}}(t) = e^{i\frac{H_0 t}{\hbar}}A_{\mathrm{S}}(t)e^{-i\frac{H_0 t}{\hbar}}, \tag{5.18}$$

which are the operators in the interaction representation. Dyson's perturbation expansion relies on the interaction representation.

The equation $i\hbar\frac{\partial}{\partial t}|\psi_{\mathrm{I}}(t)\rangle = V_{\mathrm{I}}(t)|\psi_{\mathrm{I}}(t)\rangle$ is solved iteratively in perturbation theory. We rewrite the equation in an integral form

$$\begin{aligned} |\psi_{\mathrm{I}}(t)\rangle &= U(t)|\psi_{\mathrm{I}}(0)\rangle = |\psi_{\mathrm{I}}(0)\rangle - \frac{i}{\hbar}\int_0^t V_{\mathrm{I}}(t')|\psi_{\mathrm{I}}(t')\rangle \mathrm{d}t' \\ &= |\psi_{\mathrm{I}}(0)\rangle - \frac{i}{\hbar}\int_0^t V_{\mathrm{I}}(t')U(t')|\psi_{\mathrm{I}}(0)\rangle \mathrm{d}t', \end{aligned}$$

which implies an iterative definition for $U(t)$,

$$U(t) = 1 - \frac{i}{\hbar}\int_0^t V_{\mathrm{I}}(t')U(t')\mathrm{d}t'. \tag{5.19a}$$

Iterating the equation, we obtain

$$\begin{aligned} U(t) &= 1 - \frac{i}{\hbar}\int_0^t V_{\mathrm{I}}(t_1)\mathrm{d}t_1 + \left(\frac{i}{\hbar}\right)^2\int_0^t \mathrm{d}t_1\int_0^{t_1}\mathrm{d}t_2 V_{\mathrm{I}}(t_1)V_{\mathrm{I}}(t_2) + \ldots \\ &= T\left[\exp\left(-\frac{i}{\hbar}\int_0^t V_{\mathrm{I}}(t')\mathrm{d}t'\right)\right], \end{aligned} \tag{5.19b}$$

where $T$ = time-ordering operator, defined by

$$T[V_{\mathrm{I}}(t)V_{\mathrm{I}}(t')] = \theta(t-t')V_{\mathrm{I}}(t)V_{\mathrm{I}}(t') + \theta(t'-t)V_{\mathrm{I}}(t)V_{\mathrm{I}}(t').$$

We have used the identities

$$\int_0^t \mathrm{d}t_1\int_0^{t_1}\mathrm{d}t_2 V_{\mathrm{I}}(t_1)V_{\mathrm{I}}(t_2) = \frac{1}{2}T\int_0^t \mathrm{d}t_1\int_0^{t_1}\mathrm{d}t_2 V_{\mathrm{I}}(t_1)V_{\mathrm{I}}(t_2),$$

$$\int_0^t dt_1 \int_0^{t_1} dt_2 \int_0^{t_2} dt_3\, V_I(t_1)V_I(t_2)V_I(t_3) = \frac{1}{3!} T \int_0^t dt_1 \int_0^t dt_2 \int_0^t dt_3\, V_I(t_1)V_I(t_2)V_I(t_3),$$

and so on in writing down the last result.

Note that in general $V_I(t_1)V_I(t_2) \neq V_I(t_2)V_I(t_1)$ since these are complicated operators combining $V$and $H_0$ which do not commute with each other. The perturbation theory is built up on this expansion. The non-commutation of operators at different times makes it complicated to analyze this series when compared with the path integral approach. However, the end results are the same. Readers can look at, for example, Ref. [1] for the technical details.

Note that the thermal time-evolution operator $U(\tau)$ satisfies a similar operator equation,

$$\frac{\partial}{\partial\tau} U(\tau) = -e^{\tau K_0} V e^{-\tau K} = -(e^{\tau K_0} V e^{-\tau K_0})e^{\tau K_0} e^{-\tau K} = -V(\tau)U(\tau), \tag{5.20a}$$

in the (imaginary-time) interaction picture where $K(K_0) = H(H_0) - \mu N$ (exercise) $U(\tau)$ has the formal solution

$$\begin{aligned} U(\tau) &= 1 - \int_0^\tau d\tau_1\, V(\tau_1) + \int_0^\tau d\tau_1 \int_0^{\tau_1} d\tau_2\, V(\tau_1)V(\tau_2) - \ldots \\ &= T_\tau \exp\left(-\int_0^\tau V(\tau')d\tau'\right), \end{aligned} \tag{5.20b}$$

which has the same form as the perturbative solution for $U(t)$ at real time. Because of this formal resemblance, a perturbation theory which has the same form as the real-time perturbation theory can be developed to compute physical quantities at thermal equilibrium [1, 2].

#### 5.1.2.3 Wick's Theorem in Dyson's Approach

As we shall see in the following, in Dyson's perturbation theory we also have to evaluate correlation functions of the kind

$$G_{nijm} = \langle T_\tau C_n(\tau) C_i^+(\tau') C_j(\tau') C_m^+(0)\rangle_0, \text{ etc.},$$

where we have considered a correlation function at imaginary time. A similar consideration applies at real time. $\langle\cdots\rangle_0$ denotes an average over eigenstates of the unperturbed Hamiltonian $H_0$. We can apply Wicks' theorem to evaluate this correlation function, with

$$\begin{aligned} G_{nijm} &= \langle T_\tau C_n(\tau) C_i^+(\tau') C_j(\tau') C_m^+(0)\rangle_0 \\ &= \langle T_\tau C_n(\tau) C_i^+(\tau')\rangle_0 \langle T_\tau C_j(\tau') C_m^+(0)\rangle_0 \\ &\quad \pm \langle T_\tau C_n(\tau) C_m^+(0)\rangle_0 \langle T_\tau C_j(\tau') C_i^+(\tau')\rangle_0 \end{aligned}$$

(+ for boson, − for fermion)

The physical origin of Wick's theorem can be seen easily by noting that the correlation function

$$G_{nijm} = \sum_N e^{-\beta E_N} \langle N|T_\tau C_n(\tau) C_i^+(\tau') C_j(\tau') C_m^+(0)|N\rangle$$

is evaluated over the unperturbed states $|N\rangle$, which are eigenstates of the particle number operators $n_k = C_k^+ C_k$ (for all $k$). To make sure that the expectation value $\langle N|T_\tau C_n(\tau) C_i^+(\tau') C_j(\tau') C_m^+(0)|N\rangle$ is non-zero, the final state $C_n(\tau) C_i^+(\tau') \times C_j(\tau') C_m^+(0)|N\rangle$ must be the same (aside from a multiplicative factor) as the initial state $|N\rangle$, and this is possible only if $i=n, j=m$ or $n=m, i=j$. This is exactly what happens in Wick's theorem; the time ordering is needed to keep track of the commutation rule of the $C$ operators.

#### 5.1.2.4 Example: One-Particle Green's Function

An important quantity in perturbation theory is the one-particle Green's function. One-particle Green's functions are usually defined with reference to a coherent state representation of a quantum field theory, that is when the action is represented in the form of a boson or fermion many-body system. They are important because they are the simplest correlation function of the kind $\langle B(\tau)A(\tau')\rangle$ and are building blocks in calculations of more complicated correlation functions which we shall encounter later in this chapter. In the imaginary-time formalism, the one-particle Green's functions are defined as

$$G_n(\tau-\tau') = -\langle T_\tau C_n(\tau) C_n^+(\tau')\rangle, \tag{5.21a}$$

where

$$\langle T_\tau C_n(\tau) C_n^+(\tau')\rangle = \begin{matrix} \langle C_n(\tau) C_n^+(\tau')\rangle, & \tau > \tau', \\ \pm\langle C_n^+(\tau') C_n(\tau)\rangle, & \tau \le \tau', \end{matrix} \tag{5.21b}$$

where the $+$ $(-)$ sign applies for bosons (fermions). $n$ represents chosen one-particle eigenstates which are usually the eigenstates for $H_0$ in perturbation theory. Note that we have assumed that the system is translationally invariant in time in Eq. (5.21a).

The average $\langle\cdots\rangle$ represents a thermal average, that is

$$\langle T_\tau C_n(\tau) C_n^+(\tau')\rangle = \frac{\mathrm{Tr}[e^{-\beta(H-\mu N)} T_\tau((e^{\tau(H-\mu N)} C_n e^{-\tau(H-\mu N)})(e^{-\tau'(H-\mu N)} C_n^+ e^{\tau'(H-\mu N)}))]}{\mathrm{Tr}[e^{-\beta(H-\mu N)}]}, \tag{5.22}$$

where by definition $C_n(\tau) = e^{\tau(H-\mu N)} C_n e^{-\tau(H-\mu N)}$, and represents the Heisenberg time-dependent operator at imaginary time $t \to -i\tau$. $\rho = e^{-\beta(H-\mu N)}$ is the equilibrium density matrix for the system. An important property of the Green's function following from the above definition is that (exercise)

$$G_n(\tau) = \pm G_n(\tau+\beta) \; (+(-) \text{ for boson (fermion)}),$$

which shows that the Green's functions are periodic (anti-periodic) functions at imaginary time.

In common cases of applications the system is translationally invariant in space and the eigenstates $n$ are chosen to be momentum eigenstates. In this case, $G_n(\tau-\tau') \to G(\vec{p},\tau-\tau')$. In the following we shall evaluate the Green's functions for systems of free bosons and fermions with Hamiltonian $H=\sum_{\vec{p}}\varepsilon_{\vec{p}}C^+_{\vec{p}}C_{\vec{p}}$.

The evaluation of the Green's function in this case is straightforward. For non-interacting particles,

$$C_{\vec{p}}(\tau) = e^{\tau(H-\mu N)}C_{\vec{p}}e^{-\tau(H-\mu N)} = e^{-\tau\xi_{\vec{p}}}C_{\vec{p}} \tag{5.23a}$$

and

$$C^+_{\vec{p}}(\tau) = e^{-\tau(H-\mu N)}C^+_{\vec{p}}e^{\tau(H-\mu N)} = e^{\tau\xi_{\vec{p}}}C^+_{\vec{p}}, \tag{5.23b}$$

where $\xi_{\vec{p}}=\varepsilon_{\vec{p}}-\mu$.

**Homework Q5a** Derive Equation (5.23a,b).

Therefore,

$$G(\vec{p},\tau-\tau') = -\theta(\tau)\frac{\mathrm{Tr}[\rho e^{-(\tau-\tau')\xi_{\vec{p}}}C_{\vec{p}}C^+_{\vec{p}}]}{Z} \mp \theta(-\tau)\frac{\mathrm{Tr}[\rho e^{-(\tau-\tau')\xi_{\vec{p}}}C^+_{\vec{p}}C_{\vec{p}}]}{Z}$$

$$= -e^{-(\tau-\tau')\xi_{\vec{p}}}[\theta(\tau)\langle C_{\vec{p}}C^+_{\vec{p}}\rangle \pm \theta(-\tau)\langle C^+_{\vec{p}}C_{\vec{p}}\rangle]$$

$$= -e^{-(\tau-\tau')\xi_{\vec{p}}}[\theta(\tau)(1\pm n(\xi_{\vec{p}})) \pm \theta(-\tau)n(\xi_{\vec{p}})]$$

or

$$G(\vec{p},\tau-\tau') = -e^{-(\tau-\tau')\xi_{\vec{p}}}[\theta(\tau) \pm n(\xi_{\vec{p}})], \tag{5.24}$$

where $n(\xi_{\vec{p}}) = \dfrac{1}{e^{\beta\xi_{\vec{p}}}\mp 1}$ are boson (−) and fermion (+) occupation numbers.

**Homework Q5b** Derive Equation (5.24).

It is easy to obtain the frequency Fourier transform of $G$,

$$G(\vec{p},i\omega_n) = \int_0^\beta d\tau e^{i\omega_n\tau}G(\vec{p},\tau) = -(1\pm n(\xi_{\vec{p}}))\left[\frac{e^{\beta(i\omega_n-\xi_{\vec{p}})}-1}{i\omega_n-\xi_{\vec{p}}}\right]. \tag{5.25a}$$

Note that, because of the periodicity (anti-periodicity) of the Green's functions in $\tau$,

$$\omega_n = \begin{cases} \dfrac{2n\pi}{\beta} & \text{(boson)}, \\ \dfrac{(2n+1)\pi}{\beta} & \text{(fermion)}. \end{cases}$$

Therefore, $e^{i\beta\omega_n} = \pm 1$, $(+)$ for boson, $(-)$ for fermion, and

$$G(\vec{p}, i\omega_n) = \frac{1}{i\omega_n - \xi_{\vec{p}}}, \tag{5.25b}$$

valid for both bosons and fermions.

**Homework Q5c** Derive Equation (5.25a, b).

Note that the frequency-transformed Green's function thus obtained is identical to the two-point Green's function obtained from the path integral approach, although no time ordering is mentioned at all in the path integral approach. This is not an accident. It can be shown that the frequency-transformed Green's function obtained in Dyson's approach is mathematically identical to the two-point Green's function obtained in the path integral approach, even when interaction is included in the action. You can look at for example Ref. [2] for more details.

#### 5.1.2.5 Perturbation Expansion for One-Particle Green's Function

For interacting systems, a perturbation theory based on Dyson's $U$-matrix approach can be derived. In this case, we write $H = H_0 + V$, $K = H - \mu N$, and $K_0 = H_0 - \mu N$. The Green's function can be written as

$$\begin{aligned} G_n(\tau) = \langle T_\tau C_n(\tau) C_n^+(0)\rangle &= \frac{\mathrm{Tr}[e^{-\beta K} e^{\tau K} C_n e^{-\tau K} C_n^+)]}{\mathrm{Tr}[e^{-\beta K}]} \qquad (\tau > 0) \\ &= \frac{\mathrm{Tr}[e^{-\beta K_0}(e^{\beta K_0} e^{-\beta K})(e^{\tau K} e^{-\tau K_0})(e^{\tau K_0} C_n e^{-\tau K_0})(e^{\tau K_0} e^{-\tau K}) C_n^+)]}{\mathrm{Tr}[e^{-\beta K_0}(e^{\beta K_0} e^{-\beta K})]} \\ &= \frac{\mathrm{Tr}[e^{-\beta K_0} U(\beta) U^{-1}(\tau)(e^{\tau K_0} C_n e^{-\tau K_0}) U(\tau) C_n^+)]}{\mathrm{Tr}[e^{-\beta K_0} U(\beta)]} \\ &= \frac{\langle T_\tau[U(\beta) C_n(\tau) C_n^+(0))]\rangle_0}{\langle U(\beta)\rangle_0}, \end{aligned} \tag{5.26a}$$

where $U(\tau) = e^{\tau K_0} e^{-\tau K}$, $U^{-1}(\tau) = e^{\tau K} e^{-\tau K_0}$, and $\langle T_\tau \cdots\rangle_0 = \mathrm{Tr}[e^{-\beta K_0} T_\tau \cdots]$. A formal perturbation theory for the Green's function can be set up by expanding $U(\beta)$ in a series of the strength of interaction. First, we consider the numerator in Equation (5.26a). From Equation (5.20b), we obtain

$$\langle T_\tau U(\beta) C_n(\tau) C_n^+(0))\rangle_0 = \sum_{n=0}^{\infty} \frac{(-1)^n}{n!} \int_0^\beta d\tau_1 \cdots \int_0^\beta d\tau_n \langle T_\tau C_n(\tau) \widehat{V}(\tau_1) \cdots \widehat{V}(\tau_n) C_n^+(0))\rangle_0, \tag{5.26b}$$

where $\langle T_\tau \cdots\rangle_0$ can be evaluated by Wick's theorem.

As an example we shall consider an interacting fermion gas with $\widehat{V} = \frac{1}{2}\sum_{i,j,k,l} V_{ijkl} C_i^+ C_j^+ C_k C_l$. We first consider Equation (5.26a) to first order.

In this case, Wick's theorem gives

$$
\begin{aligned}
&\langle T_\tau C_n(\tau)C_i^+(\tau')C_j^+(\tau')C_k(\tau')C_l(\tau')C_n^+(0)\rangle_0\\
&= -\langle T_\tau C_n(\tau)C_i^+(\tau')\rangle_0\langle T_\tau C_k(\tau')C_j^+(\tau')\rangle_0\langle T_\tau C_l(\tau')C_n^+(0)\rangle_0\\
&\quad + \langle T_\tau C_n(\tau)C_i^+(\tau')\rangle_0\langle T_\tau C_k(\tau')C_n^+(0)\rangle_0\langle T_\tau C_l(\tau')C_j^+(\tau')\rangle_0\\
&\quad + \langle T_\tau C_n(\tau)C_j^+(\tau')\rangle_0\langle T_\tau C_k(\tau')C_i^+(\tau')\rangle_0\langle T_\tau C_l(\tau')C_n^+(0)\rangle_0\\
&\quad - \langle T_\tau C_n(\tau)C_j^+(\tau')\rangle_0\langle T_\tau C_k(\tau')C_n^+(0)\rangle_0\langle T_\tau C_l(\tau')C_i^+(\tau')\rangle_0\\
&\quad - \langle T_\tau C_n(\tau)C_n^+(0)\rangle_0\langle T_\tau C_k(\tau')C_i^+(\tau')\rangle_0\langle T_\tau C_l(\tau')C_j^+(\tau')\rangle_0\\
&\quad + \langle T_\tau C_n(\tau)C_n^+(0)\rangle_0\langle T_\tau C_k(\tau')C_j^+(\tau')\rangle_0\langle T_\tau C_l(\tau')C_i^+(\tau')\rangle_0
\end{aligned}
\tag{5.27a}
$$

and $-\int_0^\beta d\tau_1 \langle T_\tau C_n(\tau)\widehat{V}(\tau_1)C_n^+(0))\rangle_0$

$$
= \frac{1}{2}\int_0^\beta d\tau' \sum_k \left\{
\begin{aligned}
&V_{nkkn}g_n(\tau-\tau')g_k(0)g_n(\tau'-0)\\
&-V_{nknk}g_n(\tau-\tau')g_n(\tau'-0)g_k(0)\\
&-V_{knkn}g_n(\tau-\tau')g_k(0)g_n(\tau'-0)\\
&+V_{knnk}g_n(\tau-\tau')g_n(\tau'-0)g_k(0)\\
&-V_{ikik}g_n(\tau-0)g_i(0)g_k(0)\\
&+V_{ikki}g_n(\tau-0)g_i(0)g_k(0)
\end{aligned}
\right\},
\tag{5.27b}
$$

where $g_n(\tau) = \langle T_\tau C_n(\tau)C_n^+(0)\rangle_0$.

For most cases of applications $V_{ijkl} = V_{jilk}$. In this case, the right-hand side of Equation (5.27b) becomes

$$
\frac{1}{2}\int_0^\beta d\tau' \sum_k \left\{
\begin{aligned}
&2V_{nkkn}g_n(\tau-\tau')g_k(0)g_n(\tau'-0)\\
&-2V_{nknk}g_n(\tau-\tau')g_n(\tau'-0)g_k(0)\\
&-V_{ikik}g_n(\tau-0)g_i(0)g_k(0)\\
&+V_{ikki}g_n(\tau-0)g_i(0)g_k(0)
\end{aligned}
\right\},
\tag{5.27c}
$$

and the four terms can be represented by the following diagrams:

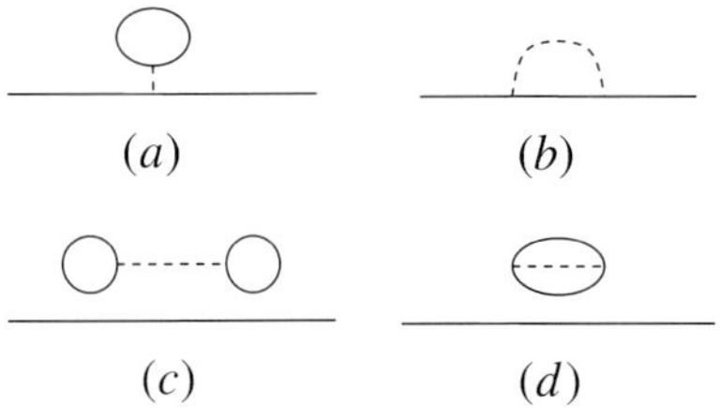

(a) (b) (c) (d)

The first and second terms (diagrams (a), (b)) can be written as $\int_0^\beta d\tau' g_n(\tau-\tau')(\Sigma_H + \Sigma_F)g_n(\tau'-0)$, where $\Sigma_{Hn} = \sum_k V_{nkkn}g_k(0)$ and $\Sigma_{Fn} = -\sum_k V_{nknk}g_k(0)$ are called the Hartree and Fock self energies of the system. Note that the third and fourth terms (diagrams (c), (d)) can be written as $g_n(\tau)\times(\cdots)$. The interaction contributes terms 'disconnected' from the one-particle Green's function.

Disconnected diagrams also appear at higher-order perturbations. The following are some of the diagrams appearing in second-order perturbation theory. (c) and (d) are 'disconnected' diagrams:

(a) (b)

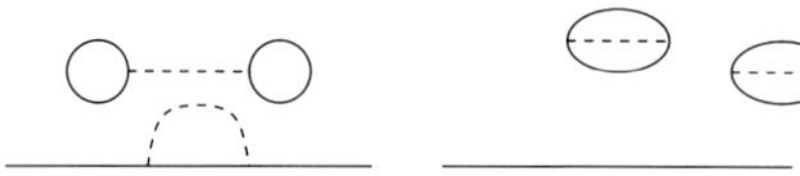

(c) (d)

Note that the 'disconnected' parts are the same as diagrams contributing to the free energy $\sim\langle U(\beta)\rangle_0$ (see previous section). It can be proved by analyzing the general structure of the diagrams (see Refs. [2, 4] for details) that

$$\langle T_\tau U(\beta) C_n(\tau) C_n^+(0)\rangle_0$$

$$= \langle T_\tau U(\beta)\rangle_0 \times \sum_{n=0}^{\infty} \frac{(-1)^n}{n!} \int_0^\beta d\tau_1 \cdots \int_0^\beta d\tau_n \langle T_\tau C_n(\tau) \widehat{V}(\tau_1) \cdots \widehat{V}(\tau_n) C_n^+(0)\rangle\rangle_{0C},$$

where $\langle T_\tau C_n(\tau) \widehat{V}(\tau_1) \cdots \widehat{V}(\tau_n) C_n^+(0)\rangle\rangle_{0C}$ includes only those diagrams which are connected to $C_n(\tau)$, $C_n^+(0)$. Therefore, from Equation (5.26a),

$$\langle T_\tau C_n(\tau) C_n^+(0)\rangle = \sum_{n=0}^{\infty} \frac{(-1)^n}{n!} \int_0^\beta d\tau_1 \cdots \int_0^\beta d\tau_n \langle T_\tau C_n(\tau) \widehat{V}(\tau_1) \cdots$$
$$\times \widehat{V}(\tau_n) C_n^+(0)\rangle\rangle_{0C} \quad (\tau > 0). \tag{5.28}$$

The disconnected diagrams cancel exactly the denominator, meaning that we just need to sum up all the 'connected' diagrams to calculate the one-particle Green's function.

The diagram (b) in the above figure is particularly interesting because it is composed of two Fock self energies linked by a *single* Green's function. The diagram is called reducible because it can be cut into two parts by cutting one fermion Green's function line ($g_k$). Mathematically, the diagram is given by the expression

$$\int d\tau' \int d\tau'' g_n(\tau-\tau') \Sigma_{Fn} g_n(\tau'-\tau'') \Sigma_{Fn} g_n(\tau''-0).$$

Together with the first-order term $\int_0^\beta d\tau' g_n(\tau-\tau') \Sigma_F g_n(\tau'-0)$, the corresponding-Fourier transform has a simple form

$$G_n(i\omega) \sim g_n(i\omega) + g_n(i\omega) \Sigma_{Fn} g_n(i\omega) + g_n(i\omega) \Sigma_{Fn} g_n(i\omega) \Sigma_{Fn} g_n(i\omega),$$

where the first term is the non-interacting fermion Green's function,

$$g_n(i\omega) = \frac{1}{i\omega - \xi_n}. \tag{5.28c}$$

More generally, one may imagine that higher-order diagrams exist corresponding to higher powers in the geometric series of the Fock self energy, that is

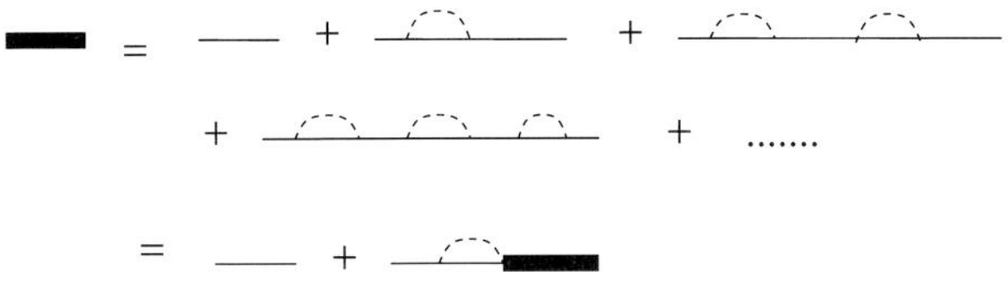

where

 = "Full" electron Green's function

Summing over the series, the corresponding Fourier-transformed Green's function has the form

$$G_n(i\omega) = \frac{1}{i\omega - \xi_n - \Sigma_{nF}}. \text{ (exercise)}$$

Most generally, it can be proved that the Green's function can be written in a general form

$$G_n(i\omega) = \frac{1}{i\omega - \xi_n - \Sigma_n(i\omega)}, \tag{5.29a}$$

where $\Sigma_n(i\omega)$ is the (total) fermion self energy which includes all diagrams that cannot be cut into two parts by cutting any single fermion Green's function line ($g_0$), that is

$$\Sigma_n(i\omega) = \Sigma_H - \Sigma_F + (a) + \ldots, \tag{5.29b}$$

where (a) = diagram (a) in second-order perturbation theory. In this way, the calculation of the Green's function reduces to the calculation of the self energy $\Sigma_n(i\omega)$ in perturbation theory.

Equation (5.29a) will be derived in an alternative way in Chapter 7, which provides a different way to understand the physical meaning of the self energy. The self energy to second order in perturbation theory is evaluated in Appendix A for a simple example.

#### 5.1.2.6 **Spectral Representation**

Spectral representation is a very useful tool to extract general properties of correlation functions. It is important because correlation functions are often difficult to compute directly. In spectral representation we write the correlation function in terms of eigenstates of the system $|m\rangle$ and look at its general behavior.

To see how it works, we consider the one-particle Green's function

$$G(\vec{p},\tau) = -\langle C_{\vec{p}}(\tau)C^+_{\vec{p}}(0)\rangle = -\frac{\mathrm{Tr}[e^{-\beta(H-\mu N)}((e^{\tau(H-\mu N)}C_{\vec{p}}e^{-\tau(H-\mu N)})C^+_{\vec{p}})]}{\mathrm{Tr}[e^{-\beta(H-\mu N)}]} \quad (\tau>0).$$

Using the completeness relation $I=\sum_m |m\rangle\langle m|$, we can write the Green's function as

$$\begin{aligned} G(\vec{p},\tau) &= -\frac{\sum_{n,m}\langle n|e^{-\beta(H-\mu N)}e^{\tau(H-\mu N)}C_{\vec{p}}|m\rangle\langle m|e^{-\tau(H-\mu N)}C^+_{\vec{p}}|n\rangle}{\sum_n e^{-\beta(E_n-\mu N)}} \\ &= -\frac{\sum_{n,m} e^{-\beta(E_n-\mu N)}e^{\tau(E_n-\mu N)}e^{-\tau(E_m-\mu(N+1))}\langle n|C_{\vec{p}}|m\rangle\langle m|C^+_{\vec{p}}|n\rangle}{\sum_n e^{-\beta(E_n-\mu N)}} \\ &= -e^{\beta\Omega}\sum_{n,m} e^{-\beta(E_n-\mu N)}e^{\tau(E_n-E_m+\mu)}|\langle n|C_{\vec{p}}|m\rangle|^2, \end{aligned} \tag{5.30a}$$

where $|n\rangle$ is an eigenstate with $N$ particles and $|m\rangle$ is an eigenstate with $N+1$ particles. $\Omega=-\frac{1}{\beta}\ln Z$.

Therefore,

$$G(\vec{p},i\omega_n) = \int_0^\beta d\tau e^{i\omega_n\tau}G(\vec{p},\tau) = e^{\beta\Omega}\sum_{n,m}\frac{(e^{-\beta(E_n-\mu N)}-e^{-\beta(E_m-\mu(N+1))})|\langle n|C_{\vec{p}}|m\rangle|^2}{i\omega_n+E_n-E_m+\mu}. \tag{5.30b}$$

We introduce the spectral function

$$A(\vec{p},\omega) = e^{\beta\Omega}\sum_{n,m}(e^{-\beta(E_n-\mu N)}-e^{-\beta(E_m-\mu(N+1))})|\langle n|C_{\vec{p}}|m\rangle|^2 2\pi\delta(\omega+E_n-E_m+\mu), \tag{5.31a}$$

so that

$$G(\vec{p},i\omega_n) = \int_{-\infty}^{\infty}\frac{d\omega'}{2\pi}\frac{A(\vec{p},\omega')}{i\omega_n-\omega'}, \tag{5.31b}$$

which is the spectral representation for the one-particle Green's function.

Note that

$$\begin{aligned} \frac{1}{2\pi}\int_{-\infty}^{\infty} d\omega A(\vec{p},\omega) &= e^{\beta\Omega}\sum_{n,m}(e^{-\beta(E_n-\mu N)}-e^{-\beta(E_m-\mu(N+1))})|\langle n|C_{\vec{p}}|m\rangle|^2 \\ &= e^{\beta\Omega}\sum_{n,m}(e^{-\beta(E_n-\mu N)}-e^{-\beta(E_m-\mu(N+1))})\langle n|C_{\vec{p}}|m\rangle\langle m|C^+_{\vec{p}}|n\rangle \\ &= e^{\beta\Omega}\sum_{n(m)}(e^{-\beta(E_n-\mu N)}\langle n|C_{\vec{p}}C^+_{\vec{p}}|n\rangle + e^{-\beta(E_m-\mu(N+1))}\langle m|C^+_{\vec{p}}C_{\vec{p}}|m\rangle) \\ &= e^{\beta\Omega}\mathrm{Tr}[e^{-\beta H}(C_{\vec{p}}C^+_{\vec{p}}+C^+_{\vec{p}}C_{\vec{p}})] = 1, \end{aligned}$$

which is an important relation (sum rule) satisfied by the spectral function.

For non-interacting particles, $G(\vec{p}, i\omega_n) = \frac{1}{i\omega_n - \xi_{\vec{p}}}$ and $A(\vec{p}, \omega) = 2\pi\delta(\omega - \xi_{\vec{p}})$. The presence of a single δ-function in the spectral function reflects that, for non-interacting systems, if $|n\rangle$ is an eigenstate of $N$ particles with energy $E_n$ then $C^+_{\vec{p}} |n\rangle = |m\rangle$ is an eigenstate of the system with $N + 1$ particles. The energy of the eigenstate $|m\rangle$ is $E_m = E_n + \varepsilon_{\vec{p}}$ and the difference $E_m - E_n$ is independent of $|n\rangle$. This is why a single δ-function appears in the spectral function.

In the presence of interaction the single δ-function structure of the spectral function will be lost because in general $C^+_{\vec{p}} |n\rangle = |m\rangle$ is no longer an eigenstate of the system. Physically, a single-particle state will be 'dressed' by the interaction and carries a 'cloud' of disturbed particles around it when it travels, much like the water wave surrounding a traveling ship. In most cases the single-particle state will decay because of resistance or dissipation from the cloud it carries. If the particle travels long enough before it decays, we expect that

$$A(\vec{p}, \omega) \sim \frac{z_{\vec{p}}}{\pi} \frac{\Gamma_{\vec{p}}}{(\omega - \xi_{\vec{p}})^2 + \Gamma^2_{\vec{p}}} + A_{\text{inc}}(\vec{p}, \omega). \tag{5.32}$$

The first term represents a particle with energy $\xi_{\vec{p}}$ and a finite lifetime $\Gamma^{-1}_{\vec{p}}$. $z_{\vec{p}} < 1$ measures the probability that the added particle represented by $C^+_{\vec{p}} |n\rangle = |m'\rangle$ is in the 'dressed' single-particle state. The second term is an incoherent part with no sharp structure representing the probability that the added particle just belongs to the 'cloud' of another single-particle state.

$A(\vec{p}, \omega)$ can be related to the fermion self energy. Note that Equation (5.31b) implies that

$$-2\text{Im}G(\vec{p}, \omega + i\delta) = A(\vec{p}, \omega). \tag{5.33a}$$

Therefore, using Equation (5.29a), we obtain

$$A(\vec{p}, \omega) = \frac{-2\text{Im}\Sigma(\vec{p}, \omega)}{(\omega - \xi_{\vec{p}} - \text{Re}\Sigma(\vec{p}, \omega))^2 + (\text{Im}\Sigma(\vec{p}, \omega))^2}. \tag{5.33b}$$

Note that, provided that $\text{Im}\Sigma(\vec{p}, \omega)$ is small, the spectral function is peaked at $\omega = E_{\vec{p}}$ satisfying $E_{\vec{p}} = \xi_{\vec{p}} + \text{Re}\Sigma(\vec{p}, E_{\vec{p}})$. For $\omega \sim E_{\vec{p}}$, we may expand the self energy to obtain (exercise)

$$A(\vec{p}, \omega) \sim \frac{-2\text{Im}\Sigma(\vec{p}, E_{\vec{p}})}{z^{-2}_{\vec{p}}(\omega - E_{\vec{p}})^2 + (\text{Im}\Sigma(E_{\vec{p}}))^2} = \frac{-2z_{\vec{p}}\Gamma_{\vec{p}}}{(\omega - E_{\vec{p}})^2 + \Gamma_{\vec{p}}^2}, \tag{5.34a}$$

where

$$z_{\vec{p}} = \left(1 - \left.\frac{\partial \text{Re}\Sigma(\vec{p}, \omega)}{\partial\omega}\right|_{\omega = E_{\vec{p}}}\right)^{-1} \quad \text{and} \quad \Gamma_{\vec{p}} = z_{\vec{p}}\text{Im}\Sigma(\vec{p}, E_{\vec{p}}), \tag{5.34b}$$

which has precisely the form (5.32). For $\omega$ far away from $E_{\vec{p}}$, the spectral function becomes a smooth function ($\sim A_{\text{inc}}(\vec{p}, \omega)$).

For *many-fermion systems* with a sharp Fermi surface, it is found that $\Gamma_{\vec{p}} \sim (|\vec{p}| - p_F)^2 \to 0$ as $|\vec{p}| \to p_F$, where $p_F$ is the Fermi momentum. This kind of system is called a Fermi liquid, and represents an important class of many-fermion systems in condensed matter physics. We shall study fermion liquids in more detail in Chapter 10.

## 5.2
## Variational Approach and Perturbation Theory

In condensed matter physics there is often no natural way to separate a Hamiltonian into two parts $H = H_0 + V$, where $H_0$ can be diagonalized exactly and $V$ is naturally small. In this situation we can only determine the ground state approximately by a variational method. We make use of the theorem that the ground-state energy is the lowest possible energy in the system and we write a trial wavefunction $|\psi_G\rangle = |\psi_G; a, b, c, \ldots\rangle$, where $a, b, c, \ldots$ are variational parameters determined by minimizing the ground-state energy with respect to these parameters, that is

$$\frac{\delta\langle\psi_G; a, b, c, \ldots|H|\psi_G; a, b, c, \ldots\rangle}{\delta a, b, c, \ldots} = 0. \tag{5.35}$$

The success of this approach depends on our ability to 'guess' a good enough trial wavefunction. The weakness of this approach is that we can only determine the ground state of the system and cannot obtain the excited states directly.

The weakness of the variational approach can be overcome by combining the variational approach with perturbation theory. The trick is to introduce a 'trial Hamiltonian' $H_{\text{trial}}\ (a, b, c, \ldots)$, where $a, b, c, \ldots$ are variational parameters. The trial Hamiltonian is chosen such that it can be diagonalized exactly and the 'trial' wavefunction is the ground-state wavefunction of $H_{\text{trial}}\ (a, b, c, \ldots)$, that is

$$H_{\text{trial}}(a, b, c, \ldots)|\psi^{(0)}_{a,b,c\ldots}\rangle = E^{(0)}_G(a, b, c, \ldots)|\psi^{(0)}_{a,b,c\ldots}\rangle. \tag{5.36a}$$

The state $|\psi^{(0)}_{a,b,c\ldots}\rangle$ is taken as a variational wavefunction for our 'real' Hamiltonian and the energy of the system is minimized with respect to H, i.e., $\frac{\delta\langle\psi^{(0)}_{a,b,c\ldots}|H|\psi^{(0)}_{a,b,c\ldots}\rangle}{\delta a,b,c,\ldots} = 0$ to determine the optimal parameters $\bar{a}, \bar{b}, \bar{c}, \ldots$.

The advantage of introducing the trial Hamiltonian is that, after optimization, we obtain an optimal trial Hamiltonian $H_{\text{trial}}(\bar{a}, \bar{b}, \bar{c}, \ldots)$, where $\bar{a}, \bar{b}, \bar{c}, \ldots$ are the optimized parameters. We can then write

$$H = H_{\text{trial}}(\bar{a}, \bar{b}, \bar{c}, \ldots) + (H - H_{\text{trial}}(\bar{a}, \bar{b}, \bar{c}, \ldots)), \tag{5.36b}$$

and the wavefunction can be improved gradually by treating $V = H - H_{\text{trial}}$ as perturbation.

This approach can also be generalized to finite temperature where the free energy $F = -\frac{1}{\beta}\ln Z$ is minimized. The free energy satisfies the minimization principle $F < F_0 + \langle H - H_0\rangle_0$, where $F_0$ is the free energy computed from the trial partition function $Z_0 = e^{-\beta H_0}$ and $\langle O\rangle_0$ is the (thermal) expectation value of the operator $O$ computed with $Z_0$. We shall discuss an example of this approach (Hartree–Fock approximation) in the following.

### 5.2.1
### Example: Hartree–Fock Approximation

Consider a system of fermions moving in an external potential $V(\vec{x})$ and interacting with each other through a potential $U(\vec{x})$. The Hamiltonian of the system is in the coherent state representation

$$\begin{aligned} H = & \int \mathrm{d}^d x \left[ \frac{\hbar^2}{2m} |\nabla \widehat{\psi}(\vec{x})|^2 + V(\vec{x}) |\widehat{\psi}(\vec{x})|^2 \right] \\ & + \frac{1}{2} \iint \mathrm{d}^d x \mathrm{d}^d x' \, U(\vec{x}-\vec{x}') |\widehat{\psi}(\vec{x})|^2 |\widehat{\psi}(\vec{x}')|^2 . \end{aligned} \tag{5.37}$$

We consider a trial ground-state wavefunction, which is the ground-state wavefunction of the trial Hamiltonian

$$H_{\text{trial}} = \int \mathrm{d}^d x \mathrm{d}^d x' \widehat{\psi}^*(\vec{x}) H_{\text{eff}}(\vec{x}, \vec{x}') \widehat{\psi}(\vec{x}'), \tag{5.38a}$$

where

$$H_{\text{eff}}(\vec{x}, \vec{x}') = \sum_n E_n \varphi_n(\vec{x}) \varphi_n^*(\vec{x}'). \tag{5.38b}$$

$H_{\text{trial}}$ is a one-particle Hamiltonian and we can always represent it in terms of its one-particle eigenstates $\varphi_n(\vec{x})$ and eigenenergies $E_n$. The many-body trial wavefunction $|\psi_{\text{trial}}\rangle$ is a Slater determinant of $\varphi_n(\vec{x})$ characterized by occupation numbers of the states $n$. The energy $E_{\text{trial}} = \langle \psi_{\text{trial}} | H | \psi_{\text{trial}} \rangle$ is

$$E_{\text{trial}} = \sum_k \varepsilon_{0k} n_k + \frac{1}{2} \sum_{kk'} (U_{kk';kk'} - U_{kk';k'k}) n_k n_{k'}, \tag{5.39a}$$

where

$$\varepsilon_{0k} = \int \mathrm{d}^d x \left[ \frac{\hbar^2}{2m} |\nabla \varphi_k(\vec{x})|^2 + V(\vec{x}) |\varphi_k(\vec{x})|^2 \right] \tag{5.39b}$$

and

$$U_{ab;cd} = \iint \mathrm{d}^d x \mathrm{d}^d x' \, U(\vec{x}-\vec{x}') \varphi_a^*(\vec{x}) \varphi_b^*(\vec{x}') \varphi_c(\vec{x}') \varphi_d(\vec{x}). \tag{5.39c}$$

**Homework Q6**

Derive Equations (5.39a)–(5.39c).

(Hint: to evaluate $E_{\text{trial}}$, first we have to express the Hamiltonian (5.37) in terms of annihilation and creation operators for the single-particle states $\varphi_n(\vec{x})$. You can use the identity $\widehat{\psi}(\vec{x}) = \sum_n \varphi_n(\vec{x})\widehat{c}$ and so on (see Chapter 3) to help. You can then use Wick's theorem to evaluate expectation values of the form $\langle \widehat{c}_n^+ \widehat{c}_m \widehat{c}_k^+ \widehat{c}_l \rangle$.)

Note that the trial Hamiltonian is completely specified by its eigenwavefunctions $\varphi_k(\vec{x})$ and eigenenergies $E_k$. Therefore, to minimize $E_{\rm trial}$ with respect to $H_{\rm trial}$, we can minimize it with respect to $E_k$ and $\varphi_k(\vec{x})$. The variation in $E_{\rm trial}$ is therefore

$$\begin{aligned}\delta E_{\rm trial} &= \sum_k \left[\varepsilon_{0k} - \mu + \sum_{k'} (U_{kk';kk'} - U_{kk';k'k}) n_{k'}\right] \frac{\delta n_k}{\delta E_k} \delta E_k \\ &+ \sum_k n_k \frac{\delta}{\delta \varphi_k} \left[\varepsilon_{0k} + \sum_{k'} (U_{kk';kk'} - U_{kk';k'k}) n_{k'}\right] \delta\varphi_k, \end{aligned} \tag{5.40}$$

where $\mu$ is a Lagrange multiplier introduced to enforce the constraint $\sum_k n_k = N$. First, we consider the variation in $\delta n_k$, neglecting $\delta\varphi_k$.

Let $\bar{E}_k = \varepsilon_{0k} + \sum_{k'} (U_{kk';kk'} - U_{kk';k'k}) n_{k'}$. For $\delta E_{\rm trial} > 0$, we must have

$$\delta n_k \begin{cases} > 0 & \bar{E}_k > \mu \\ < 0 & \bar{E}_k < \mu \end{cases} \quad \text{or} \quad n_k \begin{cases} = 0 & \bar{E}_k > \mu \\ = 1 & \bar{E}_k < \mu \end{cases} \quad \text{in the ground state,}$$

meaning that $E_k = \bar{E}_k$ and the wavefunction is the ground-state wavefunction of the trial Hamiltonian

$$H_{\rm trial} = H_{\rm HF} = \sum_k \left[\varepsilon_{0k} + \sum_{E_{k'} < \mu} (U_{kk';kk'} - U_{kk';k'k}) n_{k'}\right] n_k \tag{5.41a}$$

or, in real space,

$$\begin{aligned} H_{\rm HF} &= \int d^d x \left[\frac{\hbar^2}{2m} |\nabla \widehat{\psi}(\vec{x})|^2 + V(\vec{x}) |\widehat{\psi}(\vec{x})|^2\right] \\ &+ \frac{1}{2} \int\int d^d x d^d x' \, U(\vec{x} - \vec{x}')(\langle |\widehat{\psi}(\vec{x})|^2\rangle |\widehat{\psi}(\vec{x}')|^2 - \langle \widehat{\psi}^*(\vec{x}')\widehat{\psi}(\vec{x})\rangle \widehat{\psi}^*(\vec{x})\widehat{\psi}(\vec{x}'), \end{aligned}$$

where

$$\langle |\widehat{\psi}(\vec{x})|\rangle^2 = \sum_{E_k < 0} |\varphi_k(\vec{x})|^2 \quad \text{and} \quad \langle \widehat{\psi}^*(\vec{x}')\widehat{\psi}(\vec{x})\rangle^2 = \sum_{E_k < 0} \varphi_k^*(\vec{x}')\varphi_k(\vec{x}). \tag{5.41b}$$

Note that the second $\frac{\delta}{\delta\varphi_k}[\ldots]$ term in Equation (5.40) is automatically zero if $\varphi_k(\vec{x})$ is an eigenstate of the Hartree–Fock Hamiltonian $H_{\rm HF}$. $H_{\rm trial}$ and the $\varphi_k$ have to be determined self consistently; this scheme is called the Hartree–Fock approximation.

The approach can be generalized straightforwardly to finite temperature by minimizing $F_0 + \langle H - H_{\rm trial}\rangle_0$. In the Hartree–Fock approximation,

$$F_0 = -\frac{1}{\beta} \sum_k \ln(1 + e^{-\beta(E_k - \mu)}) \tag{5.42a}$$

and

$$\langle H - H_{\rm trial}\rangle_0 = \sum_k (\varepsilon_{0k} - E_k) n_k + \frac{1}{2} \sum_{kk'} (U_{kk';kk'} - U_{kk';k'k}) n_k n_{k'}, \tag{5.42b}$$

where $n_k = \frac{1}{1 + e^{\beta(E_k - \mu)}}$.

Minimizing $F_0 + \langle H - H_{\text{trial}}\rangle_0$ with respect to $E_k$ and $\varphi_k(\vec{x})$, we obtain

$$0 = \sum_k \left[\varepsilon_{0k} - E_k + \sum_{k'} (U_{kk';kk'} - U_{kk';k'k}) n_{k'}\right] \frac{\delta n_k}{\delta E_k} \delta E_k + \sum_k n_k \frac{\delta}{\delta \varphi_k} \left[\varepsilon_{0k} + \sum_{k'} (U_{kk';kk'} - U_{kk';k'k}) n_{k'}\right] \delta \varphi_k \tag{5.42c}$$

which produces again the Hartree–Fock equation

$$E_k = \varepsilon_{0k} + \sum_{k'} (U_{kk';kk'} - U_{kk';k'k}) n_{k'}.$$

The only difference is that $n_k = \frac{1}{1+e^{\beta(E_k-\mu)}}$ is now the finite-temperature Fermi–Dirac distribution.

**Homework Q7** The Green's function in Hartree–Fock approximation.

In this approximation we replace $H$ by $H_{\text{HF}}$ when evaluating the one-particle Green's function $G(\vec{x}, \vec{x}'; t-t')$. Write down an expression for $G(\vec{x}, \vec{x}'; t-t')$ in terms of the Hartree–Fock wavefunctions $\varphi_k$ and energies $E_k$. Discuss how $G(p; \omega)$ would look like if (i) $p$ represents the trial state $\varphi_k$ and (ii) if $p$ represents an eigenstate of the non-interacting Hamiltonian $H_0 = \int d^d x \left[\frac{\hbar^2}{2m} |\nabla \psi(\vec{x})|^2 + V(\vec{x}) |\psi(\vec{x})|^2\right]$. (Hint: the spectral representation is useful in discussing (ii).) Identify the self energy in each case and compare it with the Hartree–Fock self energy we derived in the last section.

## 5.3 Some General Properties of Correlation Functions

### 5.3.1 Linear Response Theory

As we discussed before, in quantum field theory it is often not useful to compute the ground state and/or excited state wavefunctions of the system since they are not direct physical observables. The useful quantities that are computed are often two-particle Green's functions obtained from linear response theory, which is a systematic mathematical procedure to relate physical observables in experiments to quantities we should calculate when the experiment introduces only a small perturbation to the system.

The problem can be formulated in the following way. We are interested in making observation on a system which is described by an equilibrium density matrix $\rho_0 = \sum_m e^{-\beta E_m} |m\rangle\langle m|$, with Hamiltonian $H_0$. Obviously $\langle A\rangle_{\text{eq}} = \text{Tr}(\rho_0 A)$ is the expectation value of our physical observable. $A$ can be the current operator (transport measurement), density operator (X-ray or neutron scattering experiment), spin operator (magnetization measurement), and so on.

To obtain more information about the system we may disturb our system a little by adding a voltage source, a magnetic field, putting on a beam of neutrons or a beam of microwaves, and so on. The effect of this disturbance can be described by an additional time-dependent term in the Hamiltonian,

$$H'(t) = f(t)B,$$

where $f(t)$ is a time-dependent number function and $B$ is an operator describing the (quantum) effect of the perturbation on the system. As a result of the perturbation, $\langle A\rangle \rightarrow \langle A\rangle_{\text{eq}} + \delta\langle A(t)\rangle$. The goal of linear response theory is to relate $\delta\langle A(t)\rangle$ to $f(t)$ in first order.

In general, the change of the system in the presence of $H'(t)$ is described by a change of the density matrix, $\rho_0 \rightarrow \rho_0 + \delta\rho$, and correspondingly

$$\delta\langle A(t)\rangle = \text{Tr}(\delta\rho A) \quad (\text{note that Tr}(A+B) = \text{Tr}(A) + \text{Tr}(B)).$$

Thus, to calculate $\delta\langle A(t)\rangle$, we need to determine how the density matrix changes to first order in $H'(t)$.

The density matrix can in general be changed in two ways,

$$\delta\rho = \sum_m \delta(e^{-\beta E_m})|m\rangle\langle m| + \sum_m e^{-\beta E_m}(|\delta m\rangle\langle m| + |m\rangle\langle\delta m). \tag{5.43}$$

The first term describes changes in the weighting factor (governed by thermodynamics) and is important if the system evolves to a new equilibrium state under the external perturbation. Note that the Schrödinger equation describes only changes in the second term and there is no general theory that describes how the weighting factor should change at present because of lack of a sound microscopic foundation for out-of-equilibrium thermodynamics. The second term describes changes in the wavefunction (or dynamical changes) and can be described by the Schrödinger equation in the interaction picture,

$$i\hbar\frac{\partial}{\partial t}|m(t)\rangle = H'_{\text{I}}(t)|m(t)\rangle,$$

and

$$|\delta m(t)\rangle \sim -\frac{i}{\hbar}\int_{-\infty}^{t} H'_{\text{I}}(t')\text{d}t'|m\rangle \tag{5.44}$$

to first order.

It turns out that in linear response theory we need to consider only the second, dynamical, change, that is we write

$$\begin{aligned}\delta\rho &= \sum_m e^{-\beta E_m}(|\delta m\rangle\langle m| + |m\rangle\langle\delta m|)\\ &= -\sum_m e^{-\beta E_m}\frac{i}{\hbar}\int_{-\infty}^{t}\text{d}t'(H'_{\text{I}}(t')|m\rangle\langle m| - |m\rangle\langle m|H'_{\text{I}}(t'))\\ &= \frac{i}{\hbar}\int_{-\infty}^{t} f(t')\text{d}t'[\rho, B_{\text{I}}(t')],\end{aligned} \tag{5.45a}$$

where $B_I(t) = e^{iH_0t}Be^{-iH_0t}$. Therefore,

$$\delta\langle A(t)\rangle = \mathrm{Tr}(\delta\rho A) = \frac{i}{\hbar}\int_{-\infty}^{t} f(t')\mathrm{d}t'\mathrm{Tr}[\rho, B_I(t')]A(t). \tag{5.45b}$$

Using the property Tr($ABC$) = Tr($CAB$), we may rewrite Equation (5.45b) as

$$\delta\langle A(t)\rangle = \frac{i}{\hbar}\int_{-\infty}^{t} f(t')\mathrm{d}t'\mathrm{Tr}\rho[B_I(t'), A_I(t)] = \frac{i}{\hbar}\int_{-\infty}^{t} f(t')\mathrm{d}t' K_{AB}(t-t'). \tag{5.45c}$$

Equation (5.45c) is often written as

$$\delta\langle A(t)\rangle = \frac{i}{\hbar}\int_{-\infty}^{\infty} f(t')\mathrm{d}t' K^{\mathrm{R}}_{AB}(t-t'), \tag{5.45d}$$

where $K^{\mathrm{R}}_{AB}(t-t') = \theta(t-t')K_{AB}(t-t')$ is called the retarded (linear) response function of the system describing response $A$ to perturbation $B$.

To compare with experiments, we have to compute response functions for the system with the appropriate $A$'s and $B$'s.

Some common response functions we encounter in condensed matter physics include

(a) Current–current response function

In this case we consider the current response of the system to an external applied (transverse) electric field. The electric field is expressed in terms of a vector potential $\vec{E} = \frac{1}{c}\frac{\partial\vec{A}}{\partial t}$ and the perturbation on the system to linear order in $\vec{A}$ is

$$H' = -\frac{1}{c}\int \mathrm{d}^d x \hat{\vec{j}}(\vec{x}).\vec{A}(\vec{x},t), \tag{5.46a}$$

where $\hat{\vec{j}} = \frac{e}{2mi}(\bar{\psi}\nabla\psi-(\nabla\bar{\psi})\psi)$ is the paramagnetic current operator. The current response of the system is given by

$$\langle j_\mu(\vec{x},t)\rangle = \int \mathrm{d}^d x' \sigma_{\mu\nu}(\vec{x},\vec{x}';t-t')E_\nu(\vec{x}',t') - \frac{e^2}{mc}n(\vec{x})A_\mu(\vec{x},t),$$

where the first term describes paramagnetic responses and the second term represents diamagnetic current. For a harmonic field $\vec{A}(\vec{x},t) = \vec{A}(\vec{x})e^{i\omega t}$, we obtain

$$\langle j_\mu(\vec{x},\omega)\rangle = \int \mathrm{d}^d x' \sigma^{\mathrm{R}}_{\mu\nu}(\vec{x},\vec{x}';\omega)E_\nu(\vec{x}',\omega) + i\frac{e^2}{mc\omega}n(\vec{x})E_\mu(\vec{x},\omega), \tag{5.46b}$$

where

$$\sigma^{\mathrm{R}}_{\mu\nu}(\vec{x},\vec{x}';\omega) = \frac{i}{\hbar\omega}\int \mathrm{d}t'\mathrm{Tr}(\rho[j_\mu(\vec{x}t), j_\nu(\vec{x}'t')])\theta(t-t')e^{i\omega(t-t')}. \tag{5.46c}$$

For a translationally invariant system, we can write

$$\langle j_\mu(\vec{q},\omega)\rangle = \sum_\nu \sigma_{\mu\nu}(\vec{q},\omega)E_\nu(\vec{q},\omega), \tag{5.46d}$$

where

$$\sigma_{\mu\nu}(\vec{q},\omega) = \sigma^{\mathrm{R}}_{\mu\nu}(\vec{q},\omega) + i\frac{ne^2}{mc\omega}\delta_{\mu\nu} \tag{5.46e}$$

and

$$\sigma^{\mathrm{R}}_{\mu\nu}(\vec{q},\omega) = \frac{i}{\hbar\omega}\int \mathrm{d}^d\vec{x}\int \mathrm{d}t'\mathrm{Tr}(\rho[j_\mu(\vec{x}t), j_\nu(0t')])\theta(t-t')e^{i\omega(t-t')}e^{i\vec{q}.\vec{x}}. \tag{5.46f}$$

(b) Density–density response function

In this case we consider the density response of the system to an external scalar field $\Phi$. The perturbation is

$$H' = \int \mathrm{d}^d x\,\widehat{n}(\vec{x})\Phi(\vec{x},t), \tag{5.47a}$$

where $\widehat{n} = \bar{\psi}\psi$ is the density operator. For harmonic perturbations the induced density fluctuation is given by

$$\langle n(\vec{x},\omega)\rangle = \int \mathrm{d}^d x'\chi^{\mathrm{R}}(\vec{x},\vec{x}';\omega)\Phi(\vec{x}',\omega), \tag{5.47b}$$

where

$$\chi^{\mathrm{R}}(\vec{x},\vec{x}';t) = \frac{i}{\hbar}\mathrm{Tr}(\rho[n(\vec{x}t), n(\vec{x}'0)])\theta(t). \tag{5.47c}$$

(c) Spin–spin response function

We consider a system of spins under an external magnetic field $\vec{H}$. The perturbation is

$$H' = \int \mathrm{d}^d x\widehat{\vec{S}}(\vec{x}).\vec{H}(\vec{x},t), \tag{5.48a}$$

where $\vec{S}(x)$ is the spin operator. For a harmonic perturbation the induced change in magnetization is given by

$$\langle m_\mu(\vec{x},\omega)\rangle = \int \mathrm{d}^d x'\chi^{\mathrm{R}}_{S\mu\nu}(\vec{x},\vec{x}';\omega)H_\nu(\vec{x}',\omega), \tag{5.48b}$$

where

$$\chi^{\mathrm{R}}_{S\mu\nu}(\vec{x},\vec{x}';t) = \frac{i}{\hbar}\mathrm{Tr}(\rho\left[S_\mu(\vec{x}t), S_\nu(\vec{x}'0)\right])\theta(t). \tag{5.48c}$$

For electrons, $\vec{S}(x) = \frac{1}{2}(\bar{\psi}(x)\vec{\sigma}\psi(\vec{x}))$, (5.48d)

where $\psi = \begin{pmatrix}\psi_\uparrow\\ \psi_\downarrow\end{pmatrix}$ $(\bar{\psi} = (\psi^+_\uparrow, \psi^+_\downarrow))$ and the $\vec{\sigma}$ are the Pauli matrices.

Note that, in the coherent state representation, the response functions are all linear combinations of two-particle Green's functions of the form

$$K(x,x') \sim \mathrm{Tr}\rho[\bar{\psi}(x)\psi(x)\bar{\psi}(x')\psi(x')].$$

These four-point Green's functions can be evaluated in perturbation theory as two-point Green's functions. We shall encounter some of these response functions in later chapters.

### 5.3.2
### Temperature and Causal Correlation Functions

So far, we have used the phrase 'correlation function' or 'Green's function' rather loosely. They represent different objects on different occasions. We now formally introduce and establish the general relation between three different types of Green's functions or correlation functions. We have the temperature Green's functions, which are time-ordered correlation functions calculated at imaginary time, and are the same as the correlation functions we evaluate in the imaginary-time path integral. We also have the retarded Green's functions or response functions which appear in linear response theory and the advanced Green's functions which are the complex conjugate to the retarded Green's functions, as we shall see below. The relation between these Green's functions can be seen most easily in the spectral representation we introduced in Section 1.

Let $\widehat{U}$ be an operator we are interested in. The temperature Green's function for $\widehat{U}$ is defined generally as

$$G_{\mathrm{T}}(\tau) = -\langle T_\tau \widehat{U}(\tau)\widehat{U}^+(0)\rangle, \tag{5.49a}$$

where

$$\langle T_\tau \widehat{U}(\tau)\widehat{U}^+(0)\rangle = \begin{cases} \langle \widehat{U}(\tau)\widehat{U}^+(0)\rangle, \tau > \tau', \\ \pm\langle \widehat{U}^+(0)\widehat{U}(\tau)\rangle, \tau \le \tau'. \end{cases} \tag{5.49b}$$

The $+$ $(-)$ sign applies for a bosonic (fermionic) operator $\widehat{U}$. Examples of $\widehat{U}$ include the single boson (fermion) annihilation operator $C_{\vec{p}}$ and the density, current, or spin operators we see in linear response theory. The Fourier transform of the Green's function is defined as

$$G_{\mathrm{T}}(i\omega_n) = \int_0^\beta d\tau e^{i\omega_n\tau} G_{\mathrm{T}}(\tau). \tag{5.49c}$$

A spectral representation for the Green's function $G_{\mathrm{T}}(i\omega_n)$ can be obtained following the same mathematical procedure for deriving the spectral function for the one-particle Green's function. After some straightforward algebra, we obtain

$$G_{\mathrm{T}}(i\omega_n) = e^{\beta\Omega}\sum_{n,m}\frac{(e^{-\beta E_n} \mp e^{-\beta E_m})|\langle n|\widehat{U}|m\rangle|^2}{i\omega_n + E_n - E_m}, \tag{5.50}$$

where we assume here that $\widehat{U}$ does not change the number of particles in the system (and therefore they must be bosonic). The chemical potential $\mu$ will appear as in the one-particle Green's function if $\hat{U}$ changes the particle number by one (see Section 1).

**Homework Q8a** Derive Equation (5.50).

Similarly, we define the retarded and advanced Green's functions of the operator $\hat{U}$ as

$$G_R(t) = -i\theta(t)\langle[\hat{U}(t)\hat{U}^+(0) \mp \hat{U}^+(0)\hat{U}(t)]\rangle \tag{5.51a}$$

and

$$G_A(t) = i\theta(-t)\langle[\hat{U}(t)\hat{U}^+(0) \mp \hat{U}^+(0)\hat{U}(t)]\rangle, \tag{5.51b}$$

where the $-$ $(+)$ sign applies for a bosonic (fermionic) $\hat{U}$. Note that, for (bosonic) operators which conserve particle number, $G_R(t)$ has the same form as response functions in linear response theory. The corresponding Fourier transforms are defined as

$$G_R(\omega) = \int_{-\infty}^{\infty} dte^{i\omega t}G_R(t) \tag{5.52a}$$

and

$$G_A(\omega) = \int_{-\infty}^{\infty} dte^{i\omega t}G_A(t) = G_R(\omega)^*. \tag{5.52b}$$

**Homework Q8b** Derive Equation (5.52b).

Inserting a complete set of states, that is $I = \sum_m |m\rangle\langle m|$, in $G_R(t)$ as before, we obtain

$$\begin{aligned}
G_R(t) &= -i\theta(t)e^{\beta\Omega}\textstyle\sum_{n,m} e^{-\beta E_n}[\langle n|\hat{U}(t)|m\rangle\langle m|\hat{U}^+(0)|n\rangle \mp \langle n|\hat{U}^+(0)|m\rangle\langle m|\hat{U}(t)|n\rangle]\rangle \\
&= -i\theta(t)e^{\beta\Omega}\sum_{n,m} e^{-\beta E_n}[e^{it(E_n-E_m)} \mp e^{-it(E_n-E_m)}]|\langle n|\hat{U}|m\rangle|^2 \\
&= -i\theta(t)e^{\beta\Omega}\sum_{n,m}(e^{-\beta E_n} \mp e^{-\beta E_m})|\langle n|\hat{U}|m\rangle|^2 e^{it(E_n-E_m)}.
\end{aligned} \tag{5.53}$$

Fourier transforming Equation (5.53), we obtain

$$\begin{aligned}
G_R(\omega) &= -i\int_0^{\infty} dte^{i(\omega+i\delta)t}e^{\beta\Omega}\sum_{n,m}(e^{-\beta E_n} \mp e^{-\beta E_m})|\langle n|\hat{U}|m\rangle|^2 e^{it(E_n-E_m)} \\
&= e^{\beta\Omega}\sum_{n,m}(e^{-\beta E_n} \mp e^{-\beta E_m})\frac{|\langle n|\hat{U}|m\rangle|^2}{\omega + E_n - E_m + i\delta}.
\end{aligned} \tag{5.54}$$

Note that we have to put $\omega \to \omega + i\delta$ to ensure that the integral converges. Comparing Equations (5.50), (5.52b), and (5.54), we obtain

$$G_R(\omega) = G_T(i\omega_n \to \omega + i\delta) = G_A(\omega)^*, \tag{5.55}$$

which is a general relation between the temperature and the retarded and advanced Green's functions. The relation implies that to compute retarded/advanced response functions, we may simply compute the corresponding temperature Green's functions and replace $i\omega_n \to \omega \pm i\delta$ at the end of the calculation. The temperature Green's functions are easier to calculate because they can be obtained directly from perturbation theory.

**Homework Q9** Current–current response for non-interacting electron gas

The (paramagnetic) current–current response function for a non-interacting electron gas can be computed through the following steps:

(1) We start from the corresponding (imaginary)-time-ordered current–current correlation function

$$\sigma^T_{\mu\nu}(\vec{q}, i\omega) = \frac{1}{\hbar\omega}\int d^d x \int_0^\beta d\tau \langle T_\tau \hat{j}_\mu(\vec{x}\tau)\hat{j}_\nu(00)\rangle e^{i\omega\tau} e^{i\vec{q}.\vec{x}},$$

and write down $j_\mu(\vec{x})$ in terms of fermion creation and annihilation operators in Fourier space, that is we write $\hat{j}_\mu(\vec{x}) = \frac{1}{V}\sum_{\vec{k},\vec{q}} j_{\mu\vec{k}}(\vec{q}) e^{i\vec{q}.\vec{x}} c^+_{\vec{k}-\vec{q}/2} c_{\vec{k}+\vec{q}/2}$. Determine $j_{\mu\vec{k}}(\vec{q})$.

(2) Show that

$$\sigma^T_{\mu\nu}(\vec{q}, i\omega) = \frac{1}{\hbar\omega V}\sum_{\vec{k},\vec{k}'} j_{\mu\vec{k}}(\vec{q}) j_{\nu\vec{k}'}(\vec{q}) \int_0^\beta d\tau \langle T_\tau c^+_{\vec{k}-\vec{q}/2}(\tau) c_{\vec{k}+\vec{q}/2}(\tau) c^+_{\vec{k}'+\vec{q}/2}(0) c_{\vec{k}'-\vec{q}/2}(0)\rangle e^{i\omega\tau}.$$

(3) Use Wick's theorem for fermion Green's functions to show that

$$\sigma^T_{\mu\nu}(\vec{q}, i\omega) = \frac{-1}{\hbar\omega V\beta}\sum_{\vec{k}, i\Omega_n} j_{\mu\vec{k}}(\vec{q}) j_{\nu\vec{k}}(\vec{q}) g(\vec{k}+\vec{q}/2, i\omega + i\Omega_n) g(\vec{k}-\vec{q}/2, i\Omega_n).$$

(4) Using Equation (5.25b) for the g, evaluate the sum over $i\Omega_n$.
(Caution: note that $\Omega_n = \frac{(2n+1)\pi}{\beta}$. How about $\omega$?)

The (retarded) current–current response function is obtained by taking $i\omega \to \omega + i\delta$ at the end of the calculation.

### 5.3.3 Fluctuation–Dissipation Theorem

Imagine measuring the current flowing in a macroscopic system at thermal equilibrium. You will find that, although the current is zero on average, it does not stay constant at zero all the time. There is always a weak, fluctuating current in the system, that is you will find that

$$\langle \vec{j}(t)\rangle = 0, \quad \langle \vec{j}(t)\cdot\vec{j}(t')\rangle \neq 0,$$

where $\langle \ldots \rangle$ denotes time average. Similar results are found for many other measurable macroscopic quantities. The presence of these small fluctuations in classical systems can be understood from thermodynamics where the system has a finite probability to stay in any state at temperature $T \neq 0$. In quantum systems fluctuations exist even at $T = 0$ because of intrinsic quantum fluctuations. The spectra of these fluctuations are measured by Fourier transformation of correlation functions of the kind

$$\langle A(t)A(t')\rangle = \frac{1}{Z}\sum_n e^{-\beta E_n}\langle n|A(t)A(t')|n\rangle, \tag{5.56}$$

which can be computed (at least perturbatively) using the tools we have developed. Note that the fluctuation spectrum reflects the *equilibrium* density matrix as well as the dynamics of the system.

Next, we imagine the following experiment. We put on adiabatically a small external, say magnetic, field $\vec{H}(\vec{x},t) = \lim_{\eta\to 0} H(\vec{x})\,\widehat{z}\,e^{\eta t}$ such that, at time $t = 0$, a finite magnetization is built up in the system given by $\vec{m}(\vec{x}) = m(\vec{x})\,\widehat{z}$, where

$$m(\vec{x}) = \int d^d x' \int_{-\infty}^{0} \chi_s(\vec{x}-\vec{x}'; -t')H(\vec{x}',t')dt' = \int d^d x' \bar{\chi}_s(\vec{x}-\vec{x}')H(\vec{x}') \tag{5.57a}$$

or, after Fourier transformation,

$$m(\vec{q}) = \bar{\chi}_s(\vec{q})H(\vec{q}), \tag{5.57b}$$

where $\chi_s(\vec{q})$ is the static spin–spin response function.

**Exercise**

Use the relation between the causal and temperature Green's functions to show that

$$\bar{\chi}_s(\vec{q}) = \int_{-\infty}^{0} \chi_s(\vec{q}, -t')dt' = \beta\langle m(\vec{q}, \tau = 0)m(-\vec{q}, 0)\rangle.$$

We have assumed space isotropy and translational invariance in writing down Equation (5.57b). Now, let us turn off the magnetic field at $t = 0$ so that the system relaxes to equilibrium, $\vec{m}(\vec{x},t) \to 0$ as $t \to \infty$. The magnetization at $t > 0$ can be determined in two ways. First, it is determined by $\vec{H}(\vec{q},t)$ via linear response theory, that is

$$m(\vec{q},t) = \int_{-\infty}^{0} dt' \chi_s(\vec{q}; t-t')H(\vec{q})e^{\eta t'}. \tag{5.58}$$

However, it seems physically reasonable that the decay of the magnetization should also be related to the correlation function $C(\vec{q};t) = \langle m(\vec{q},t)m(-\vec{q},0)\rangle$, which measures the decay of *spontaneous* small magnetization fluctuations away from equilibrium, that is we expect that

$$m(\vec{q},t) = R_s(\vec{q};t)m(\vec{q},t=0) \sim C(\vec{q};t)m(\vec{q}). \tag{5.59}$$

The function $R_s(\vec{q};t)$ is called the relaxation function and describes the relaxation of the system towards equilibrium. The above two ways of looking at how $m(\vec{q},t)$ decays implies that there exists a deep relation between the responses of the system to weak external perturbations and the corresponding (spontaneous) fluctuation spectrum of the system. The relation between these two aspects of a macroscopic system is the fluctuation–dissipation theorem.

The relation between $R_s(\vec{q},t)$ and $C(\vec{q};t)$ can be made precise in the hydrodynamic regime where, for small fluctuations, the magnetization can be treated as a classical variable satisfying linearized, hydrodynamic equations of the form

$$\widehat{L}\left(\frac{\partial}{\partial t},\nabla\right)m(\vec{x},t) = H(\vec{x},t), \tag{5.60}$$

where $\widehat{L}$ is a linear operator and $H(\vec{x},t)$ represents an external force acting on $m$. By definition it is obvious that $R_s(\vec{q};t)m(\vec{q})$ is a solution of the corresponding homogeneous equation $\widehat{L}\left(\frac{\partial}{\partial t},\nabla\right)m(\vec{x},t) = 0$, with initial condition $m(\vec{q},t=0) = m(\vec{q})$.

Similarly, the correlation function $C(\vec{q};t) = \langle m(\vec{q},t)m(-\vec{q},0)\rangle$ should also be a solution of the same homogeneous equation with initial condition

$$C(\vec{q},t=0) = \langle m(\vec{q},0)m(-\vec{q},0)\rangle = \beta^{-1}\chi_s(\vec{q}).$$

Comparing the two, we obtain

$$R_s(\vec{q};t) = \frac{\beta C(\vec{q};t)}{\chi_s(\vec{q})}. \tag{5.61}$$

$R_s(\vec{q};t)$ can be related to the response function $\chi_s$ using Equations (5.58) and (5.59). We obtain, after Fourier transforming and using Equation (5.61),

$$m(\vec{q},\omega) = \frac{(2\mathrm{Im}\chi_s(\vec{q};\omega))H(\vec{q})}{\omega} = \beta C(\vec{q},\omega)H(\vec{q})$$

or

$$\mathrm{Im}\chi_s(\vec{q};\omega) = \frac{\beta\omega}{2}C(\vec{q},\omega). \tag{5.62}$$

The Fourier transforms of the response functions are defined as

$$\chi_s(\vec{q};\omega) = \int dt e^{i\omega t}\chi_s(\vec{q},t) = \int \frac{d\omega'}{\pi}\frac{\mathrm{Im}\chi_s(\vec{q},\omega')}{\omega-\omega'+i\delta}$$

and $C(\vec{q},\omega) = \int dt e^{i\omega t}C(\vec{q},t)$.

Equation (5.62) is the fluctuation–dissipation theorem in the hydrodynamic regime. Note that the theorem was derived *without* assuming any form of $\chi_s(\vec{q};\omega)$ or $\widehat{L}$ but is based on a pure phenomenological argument that *the thermodynamic state with a small magnetization $m(\vec{q})$ generated by an external field is the same as* *the thermodynamic state $m(\vec{q})$ arising from pure (spontaneous) fluctuations in the system.* As a result, a single equation governing $m(\vec{q}, t)$ can be derived in the hydrodynamic regime, independent of the microscopic details of the system. The validity of this assumption is unclear in general. However, if valid, the theorem will certainly not be restricted to magnetization and magnetic fields but can be generalized to any hydrodynamic variable $\phi$ coupling to an external force $F$. The theorem imposes a strong constraint on microscopic theories which describe properties of macroscopic systems away from (but close to) equilibrium.

The fluctuation–dissipation theorem can be checked directly in linear response theory, which provides a general microscopic expression for any response function $\chi(\vec{q};\omega)$. In particular, relation(s) between arbitrary response functions $\mathrm{Im}\chi(\vec{q};\omega)$ and corresponding correlation functions $C(\vec{q},\omega)$ which are valid beyond the hydrodynamic regime can be derived. We start with the exact expression for the response function

$$\begin{aligned}\chi_R(\vec{q};t) &= \frac{i}{\hbar}\mathrm{Tr}(\rho[\phi(\vec{q},t),\phi(-\vec{q},0)]\theta(t))\\ &= \frac{i}{\hbar}[\langle\phi(\vec{q},t)\phi(-\vec{q},0)\rangle - \langle\phi(-\vec{q},0)\phi(\vec{q},t)\rangle]\theta(t)\\ &= \frac{i}{\hbar}[C(\vec{q},t)-C(-\vec{q},-t)]\theta(t) = \frac{i}{\hbar}[C(\vec{q},t)-C(\vec{q},-t)]\theta(t),\end{aligned} \tag{5.63}$$

where we assume space isotropy in writing down the last equality. Using the spectral representation

$$C(\vec{q},t) = \frac{1}{Z}\sum_{n,k} e^{-\beta E_k}\langle k|\phi(\vec{q})|n\rangle\langle n|\phi(\vec{q})|k\rangle e^{i(E_k-E_n)t},$$

it is easy to show that $C(\vec{q},-\omega) = C(\vec{q},\omega)e^{-\beta\hbar\omega}$

and correspondingly $\mathrm{Im}\chi_R(\vec{q};\omega) = \frac{1}{2\hbar}(1-e^{-\beta\hbar\omega})C(\vec{q},\omega)$, (5.64)

which is called the (general) fluctuation–dissipation theorem. It is easy to show that Equation (5.64) reduces to Equation (5.62) in the limit $\beta\hbar\omega \ll 1$, indicating the correctness of linear response theory and the physical assumption underlying the fluctuation–dissipation theorem.

**Homework Q10**

Derive Equation (5.64). Using the spectral representation, show that in general $\omega\mathrm{Im}\chi_R(\vec{q};\omega) \geq 0$, $\mathrm{Im}\chi_R(\vec{q};\omega) = -\mathrm{Im}\chi_R(\vec{q};-\omega)$, and $\mathrm{Re}\chi_R(\vec{q};\omega) = \mathrm{Re}\chi_R(\vec{q};-\omega)$.

The result $\omega\mathrm{Im}\chi_R(\vec{q};\omega) \geq 0$ is often expressed as a criterion for stability of the system. To see why this is the case we examine the work done $W$ by the external field $F(\vec{x},t) \sim F(\vec{q},\omega)e^{i\vec{q}.\vec{x}}\cos\omega t$ on the system,

$$W = \int \mathrm{d}x \int F\mathrm{d}\phi = \int \mathrm{d}x \int_{-\infty}^{\infty} F\frac{\mathrm{d}\phi}{\mathrm{d}t}\mathrm{d}t = \omega\mathrm{Im}\chi_R(\vec{q},\omega)|F(\vec{q},\omega)|^2 + (\omega\to-\omega). \tag{5.65}$$

Requiring that $W > 0$ (we always need to do work to drive the system away from equilibrium) implies that $\omega \mathrm{Im}\chi_R(\vec{q};\omega) \geq 0$.

The linear response theory and fluctuation–dissipation theorem are general theories describing the behavior of systems away from, but close to, equilibrium. They represent important steps in our effort to understand non-equilibrium systems that are not far from equilibrium. General quantum/classical statistical theories applicable to systems far away from equilibrium are still at a very preliminary stage nowadays and this is a wide-open area of research in physics.

## References


1 Mahan, G.D. (1990) Many Particle Physics, Plenum Press (N.Y. and London).

2 Negele, J.W. and Orland, H. (1987) Quantum Many-Particle Systems, Addison-Wesley (The Advanced Book Program).

3 Wen, X.G. (2004) Theory of Quantum Many-Body Systems, Oxford University Press.

4 Weinberg, S. (1995) The Quantum Theory of Fields, Vol. 1; Cambridge University Press.

5 Sachdev, S. (1999) Quantum Phase Transitions, Cambridge University Press.

6 Forster, D. (1975) Hydrodynamic Fluctuations, Broken Symmetry, and Correlation Function, Addison-Wesley (The Advanced Book Program).

# 6
# Introduction to Berry Phase and Gauge Theory

## 6.1
## Introduction to Berry Phase

Much modern advancement in quantum field theory is tied with the understanding of non-trivial phase structures (and their generalizations) in quantum systems. In this chapter we shall give an introduction to Berry phase and gauge theory hoping that it can help the readers to start reading texts and papers in this very active area of research. A more in depth discussion of gauge theory will be presented in Chapter 12.

In quantum mechanics, two wavefunctions $\psi(x)$ and $\psi(x)e^{i\vartheta}$ that differ by a constant phase factor are considered to represent the same state vector in Hilbert space. However, a question remains: does the phase $\theta$ introduce any physical (observable) effect?

The answer is 'yes' if we consider an interference effect of two wavefunctions that differ only by $\theta$:

$$|\psi(x) + \psi(x)e^{i\theta}|^2 = |\psi(x)|^2\cos(\theta/2).$$

The result of an interference experiment provides a measure for $\theta$. You can look at, for example, Ref. [1] for a discussion of experiments of this kind. Note that this is a pure quantum mechanical effect which has no classical analogue.

The first person who systematically examined how $\theta$ can enter physical phenomena was Michael Berry. In the following we shall reproduce his analysis for a simple quantum system.

### 6.1.1
### Berry Phase for a Simple Quantum System

Imagine an electron (or whatever particle) trapped by a potential centered at a controllable position $\vec{R}$. The potential may be the electrostatic potential coming from an atomic nucleus, the confining potential of a quantum dot, or whatever you can imagine. $\vec{R}$ can be changed by moving the trapping potential. We shall assume that initially the electron wavefunction is at the ground state of the potential, $\psi(\vec{x}) = \psi_0(\vec{x}-\vec{R})$, where $\vec{x}$ is the electron coordinate.

*Introduction to Classical and Quantum Field Theory.* Tai-Kai Ng

ISBN: 978-3-527-40726-2

Now consider moving $\vec{R}$ very slowly in time such that, at each instant of time, the electron is still located at the ground state, that is

$$\hat{H}[\vec{R}(t)]\psi_0(\vec{x}-\vec{R}(t)) = E_G\psi_0(\vec{x}-\vec{R}(t)). \quad (6.1)$$

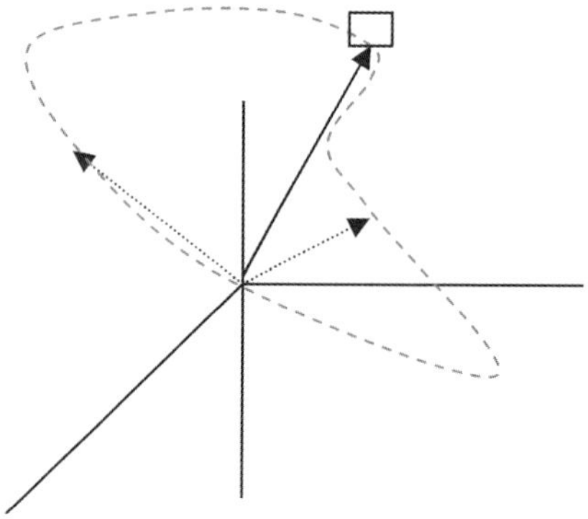

We consider in particular a closed path $\vec{R}(t)$ with $\vec{R}(T) = \vec{R}(0)$ (see figure).

In general, the wavefunction at time $T$ will be the 'same' as the original wavefunction except that the overall phase of the wavefunction is allowed to change according to quantum mechanics, that is we have in general

$$\psi(\vec{x}, T) = \psi_0(\vec{x}, \vec{R}(T)) = e^{i\gamma(T,\vec{R})}\psi_0(\vec{x}, \vec{R}(0)). \quad (6.2)$$

Therefore, a quantum interference effect can be observed between a wavefunction representing the electron staying at $\vec{R}(0)$ all the time and another wavefunction representing the electron going around a closed path.

### Question

How about the dynamic phase $e^{-i\frac{E_G}{\hbar}T}$ which should be present in the experiment? Is this observable in the interference experiment?

> Note: the requirement of *no* transition of the wavefunction to an excited state when $\vec{R}$ changes means essentially that $T \gg \Delta E/\hbar$, where $\Delta E$ is the excitation energy for the first excited state.

The above argument only suggests that the Berry phase $\gamma(T, \vec{R})$ is observable if it is non-zero. To show that $\gamma(T, \vec{R})$ is really non-zero, we have to analyze the time-dependent Schrödinger equation

$$i\hbar\frac{\partial}{\partial t}|\psi(t)\rangle = \hat{H}[\vec{R}(t)]|\psi(t)\rangle. \quad (6.3a)$$

Let $|\psi_0[\vec{R}(t)]\rangle$ be the (normalized) ground-state wavefunction of $\widehat{H}[\vec{R}(t)]$, that is

$$\widehat{H}[\vec{R}(t)]|\psi_0[\vec{R}(t)]\rangle = E_G|\psi_0[\vec{R}(t)]\rangle \tag{6.3b}$$

and $\langle\psi_0[\vec{R}(t)]|\psi_0[\vec{R}(t)]\rangle = 1$. Note that, if $\vec{R}$ is time independent, then $|\psi(t)\rangle = e^{-i\frac{E_G}{\hbar}t}|\psi_0[\vec{R}]\rangle$.

For time-dependent $\vec{R}(t)$, we can write the wavefunction as

$$|\psi(t)\rangle = e^{-i\frac{E_G}{\hbar}t}e^{i\gamma(t)}|\psi_0[\vec{R}(t)]\rangle. \tag{6.3c}$$

Note that, since $\langle\psi_0[\vec{R}(t)]|\psi_0[\vec{R}(t)]\rangle = 1$, $\gamma(t)$ must be a real number, that is $e^{i\gamma(t)}$ is a phase factor.

Substituting Equation (6.3c) into the Schrödinger equation (6.3a), and using Equation (6.3b), we obtain

$$\begin{aligned} i\hbar\frac{\partial}{\partial t}|\psi(t)\rangle &= E_G|\psi(t)\rangle - \hbar\dot{\gamma}(t)|\psi(t)\rangle + e^{-i\frac{E_G}{\hbar}t}e^{-i\gamma(t)}(i\hbar\dot{\vec{R}}(t)\cdot\nabla_{\vec{R}}|\psi_0[\vec{R}(t)]\rangle \\ &= \widehat{H}[\vec{R}(t)]|\psi(t)\rangle = E_G|\psi(t)\rangle. \end{aligned} \tag{6.4a}$$

Therefore,

$$\dot{\gamma}(t) = i\langle\psi_0[\vec{R}(t)]|\nabla_{\vec{R}}\psi_0[\vec{R}(t)]\rangle\cdot\dot{\vec{R}}(t), \tag{6.4b}$$

and the total phase change going from $\vec{R}(0)$ to $\vec{R}(0)$ is

$$\gamma(T) = \int_0^T \dot{\gamma}(t)\mathrm{d}t = i\int_{\vec{R}(0)}^{\vec{R}(T)} \langle\psi_0[\vec{R}]|\nabla\vec{R}\psi_0[\vec{R}]\rangle\cdot \mathrm{d}\vec{R}. \tag{6.5a}$$

For a closed loop $C(\vec{R}(0) = \vec{R}(0))$, we obtain

$$\gamma(C) = i\oint_C\langle\psi_0[\vec{R}]|\nabla_{\vec{R}}\psi_0[\vec{R}]\rangle\cdot \mathrm{d}\vec{R}. \tag{6.5b}$$

Note that the phase factor depends on the path $\vec{R}$ only and time drops out from Equation (6.5a), that is the phase is a property that depends only on geometry of the path and has no dynamical effect. For this reason the Berry phase is often called a geometrical phase.

**Homework Q1a**

Prove that $\langle\psi_0[\vec{R}]|\nabla\vec{R}\psi_0[\vec{R}]\rangle$ is a pure imaginary number.

**Question**

Under what condition is $\gamma(C) \neq 0$?

To analyze $\gamma(C)$, we introduce a vector field

$$\vec{A}(\vec{R}) = i\langle\psi_0[\vec{R}]|\nabla\vec{R}\psi_0[\vec{R}]\rangle. \tag{6.6a}$$

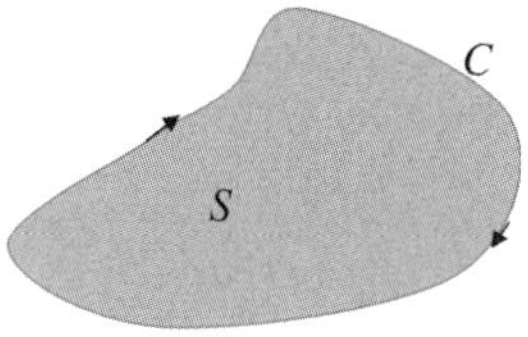


With $\vec{A}(\vec{R})$, we can write

$$\gamma(C) = \oint_C \vec{A}(\vec{R}) \cdot d\vec{R} = \oint_S (\nabla \times \vec{A}) \cdot d\vec{S}, \tag{6.6b}$$

where $S$ is a surface enclosed by the loop $C$ (see above figure). Therefore, we need $\nabla \times \vec{A} \neq 0$ for at least some region in space if $\gamma(C) \neq 0$.

To see what it means to the phase of the wavefunction, we examine what happens if $\nabla \times \vec{A} \equiv 0$. In this case, we may write $\vec{A} = \nabla\phi$.

Therefore, $\gamma(T) = \int_{\vec{R}(0)}^{\vec{R}(T)} \nabla\phi(\vec{R}).d\vec{R} = \phi(\vec{R}(T)) - \phi(\vec{R}(0))$, and the Berry phase can be absorbed into the wavefunction as

$$|\psi[\vec{R}]\rangle = e^{i\phi(\vec{R})}|\psi_0[\vec{R}]\rangle. \tag{6.7}$$

In this case the phase of the wavefunction can be defined unambiguously at each point $\vec{R}$ in space up to an overall constant (like an overall constant in a static electrostatic potential). What is most interesting is that in this case the phase is *not* observable in experiments since the interference effect involves $\gamma(C) = \oint_S (\nabla \times \vec{A}).d\vec{S}$, which is non-zero only if $\nabla \times \vec{A} \neq 0$. The Berry phase has a physical effect only if we *cannot* define a unique phase for the wavefunction $|\psi[\vec{R}]\rangle$ over the whole space $\vec{R}$ – therefore, it is called an *irreducible phase.*

**Example**

Berry phase of a single spin $S = {}^1/_2$ in a slowly varying magnetic field that guides the direction of the spin.

First, we introduce a general formula to compute

$$\nabla \times \vec{A}(\vec{R}) = i\nabla_{\vec{R}} \times \langle\psi_0[\vec{R}]|\nabla_{\vec{R}}\psi_0[\vec{R}]\rangle$$
$$= i\sum_m \nabla_{\vec{R}} \times (\langle\psi_0[\vec{R}]|m\rangle\langle m|\nabla_{\vec{R}}\psi_0[\vec{R}]\rangle), \tag{6.8a}$$

where $I = \sum_m |m(\vec{R})\rangle\langle m(\vec{R})|$ (completeness). Using $\nabla \times \nabla f(\vec{R}) = 0$ for any function $f(\vec{R})$, we obtain

$$\nabla \times \vec{A}(\vec{R}) = i\sum_{m\neq 0} \langle\nabla_{\vec{R}}\psi_0[\vec{R}]|m(\vec{R})\rangle \times \langle m(\vec{R})|\nabla_{\vec{R}}\psi_0[\vec{R}]\rangle. \tag{6.8b}$$

Note the absence of the $m = 0$ term in the sum (*question: why is this correct?*).

Using $\widehat{H}[\vec{R}+\vec{a}]|\psi_0(\vec{R}+\vec{a})\rangle = E_G|\psi_0(\vec{R}+\vec{a})\rangle$ and applying first-order perturbation theory to $\widehat{H}[\vec{R}+\vec{a}] \sim \widehat{H}[\vec{R}] + \vec{a}.\nabla_{\vec{R}}\widehat{H}[\vec{R}]$ for small $\vec{a}$, we obtain

$$\langle m(\vec{R})|\nabla_{\vec{R}}\psi_0(\vec{R})\rangle = \frac{\langle m(\vec{R})|(\nabla_{\vec{R}}\widehat{H}[\vec{R}])|\psi_0(\vec{R})\rangle}{E_G - E_m}, \tag{6.8c}$$

where $H[\vec{R}]|m(\vec{R})\rangle = E_m|m(\vec{R})\rangle$.

**Homework Q1b**

Derive Equations (6.8b) and (6.8c).

Therefore,

$$\nabla \times \vec{A}(\vec{R}) = i\sum_{m\neq 0}\frac{\langle\psi_0[\vec{R}]|\nabla_{\vec{R}}\widehat{H}[\vec{R}]|m(\vec{R})\rangle \times \langle m(\vec{R})|\nabla_{\vec{R}}\widehat{H}[\vec{R}]|\psi_0[\vec{R}]\rangle}{(E_G - E_m)^2}. \tag{6.8d}$$

For spin $^1/_2$ in a magnetic field, the continuous variable $\vec{R} \to \widehat{n}$ (unit vector) represents the direction of the external magnetic field and

$$\widehat{H}[\widehat{n}] = -B\widehat{n}\cdot\vec{S} \quad \text{and} \quad \nabla_{\widehat{n}}\widehat{H} = -B\vec{S}. \tag{6.9}$$

Therefore,

$$\begin{aligned}\nabla \times \vec{A} &\to iB^2 \sum_{\sigma=-1/2}\frac{\langle 1/2|\vec{S}|\sigma\rangle \times \langle\sigma|\vec{S}|1/2\rangle}{(\hbar B)^2} = \frac{i}{\hbar^2}\langle 1/2|\vec{S}\times\vec{S}|1/2\rangle \\ &= -\frac{1}{\hbar}\langle\vec{S}\rangle = -\frac{\widehat{n}}{2},\end{aligned} \tag{6.10}$$

which represents the 'magnetic field' of a magnetic monopole with strength '$^1/_2$' located at the origin (see also Ref. [1]).

Another commonly quoted example of non-trivial Berry phase is the Aharonov–Bohm effect, corresponding to charges moving around an enclosed flux tube. We shall study this example in detail later.

### 6.1.2 Berry Phase and Particle Statistics

As mentioned in Chapter 3, Berry phase has an intimate relation with particle statistics. We consider wavefunctions of $N$-particle systems $\psi(\vec{r}_1,\vec{r}_2,\ldots,\vec{r}_N)$, where $(\vec{r}_1,\vec{r}_2,\ldots,\vec{r}_N)$ are the coordinates of the particles.

The statistics of the particles can be defined by asking how the phase of $\psi(\vec{r}_1,\vec{r}_2,\ldots,\vec{r}_N)$ changes when we exchange adiabatically two of the coordinates $(\vec{r}_i,\vec{r}_j)$. The change in phase can be calculated in a Berry phase calculation where the particles move from their initial positions $(\vec{r}_i,\vec{r}_j)$ to final positions $(\vec{r}_j,\vec{r}_i)$ in space. Note that in this sense statistics is not a fundamental property of particles but is a property of the many-body wavefunction. The phase can be quite 'arbitrary' and is not necessarily restricted to fermions or bosons where the phases are $\pm 1$ upon exchange. We shall discuss the possible values of exchange phases in more detail later in this chapter.

### 6.1.3 Berry Phase and U(1) Gauge Theory

We have seen that, in quantum mechanics, two wavefunctions $\psi(x)$ and $\psi(x)e^{i\vartheta}$ differing by a constant phase factor $\vartheta$ represent the same state vector (*global gauge invariance*). However, through Berry phase analysis, we know that the phase factor has a physical effect if it is an irreducible phase.

Inspired by the quantum theory of electromagnetism, physicists asked the following question: how about two wavefunctions differing by a space–time-dependent phase factor, that is $\psi(x)$ and $\psi e^{i\vartheta(x)}(x = (\vec{x}, t))$? What constraint would it impose on quantum mechanics if we require that they represent the same physical state, or if quantum mechanics is locally not sensitive to the absolute value of phase of a wavefunction? Is there also an associated Berry phase effect in this case?

The requirement that quantum mechanics is locally insensitive to the absolute value of the phase of a wavefunction is called *local gauge invariance*. In the following, we shall address the above two questions in the simplest case of U(1) gauge theory.

Let us start with the (non-relativistic) free-particle Schrödinger equation

$$i\hbar\frac{\partial}{\partial t}\psi(\vec{x},t) = -\frac{\hbar^2}{2m}\nabla^2\psi(\vec{x},t).$$

**Question**

Suppose that $\psi_0(\vec{x},t)$ is a solution of the above equation, what is the equation that $\psi_1(\vec{x},t) = \psi_0(\vec{x},t)e^{i\vartheta(\vec{x},t)}$ satisfies?

It is easy to show by direct substitution that the equation that $\psi_1(\vec{x},t)$ satisfies is

$$i\hbar\left[\frac{\partial}{\partial t} - i\dot{\vartheta}(\vec{x},t)\right]\psi_1(\vec{x},t) = -\frac{\hbar^2}{2m}[\nabla - i\nabla\vartheta(\vec{x},t)]^2\psi_1(\vec{x},t). \tag{6.11}$$

In fact, it is quite easy to see that in general that if $\psi_0(\vec{x},t)$ satisfies an equation of the form $L[\frac{\partial}{\partial t},\nabla]\psi(\vec{x},t) = 0$, then $\psi_1(\vec{x},t) = \psi_0(\vec{x},t)e^{i\vartheta(\vec{x},t)}$ satisfies $L[\frac{\partial}{\partial t} - i\dot{\vartheta},\ \nabla - i\nabla\vartheta]\psi_1(\vec{x},t) = 0$.

What the above correspondence means is that if $\psi_0(\vec{x},t)$ is an eigenstate of the original Schrödinger equation with energy $E_n$, then $\psi_1(\vec{x},t)$ is an eigenstate of the 'new' Schrödinger equation with the *same* energy, that is there exists a one-to-one correspondence between the wavefunctions of the two 'Schrödinger' equations.

Well, this is encouraging, but the real question is, does this mean that the two wavefunctions represent the *same physical state*? That is, can $\langle\psi_0|\hat{O}|\psi_0\rangle = \langle\psi_1|\hat{O}|\psi_1\rangle$ for all physical observables $A$?

The answer is 'yes' if the operator $O$ is of the form $\hat{O} = f(\vec{x},t)$, since the difference in phase between $\psi_0$ and $\psi_1$ will not affect the expectation value of $\int d^d x|\psi(\vec{x},t)|^2 f(\vec{x},t)$. However, if the operator $A$ involves derivatives $\hat{O}_\mu \sim \partial_\mu$, then $\langle\psi_0|\hat{O}_\mu|\psi_0\rangle \neq \langle\psi_1|\hat{O}_\mu|\psi_1\rangle$ in general.

This difficulty can be resolved by replacing the ordinary derivatives by the so-called *covariant derivatives*,

$$\frac{D}{Dx_\mu} = \frac{\partial}{\partial x_\mu} - iA_\mu(\vec{x},t) \quad (\mu = 0,1,2,3), \tag{6.12a}$$

where $A_\mu$ is a vector field introduced to keep track of the local phases of the wavefunctions. In particular, we require that

$$A_\mu \to A_\mu + \partial_\mu \vartheta \quad \text{when} \quad \psi(\vec{x},t) \to \psi(\vec{x},t)e^{i\vartheta(\vec{x},t)}. \tag{6.12b}$$

It is easy to see that if we require that in quantum mechanics all physical processes are expressed in terms of covariant derivatives rather than ordinary derivatives, then

$$\langle\psi_0|\,\widehat{O}(D_\mu)|\psi_0\rangle \equiv \langle\psi_1|\,\widehat{O}(D_\mu)|\psi_1\rangle \tag{6.12c}$$

and our problem is solved!

For example, the 'free-particle' Schrödinger equation should be replaced by

$$i\hbar\frac{D}{Dt}\psi(\vec{x},t) = -\frac{\hbar^2}{2m}\vec{D}^2\psi(\vec{x},t), \quad \vec{D}^2 = \sum_{\mu=1,2,3}\left(\frac{D}{Dx_\mu}\right)^2 \tag{6.13a}$$

or

$$i\hbar\left(\frac{\partial}{\partial t} - iA_0(\vec{x},t)\right)\psi(\vec{x},t) = -\frac{\hbar^2}{2m}(\nabla - i\vec{A}(\vec{x},t))^2\psi(\vec{x},t), \tag{6.13b}$$

which, aside from some numerical factors, is the Schrödinger equation for (non-relativistic) charged particles in an electromagnetic field ($A_0.\vec{A}$).

**Question**

What are the corresponding relativistic Klein–Gordon and Dirac equations?

#### 6.1.3.1 Relation Between $A_\mu$ and Berry Phase

To see the relation of $A_\mu$ to Berry phase we first consider the translation operator $T(x)$ on a wavefunction defined by

$$T(x)\psi(x') = \psi(x + x') \quad (x = (\vec{x},t), \text{ etc.}). \tag{6.14a}$$

It can be shown (from group theory) that the operator $T(x)$ is characterized completely by what it is for an infinitesimal translation $T(\delta x) = 1 - i\pi_\mu \delta x_\mu$, where $\pi_\mu$ is called the generator of the translation (along the direction $\mu$). Naively, we expect that

$$T(\delta x)\psi(x') = \psi(\delta x + x') \sim \left[1 + \delta x_\mu \frac{\partial}{\partial x'_\mu}\right]\psi(x')$$

and the generator of the translation is $\pi_\mu = -i\frac{\partial}{\partial x_\mu}$, which is nothing but the usual canonical momentum (setting $\hbar = 1$).

However, recall that, in general, a quantum wavefunction may acquire a Berry phase upon translation, that is, in general,

$$T(x)\psi(x') = \psi(x + x')e^{i(\vartheta(x'+x) - \vartheta(x'))}.$$

Therefore, in general,

$$T(\delta x)\psi(x') = \psi(\delta x + x')e^{i(\vartheta(\delta x + x') - \vartheta(x'))} = \left[1 + \delta x_\mu \frac{\partial}{\partial x'_\mu} + i\delta x_\mu \partial_\mu \vartheta(x')\right]\psi(x')$$

and the generator of the translation becomes

$$\pi_\mu = -i\left(\frac{\partial}{\partial x_\mu} + i\partial_\mu \vartheta\right) \rightarrow -i\left(\frac{\partial}{\partial x_\mu} - iA_\mu\right). \tag{6.14b}$$

Therefore, the covariant derivative $\frac{D}{Dx_\mu} = \frac{\partial}{\partial x_\mu} - iA_\mu(\vec{x}, t)$ is essentially the generator of the translation when Berry phase is incorporated. The Berry phase change is related to $A_\mu$ by

$$\vartheta(x + x') - \vartheta(x') = -\int_{x'}^{x+x'} A_\mu(x'')\mathrm{d}x''_\mu .$$

In this regard, gauge invariance is just another way to express the fact that Berry phase has no physical effect if it is single valued ($\nabla \times \vec{A} = 0$) and the electromagnetic field represents nothing but a field that induced irreducible phases in a quantum system.

Note: the requirement of local gauge invariance can be generalized if we consider $N$-component wavefunctions ($N > 1$)

$$\psi = \begin{pmatrix} \psi_1 \\ \cdot \\ \cdot \\ \psi_N \end{pmatrix}.$$

In this case, local gauge invariance means invariance with respective to a unitary transformation $\psi(x) \rightarrow U(x)\psi(x)$, where $U(x)$ is an $N \times N$ unitary matrix. This is what Yang–Mills theory considered. Yang–Mills theory considers $U(x)$ within the next simplest group structure SU(2), which, in the simplest form, corresponds to $2 \times 2$ unitary matrices. We shall not go into these more complicated gauge theories here.

### 6.1.4 Electromagnetism as Gauge Theory

Recall that, in electromagnetism,

$$\vec{E} = -\nabla\phi - \frac{1}{c}\frac{\partial \vec{A}}{\partial t}, \quad \vec{B} = \nabla \times \vec{A}, \tag{6.15a}$$

and the single-particle Hamiltonian is $H = \frac{1}{2m}\left(\vec{p} - \frac{e\vec{A}}{c}\right)^2 + e\phi$, where $\vec{p} = -i\hbar\nabla$.

The Schrödinger equation is

$$i\hbar\frac{\partial}{\partial t}\psi(\vec{x},t) = -\frac{\hbar^2}{2m}\left(\nabla - i\frac{e\vec{A}(\vec{x},t)}{\hbar c}\right)^2\psi(\vec{x},t) + e\phi(\vec{x},t)\psi(\vec{x},t), \qquad (6.15\text{b})$$

which is precisely the U(1) gauge theory we constructed based on local gauge invariance,

$$i\hbar\left(\frac{\partial}{\partial t} - iA_0(\vec{x},t)\right)\psi(\vec{x},t) = -\frac{\hbar^2}{2m}(\nabla - i\vec{A}(\vec{x},t))^2\psi(\vec{x},t),$$

if we identify $A_0 \to -\frac{e\phi}{\hbar}$ and $\vec{A} \to \frac{e\vec{A}}{\hbar c}$.

Let us check whether physical observables are indeed expressed in terms of covariant derivatives.

*Notation:*
$\vec{p} = -i\hbar\nabla$ (canonical momentum, derived from $p_\alpha = \frac{\delta L}{\delta q_\alpha}$)
$\vec{\pi} = \vec{p} - \frac{e}{c}\vec{A}$ (kinetic momentum)

Physical observables should associate with $\vec{\pi}$ but not with $\vec{p}$ if gauge theory is self consistent. To see whether this is true, we start with the physical observable $\vec{x}$ (position of particle). We expect that $\vec{v} = \frac{d\vec{x}}{dt}$ is also a physical observable.

Using the Heisenberg equation of motion, its is easy to show that

$$\dot{\vec{x}} = \frac{d\vec{x}}{dt} = \frac{1}{m}\left(\vec{p} - \frac{e}{c}\vec{A}\right), \qquad (6.16)$$

that is the velocity (which is a physical observable) of a particle is given by $\frac{\vec{\pi}}{m}$ (not $\frac{\vec{p}}{m}$).

**Homework Q2a**
Derive the above result and derive also the commutation relations between different components of kinetic momentum.

Similarly, we can also derive the current operator through the continuity equation $\frac{\partial\rho}{\partial t} + \nabla.\vec{j} = 0$.

**Homework Q2b**
Starting with the density operator $\rho = \psi^*\psi$, show that

$$\vec{j} = \frac{\hbar}{2mi}[\psi^*\vec{\pi}\psi + (\vec{\pi}^*\psi^*)\psi] = \frac{\hbar}{2mi}[\psi^*\nabla\psi - (\nabla\psi^*)\psi] - \frac{e}{mc}\vec{A}\rho.$$

Other physical observables can be constructed similarly.

**Homework Q2c**
Show that the angular momentum operator is $\vec{L} = \vec{r}\times\vec{\pi}$.

The results can be generalized to quantum field theory if you replace the single-particle wavefunction by a coherent state representation, $\psi \to \widehat{\psi}$ (boson or fermion field operators) and $A_\mu \to \widehat{A}_\mu$. We shall study the quantum theory of electromagnetism in Chapter 12.

## 6.2 Singular Gauge Potentials and Angular Momentum Quantization

We shall study two examples of so-called singular gauge potential here. The first example we shall consider is the Aharonov–Bohm effect. The second example is a magnetic monopole. First, we consider the Aharonov–Bohm effect.

### 6.2.1 Aharonov–Bohm (AB) Effect

The ideal Aharonov–Bohm (AB) effect considers an infinitely long solenoid with impenetrable walls. A magnetic flux $\Phi$ is trapped inside the solenoid but the magnetic field outside the solenoid is zero. We consider the motion of charges (electrons) outside the solenoid. Experimentally, a situation that mimics the ideal AB effect is a trapped vortex inside a superconductor where a magnetic flux $\Phi = hc/e^*$ is automatically trapped (see Chapter 11). Electrons in a superconductor see the AB effect and their motions are affected.

Classically, a charged particle outside the solenoid will never be aware of the existence of a magnetic field inside the solenoid because of the impenetrable wall. Therefore, the motion of the charge is not affected by the presence of the magnetic flux $\Phi$. Quantum mechanically, we have to solve the corresponding Schrödinger equation

$$i\hbar\frac{\partial}{\partial t}\psi(\vec{x},t) = -\frac{\hbar^2}{2m}\left(\nabla - i\frac{e\vec{A}(\vec{x})}{\hbar c}\right)^2\psi(\vec{x},t) + \phi(\vec{x})\psi(\vec{x},t) \tag{6.17}$$

with a vector potential $\vec{A}(\vec{x})$ representing the magnetic field in the solenoid and $\phi = 0$. The first question is thus: what is $\vec{A}(\vec{x})$ outside the solenoid? Is it zero?

Using Stokes' theorem, we have for $r > r_0 =$ radius of solenoid, $\oint_C \vec{A}.d\vec{l} = \Phi$, where we have considered an infinite solenoid located along the $z$ axis, and $C$ is a circular loop centered at the origin with radius $r$. Taking advantage of cylindrical symmetry, we can write $\vec{A}(r,\vartheta) = A(r)\,\widehat{\vartheta}$ (see figure below); therefore,

$$\oint_C \vec{A}\cdot d\vec{l} = \int_0^{2\pi} r d\vartheta A(r) = \Phi$$

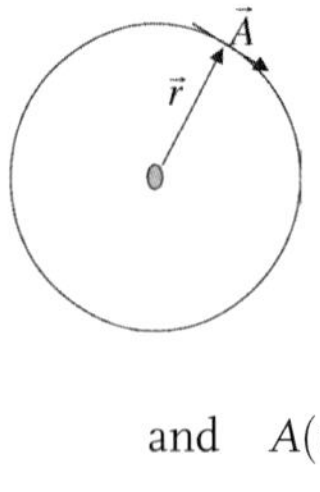


$$\text{and}\quad A(r) = \frac{\Phi}{2\pi r}. \tag{6.18}$$

**Homework Q3a**
Show that $\nabla \times \vec{A} = \vec{B} = 0$ for $r > r_0$.

Since $\nabla \times \vec{A} = \vec{B} = 0$ at $r > r_0$, for $r > r_0$, we should be able to write $\vec{A} = \nabla\varphi$, or

$$A(r) = \frac{1}{r}\frac{\partial \varphi}{\partial \vartheta} = \frac{\Phi}{2\pi r}, \tag{6.19a}$$

which suggests that

$$\varphi = \varphi(\vartheta) = \frac{\Phi\vartheta}{2\pi}, \tag{6.19b}$$

which is in fact not a well-defined function since $\vartheta$ and $\vartheta + 2n\pi$ ($n$ = integer) represent the same physical point in space. The function is said to be 'multi-valued' in this case since different values of $\varphi = \varphi(\vartheta)$ correspond to the same point in space. This is an example of a singular gauge potential.

To see the physical consequences of the singular gauge potential, we consider the Schrödinger equation for a charged particle in $\vec{A}$.

In cylindrical coordinates,

$$\left(\nabla - i\frac{e\vec{A}}{\hbar c}\right)^2 \rightarrow \frac{1}{r}\frac{\partial}{\partial r}\left(r\frac{\partial}{\partial r}\right) + \frac{\partial^2}{\partial z^2} + \frac{1}{r^2}\left(\frac{\partial}{\partial \vartheta} - i\frac{e\Phi}{hc}\right)^2 \tag{6.20}$$

and the time-independent Schrödinger equation is

$$\frac{-\hbar^2}{2m}\left(\nabla - i\frac{e\vec{A}}{hc}\right)^2 \psi(r,\vartheta,z) = E\psi(r,\vartheta,z)$$

with boundary condition $\psi(r \to r_0, \vartheta, z) \to 0$ (impenetrable solenoid).

Writing $\psi(r,\vartheta,z) = e^{ik_z z}e^{im\vartheta}R(r)$ and $k^2 = 2mE/\hbar^2$, we obtain

$$\left(-\frac{1}{r}\frac{\partial}{\partial r}r\frac{\partial}{\partial r} + \frac{1}{r^2}\left(m - \frac{e\Phi}{hc}\right)^2\right)R(r) = (k^2 - k_z^2)R(r). \tag{6.21a}$$

The solution of the equation is the Bessel function of order $m - \frac{e\Phi}{hc}$,

$$R(r) = J_{m-\frac{e\Phi}{hc}}(k'r), \quad \text{where} \quad k' = \sqrt{k^2 - k_z^2}, \tag{6.21b}$$

$$\text{with boundary condition. } J_{m-\frac{e\Phi}{hc}}(k'r_0) = 0, \tag{6.21c}$$

which determines the eigenvalues $k^2$.

We observe that

1. The eigenvalues depend on $\Phi$, implying that although the charges never 'see' the magnetic field, there is a physical effect on the particles associated with the magnetic flux through the $\vec{A}$ field.
2. There is a periodic structure in the energy spectrum. When $\Phi = \frac{hc}{e}n$, then $m - \frac{\Phi e}{hc} = m - n$. The spectrum has the same structure as $\Phi = 0$ except that the quantum number $m$ is shifted to $m - n$.

($\Phi_0 = \frac{hc}{e}$ is called the magnetic flux quantum.)

### 6.2.2
### Alternative Description of the Problem

Suppose that we now 'remove' the $\vec{A}$ field by a gauge transformation,

$$\vec{A} \to \vec{A} + \nabla\Lambda, \quad \text{where} \quad \Lambda = -\frac{\Phi\vartheta}{2\pi},$$

then $H \to -\frac{\hbar^2}{2m}\nabla^2$, which is the free-particle Hamiltonian!

**Question**
Where is the effect of magnetic flux now?

The effect of magnetic flux shows up as a change in the boundary condition of the wavefunction. The wavefunction becomes 'multi-valued' (i.e. $\psi(\vartheta) \neq \psi(\vartheta + 2\pi)$) after the gauge transformation (see (6.19b)). This is an example where the gauge field can be replaced explicitly by an irreducible phase.

**Homework Q3b**
Write down the boundary condition for $\psi(\vartheta)$ after the gauge transformation.

Note that physical observables should *not* depend on which way you describe the system.

**Homework Q3c**
Compute the $z$ component of angular momentum $L_z$ in the above two descriptions. Show that in the first description where $\vec{A} \neq 0$ (i) the usual quantization condition $L_z = m\hbar$ is modified in the presence of magnetic flux $\Phi \neq 0$ and (ii) the same value of $L_z$ is obtained in the $\vec{A} = 0$ description where the boundary condition is modified.

### 6.2.3
### Magnetic Monopole and Angular Momentum Quantization

In the previous section we saw that the $z$-component angular momentum quantization condition is modified in the presence of a magnetic flux tube. In this section, we shall study this question in more depth. We shall ask the question: what is the physical origin of angular momentum quantization (in 3D)?

First, we recall some basics of the theory of angular momentum. Angular momentum is related to rotation. A rotation can be characterized by three parameters, $\widehat{n}\,(\theta, \phi)$ which specifies the axis of rotation, and $\vartheta$ which specifies the angle of rotation.

In quantum mechanics, rotation is a symmetry operation and its effect on a state vector is represented by a unitary transformation, $\widehat{U}(\widehat{n}, \vartheta) = e^{-i\frac{\vartheta}{\hbar}\vec{J}\cdot\vec{n}}$, where $-i\frac{J_\alpha}{\hbar}(\alpha = \widehat{x}, \widehat{y}, \widehat{z})$ are the generators of rotation, and the $J_\alpha$ are the angular momentum operators.

Note that angular momentum is defined with respect to the physical process of rotation here, but not defined from $\vec{L} = \vec{x} \times \vec{p}$.

Since $\hat{n}(\theta, \phi)$ is characterized by two parameters only, the angular momentum operators cannot all be independent (or commute) with each other. The commutation relation in 3D space can be derived by requiring that the usual rotation rules for vectors be satisfied (see for example Ref. [1]) and is given by

$$[J_\alpha, J_\beta] = i\hbar\varepsilon_{\alpha\beta\gamma}J_\gamma. \tag{6.22}$$

Starting from these commutation rules, we can construct simultaneous eigenstates of operators (*algebraic theory of angular momentum*) $J^2 = J_x^2 + J_y^2 + J_z^2$ and $J_z$, denoted by $|j, m\rangle$, where

$$J^2|j,m\rangle = \hbar^2 j(j+1)|j,m\rangle \quad \text{and} \quad J_z|j,m\rangle = m\hbar|j,m\rangle, \tag{6.23a}$$

$$\text{where } j = \frac{n}{2} \ (n = \text{integer}) \quad \text{and} \quad m = -j, \ldots, j. \tag{6.23b}$$

**Question**

How about in 2D and 1D, what do you think the angular momentum operators will look like? How about their eigenvalues? Are they quantized as in 3D?

The question we want to understand is why $j$ takes the values $j = n/2$ ($n$ = integer).

#### 6.2.3.1 Geometrical Theory of Angular Momentum

Since angular momentum is tied to rotation in real space, we should be able to understand the quantization rules of angular momentum from examining directly the operation of rotation in real space. This is what we shall do in the following.

We consider an arbitrary quantum state $|\psi\rangle$. The state is, in general, a many-particle state. However, we shall consider a one-particle state for illustration in the following.

Let us construct an infinitesimal rotation on $|\psi\rangle$. We write

$$|\psi\rangle = \int d^3x'|\vec{x}'\rangle\langle\vec{x}'|\psi\rangle|\psi\rangle = \int d^3x'|\vec{x}'\rangle\langle\vec{x}'|\psi\rangle. \tag{6.24a}$$

$$\text{Upon rotation, } |\psi\rangle \to \int d^3x'|\vec{x}' + \delta\vec{x}'\rangle\langle\vec{x}'|\psi\rangle. \tag{6.24b}$$

Therefore, we should first study the effect of rotation on $|\vec{x}'\rangle$. Let us consider rotation of an amount $\delta\phi$ about the $z$ axis. (see figure below)

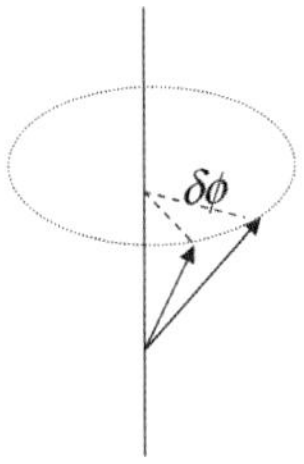

Upon rotation, it is obvious that (see figure below)

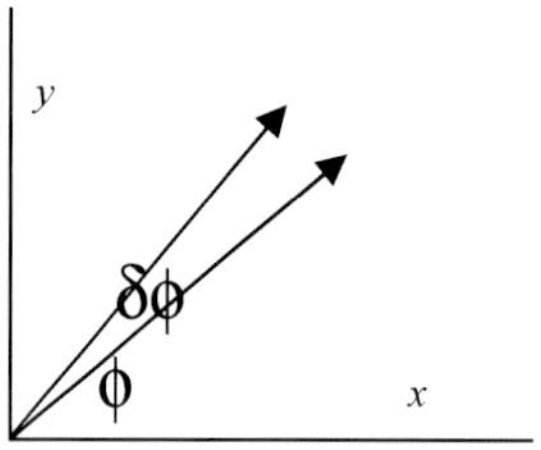


$$|\vec{x}' = x', y', z'\rangle \to |x' - y'\delta\phi, y' + x'\delta\phi, z'\rangle \tag{6.25a}$$

and

$$\begin{aligned} |x' - y'\delta\phi, y' + x'\delta\phi, z'\rangle &\sim \left[1 - (x'\delta\phi) i\frac{\widehat{p}_y}{\hbar} + (y'\delta\phi) i\frac{\widehat{p}_x}{\hbar}\right] |x', y', z'\rangle \\ &= \left[1 - \frac{i\delta\phi}{\hbar}\widehat{L}_z\right] |x', y', z'\rangle \end{aligned} \tag{6.25b}$$

to first order in $\delta\phi$. Therefore, we obtain $L_z = xp_y - yp_x$. (6.25c)

By rotational symmetry, we also expect that $L_x = yp_z - zp_y$ and $L_y = zp_x - xp_z$, or $\vec{L} = \vec{x} \times \vec{p}$, in agreement with our usual definition of angular momentum.

We can study the angular momentum eigenstates in this representation directly, that is we can solve directly the eigenvalue equations

$$\vec{L}^2 |\psi_{lm}\rangle = \lambda_l |\psi_{lm}\rangle \quad \text{and} \quad L_z |\psi_{lm}\rangle = \lambda_m |\psi_{lm}\rangle \tag{6.26a}$$

in real-space representation.

In spherical coordinates, we obtain

$$\langle \vec{x} | \vec{L}^2 | \psi_{lm}\rangle = -\hbar^2 \left[\frac{1}{\sin^2\vartheta} \frac{\partial^2}{\partial\phi^2} + \frac{1}{\sin\vartheta} \frac{\partial}{\partial\vartheta}\left(\sin\vartheta \frac{\partial}{\partial\vartheta}\right)\right] \langle \vec{x} | \psi_{lm}\rangle = \lambda_l \langle \vec{x} | \psi_{lm}\rangle \tag{6.26b}$$

and

$$\langle \vec{x} | L_z | \psi_{lm}\rangle = -i\hbar \frac{\partial}{\partial\phi} \langle \vec{x} | \psi_{lm}\rangle = \lambda_m \langle \vec{x} | \psi_{lm}\rangle, \tag{6.26c}$$

of which the solutions are spherical harmonics $\langle \vec{x} | \psi_{lm}\rangle = Y_{lm}(\vartheta, \phi)$ with eigenvalues $\lambda_l = \hbar^2 l(l+1)$ and $\lambda_m = m\hbar$, where $l =$ integer.

The spectrum differs from the algebraic theory where $j = n/2$ can be half-integers. The question is then what is wrong with the geometrical theory?

> *Why can't we obtain half-integer angular momentum values from the real-space representation of the angular momentum operator?*

#### 6.2.3.2 Berry Phase and Angular Momentum Quantization

The answer to the above question rests in Berry phase. Recall that in general when a particle is 'moved' from $|\vec{x}_1\rangle$ to $|\vec{x}_2\rangle$ via path $l$, the wavefunction can take up an overall phase $\gamma(l)$ which depends on the path $l$. $\gamma(l)$ can be written as

$$\gamma(l) = \int_{\vec{x}_1}^{\vec{x}_2} \vec{A}.d\vec{l},$$

where $\vec{A}$ is an effective vector potential representing the Berry phase. Therefore, when a particle is physically moved from a point $\vec{x}$ to $\vec{x}+\delta\vec{x}$, we have in general

$$|\vec{x}\rangle \to e^{i\gamma(\delta l)}|\vec{x}+\delta\vec{x}\rangle \sim \left(1-\frac{i}{\hbar}(\vec{p}-\hbar\vec{A})\cdot\delta\vec{x}\right)|\vec{x}\rangle, \tag{6.27}$$

or the generator of translation is in general $\vec{\pi} = \vec{p}-\vec{A}$, and $\vec{L} = \vec{r}\times\vec{\pi}$.

To understand the angular momentum quantization, let us ask the following question: suppose that we have a system with rotational symmetry (in 3D space) so that the states are eigenstates of angular momentum, what is the most general form of $\vec{A}$ (or $\vec{B} = \nabla\times\vec{A}$) that is allowed? This question was first addressed by Wilczek [2].

From rotational symmetry, we expect that $\vec{B}(\vec{r}) = B(r)\widehat{r}$, corresponding to a magnetic monopole type potential!

#### 6.2.3.3 Quantization of Magnetic Monopole

Consider the (Berry) phase picked up by a particle when translated around a circle $C$ around the monopole.

$$\text{The phase pick up is } \gamma = \oint_C \vec{A}.d\vec{l} = \oint_S \vec{B}\cdot d\vec{s} \tag{6.28a}$$

(see figure below).

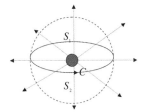

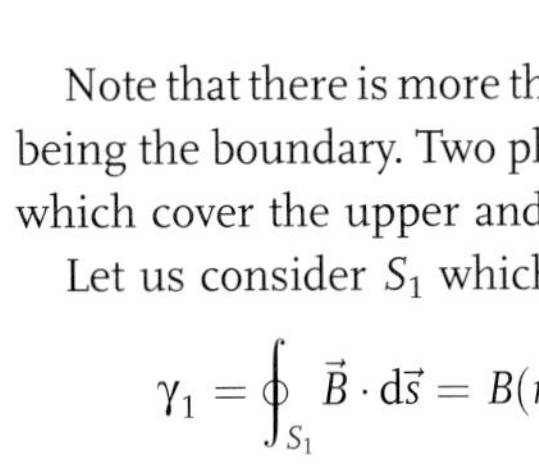


Note that there is more than one way to identify the surface $S$ with the closed loop $C$ being the boundary. Two plausible choices are shown in the above figure, $S_1$ and $S_2$, which cover the upper and lower hemispheres of the sphere, respectively.

Let us consider $S_1$ which covers the upper hemisphere. In this case,

$$\gamma_1 = \oint_{S_1} \vec{B}\cdot d\vec{s} = B(r)(2\pi r^2). \tag{6.28b}$$

Note that the line integral $\oint_C \vec{A}.d\vec{l}$ goes counterclockwise viewing from above (north pole).

However, from the lower hemisphere ($S_2$), we obtain

$$\gamma_2 = \oint_{S_2} \vec{B} \cdot d\vec{s} = -B(r)(2\pi r^2). \tag{6.28c}$$

This is because the line integral $\oint_C \vec{A}.d\vec{l}$ goes clockwise viewing from below (south pole).

However, the physical effect associated with the phases should be unique, since we are just viewing the same physical process from two different angles, that is we must have

$$e^{i\gamma_1} = e^{i\gamma_2}, \quad \text{or} \quad e^{i[4\pi r^2 B(r)]} = 1, \tag{6.29}$$

implying that

$$4\pi B(r) r^2 = 2\pi n, \quad \text{or} \quad B(r) = \frac{(n/2)}{r^2}, \tag{6.30a}$$

corresponding to the magnetic field from a magnetic monopole with quantized strength $g = n/2$.

Note that, in electromagnetism, similar considerations were applied to predict (if they exist) the strength of magnetic monopoles (see Ref. [1]). In this case, the kinetic momentum is

$$\vec{\pi} = \vec{p} - \frac{e\vec{A}}{c} = -i\hbar\left(\nabla - i\frac{e\vec{A}}{\hbar c}\right).$$

For charges to pick up a consistent Berry phase when moving around a monopole, we must have $e^{i\frac{e}{\hbar c}[4\pi r^2 B(r)]} = 1$ and correspondingly

$$B(r) = \frac{g}{r^2}, \quad \text{where } \frac{2ge}{\hbar c} = n \text{ (integer)}. \tag{6.30b}$$

This is called the Dirac quantization condition of a magnetic monopole.

Note that the above quantization result is based purely on geometrical considerations. No dynamics is involved. This is an example of topological quantization. You will see more examples of topological quantization later.

Therefore, the most general eigenvalues of the angular momentum should be determined by the Schrödinger equation

$$-\hbar^2\left[\frac{1}{\sin^2\vartheta}\left(\frac{\partial}{\partial\phi} - iA_\phi\right)^2 + \frac{1}{\sin\vartheta}\frac{\partial}{\partial\vartheta}\left(\sin\vartheta\frac{\partial}{\partial\vartheta}\right)\right]\langle\vec{x}|\psi_{lm}\rangle = \lambda_l\langle\vec{x}|\psi_{lm}\rangle,$$
$$\text{where } A_\phi = \begin{bmatrix} g\dfrac{1-\cos\vartheta}{\sin\vartheta}, & \vartheta < \pi - \varepsilon, \\ g\dfrac{1+\cos\vartheta}{\sin\vartheta}, & \vartheta > \varepsilon \end{bmatrix} \tag{6.31}$$

is the vector potential for the magnetic monopole. There is an ambiguity in $A_\phi$ because of its singular nature. You can look at Ref. [1] for more details. The above

Schrödinger equation was solved by Wu and Yang [3], where they observed that the angular momentum $j$ is an integer if $g = n/2$ is an integer, and is a half-integer if $g$ is a half-integer. We shall not go into the mathematical details of this differential equation here. The physical effect of the magnetic monopole potential on angular momentum quantization can be understood qualitatively from the Bohr–Sommerfeld quantization where stable quantum states are represented by classical closed orbits with the quantization condition $\oint_C p\,dq = nh$.

In the absence of the monopole potential ($\vec{A} = 0$), it results in the usual quantization of angular momentum $L = pr$, where $2\pi L = nh$, or $L = n\hbar$.

In our case, the angular momentum of a particle moving around a monopole potential becomes

$$L \to \frac{1}{2\pi}\oint(\vec{p} - \hbar\vec{A}) \cdot d\vec{l}. \tag{6.32}$$

For $\vec{A}$ = vector potential associated with a magnetic monopole, we obtain

$$L \to n\hbar - \frac{\hbar}{2\pi}\int_S \vec{B} \cdot d\vec{s} = (n+g)\hbar \tag{6.33}$$

and the angular momentum quantization condition is modified. In particular, for $g$ = (half) integer, angular momentum is also a (half) integer.

#### 6.2.3.4 Two Dimensions

**Homework Q4**

We can ask the same question of what is the most general Berry phase that can occur in 2D that respects rotational invariance. Show that the vector potential structure that replaces a magnetic monopole in 3D is a magnetic flux tube perpendicular to the plane and that the angular momentum can take arbitrary values (Homework Q3a-c).

#### 6.2.3.5 Angular Momentum: Statistics Theorem

Quantization of angular momentum is intimately related to Berry phase of particles picked up upon exchange. Their relation can be visualized by seeing that exchange of two particles can be viewed as 2 × (half) rotation of particle A (B) around B (A), if we view the process from the angle of particle B (A) (see figure below). The factor of two comes because the same phase is picked up by both particles.

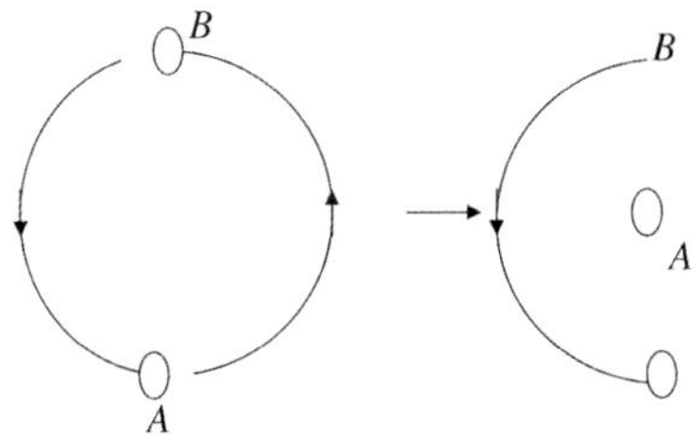

Therefore, the phase picked up by exchanging the particles should be the same as the phase taken up by rotating the particles around one another, which is represented by a unitary transformation, $\widehat{U}(\widehat{z}, 2\pi) = e^{-i\frac{2\pi}{\hbar}J_z}$, and can only be 1 (angular momentum $= n\hbar \leftrightarrow$ bosons) and $-1$ (angular momentum $= (n+1/2)\hbar \leftrightarrow$ fermions) in 3D, according to our Berry phase analysis of rotation. The above relation between particle statistics and spin (angular momentum) is the angular momentum–statistics theorem.

Since angular momentum can take arbitrary values in 2D, particles can have arbitrary statistics in 2D. They are called anyons. This is the case of quasi-particles in the fractional quantum Hall effect [4].

## 6.3
## Quantization of Electromagnetic Field

The next natural topic in the study of Berry phase would be the effect of Berry phase on quantum field theory (or quantum gauge field theory). We shall postpone this topic to Chapter 12 where we shall explore the interesting mathematical structure behind a few simple lattice gauge field theories. Here we shall lay down the groundwork by discussing the conventional ways of quantizing the electromagnetic (EM) field.

The quantization of the EM field faces some mathematical subtlety associated with gauge invariance. We shall first discuss quantization in the canonical formulation, and then path integral quantization.

### 6.3.1
### Canonical Quantization

For simplicity we consider the Lagrangian for the EM field $(\varphi(=A_0), \vec{A})$ in a fixed background charge,

$$\begin{aligned} L &= \sum_{\nu,\mu} \frac{1}{16\pi} \int d^3x F_{\mu\nu}^2(x) - \int d^3x \rho(x)\varphi(x) \\ &= \sum_{\nu,\mu} \frac{1}{16\pi} \int d^3x (\partial_\mu A_\nu(x) - \partial_\nu A_\mu(x))^2 - \int d^3x \rho(x)\varphi(x). \end{aligned} \tag{6.34a}$$

The Lagrangian can be written explicitly in terms of electric and magnetic fields,

$$L = \frac{1}{8\pi} \int d^3x [\vec{E}(x)^2 - \vec{B}(x)^2] - \int d^3x \rho(\vec{x})\varphi(x), \tag{6.34b}$$

$$\text{where } \vec{E}(x) = -\frac{1}{c}\frac{\partial \vec{A}(x)}{\partial t} - \nabla\varphi(x) \quad \text{and} \quad \vec{B}(x) = \nabla \times \vec{A}(x). \tag{6.34c}$$

The conjugate momenta are

$$\pi_\varphi = \frac{\delta L}{\delta \varphi} = 0, \quad \pi_{\vec{A}} = \frac{\delta L}{\delta \vec{A}} = -\frac{\vec{E}(x)}{4\pi c}, \tag{6.34d}$$

and the corresponding Hamiltonian is

$$H = \frac{-1}{4\pi c}\int d^3x\vec{A}\ (x).\vec{E}(x) - L$$

$$= \frac{1}{8\pi}\int d^3x[\vec{E}(x)^2 + \vec{B}(x)^2 + 2\nabla\varphi(\vec{x}).\vec{E}(x)] + \int d^3x\rho(\vec{x})\varphi(x). \qquad (6.35)$$

Note that the canonical momenta for the $\varphi$ field vanish. This situation has been discussed in Homework Q8 of Chapter 3.

In this case, the $\varphi$ field is determined by minimizing the action directly. We obtain

$$\frac{\delta L}{\delta\varphi} = \frac{1}{4\pi}\nabla\cdot\vec{E} - \rho = 0, \qquad (6.36)$$

which is Gauss' law. Note that the classical Gauss' law is *not* affected by quantization.

Therefore,

$$H = \frac{1}{8\pi}\int d^3x[\vec{E}(x)^2 + \vec{B}(x)^2 + 2\nabla\varphi(\vec{x}).\vec{E}(x)] + \int d^3x\rho(\vec{x})\varphi(x)$$

$$= \frac{1}{8\pi}\int d^3x[\vec{E}(x)^2 + \vec{B}(x)^2 - 2\varphi(\vec{x})\nabla.\vec{E}(x)] + \int d^3x\rho(\vec{x})\varphi(x) \qquad (6.37)$$

$$= \frac{1}{8\pi}\int d^3x[\vec{E}(x)^2 + \vec{B}(x)^2].$$

Before proceeding further, let us examine the consequence of gauge invariance, that is invariance of action under the transformation $\vec{A} \to \vec{A} + \nabla\Lambda,\ \varphi \to \varphi + \dot{\Lambda}$.

In this case, $\vec{E}$ and $\vec{B}$ remain invariant, and

$$\int d^3x dt\rho(\vec{x})\varphi(x) \to \int d^3x dt\rho(\vec{x})\left(\varphi(x) + \frac{\partial\Lambda(x)}{\partial t}\right)$$

$$= \int d^3x dt\left[\rho(\vec{x})\varphi(x) - \Lambda(x)\frac{\partial\rho}{\partial t}\right] = \int d^3x dt\rho(\vec{x})\varphi(x).$$

Therefore, both the action and the Hamiltonian are gauge invariant.

However, we have a mathematical problem here – the fields $\vec{A}$ and $\vec{A} + \nabla\Lambda$ have identical canonical momenta $\vec{E}(x) = -\frac{1}{c}\frac{\partial\vec{A}(x)}{\partial t} - \nabla\varphi(x)$.

This duplication imposes a mathematical problem in the canonical quantization scheme since the pair of canonical variables becomes ill defined. Note that $(\vec{A}, \varphi)$ and $(\vec{A} + \nabla\Lambda, \varphi + \dot{\Lambda})$ both describe the same EM fields physically, and the real problem is that we have too many mathematical functions representing the same physical quantity in our field theory.

A straightforward way to solve the above problem is to eliminate these redundancies by fixing the gauge, or putting constraints on the vector potentials $(\vec{A}, \varphi)$ so that the real EM fields are represented by unique choices of $(\vec{A}, \varphi)$. A common gauge that is convenient for canonical quantization is the Coulomb gauge $\nabla\cdot\vec{A} = 0$

($\varphi$ arbitrary). With this gauge the Hamiltonian can be conveniently written as

$$H = \int \frac{d^3x}{8\pi}\left[\left(\frac{1}{c}\frac{\partial \vec{A}}{\partial t} + \nabla\varphi\right)^2 + (\nabla \times \vec{A})^2\right]$$

$$= \int \frac{d^3x}{8\pi}\left[\left(\frac{1}{c}\frac{\partial \vec{A}}{\partial t}\right)^2 + (\nabla \times \vec{A})^2 + (\nabla\varphi)^2\right], \tag{6.38a}$$

since the longitudinal and transverse components of electric fields decouple in the integral, that is

$$\vec{E} = \vec{E}_l + \vec{E}_t, \quad \vec{E}_l = -\nabla\varphi, \quad \vec{E}_t = -\frac{1}{c}\frac{\partial \vec{A}}{\partial t}, \tag{6.38b}$$

$$\nabla.\vec{E}_t = 0, \quad \nabla \times \vec{E}_l = 0. \tag{6.38c}$$

$$\text{With the Coulomb gauge,} \quad \vec{\pi}_{\vec{A}} = \frac{\delta L}{\delta \vec{A}} = -\frac{\vec{E}_t(x)}{4\pi c} \tag{6.39}$$

and

$$H = \frac{1}{8\pi}\int d^3x[(4\pi c\vec{\pi}_{\vec{A}}(x))^2 + (\nabla \times \vec{A})^2 + (\nabla\varphi)^2]. \tag{6.40}$$

The EM field can be quantized by imposing the usual canonical quantization condition (with the gauge-fixing condition $\nabla.\vec{A} = 0$)

$$[A_\alpha(\vec{x}), \pi_\beta(\vec{x}')] = i\hbar\delta(\vec{x}-\vec{x}')\delta_{\alpha\beta}. \tag{6.41a}$$

In the literature, the gauge-fixing condition is often included directly in the quantization condition by writing

$$[A_\alpha(\vec{x}), \pi_\beta(\vec{x}')] = i\hbar\left(\delta_{\alpha\beta} - \frac{\partial_\alpha\partial_\beta}{\nabla^2}\right)\delta(\vec{x}-\vec{x}'), \tag{6.41b}$$

$$\text{which guarantees that } [\nabla \cdot A(\vec{x}), \pi_\beta(\vec{x}')] = 0. \tag{6.41c}$$

The spectrum and wavefunction of the EM field can be obtained from the Hamiltonian following the quantization procedure of bosonic field theories discussed in Chapter 3.

**Homework Q5**

Obtain the ground-state wavefunction and excitation spectrum for the quantized EM field. What are the effects of static charges on the wavefunction and the spectrum?

### 6.3.2
### Path Integral Quantization

In Chapters 3 and 4, we derived the path integral quantization scheme by studying the time-evolution operator in the Hamiltonian formulation. As time went on, physicists realized that one may define quantum mechanics directly from a path integral formulation, that is we assume directly that

$$\langle \psi | \hat{U}(t,t') | \psi' \rangle \sim \int D\psi^*(t'') D\psi(t'') e^{\frac{i}{\hbar}\int_{t'}^{t} L(t'')dt''}$$

for the field $\psi$ under investigation without showing its validity starting from the Hamiltonian. Note that one has to be careful that, a priori, the path integral and canonical quantization procedures may not be always the same, though for problems physicists have attacked so far, they seem to be equivalent. In the following we shall apply this procedure directly to the electromagnetic field where we shall examine the problem of gauge invariance by assuming that we can write down the Lagrangian in the path integral directly, that is we assume that

$$\langle A | \hat{U}(t,t') | A' \rangle \sim \int DA(t'') e^{\frac{i}{\hbar}\int_{t'}^{t} L(t'')dt''}, \tag{6.42}$$

where $L = \sum_{\mu,\nu} \frac{1}{16\pi} \int d^3x F_{\mu\nu}^2(x) - \int d^3x \rho(\vec{x})\varphi(x)$.

### 6.3.3
### Gauge Problem

As discussed before, the electromagnetic field Lagrangian is gauge invariant, meaning that if we write

$$A_\mu = \left(\delta_{\mu\nu} - \frac{\partial_\mu \partial_\nu}{\nabla^2}\right) A_\nu + \frac{\partial_\mu \partial_\nu}{\nabla^2} A_\nu = A_\mu^t + \partial_\mu \Lambda, \tag{6.43a}$$

then $L = L[A^t]$ is independent of the 'pure gauge' component $\Lambda$. This creates a problem in the path integral, where all plausible forms of $A_\mu$ should be 'summed over' in the evaluation of the path integral. Formally, we may write

$$\begin{aligned} \int DA(t'') e^{\frac{i}{\hbar}\int_{t'}^{t} L(t'')dt''} &= \int D\Lambda(t'') DA^t(t'') e^{\frac{i}{\hbar}\int_{t'}^{t} L(A^t;t'')dt''} \\ &= \int D\Lambda(t'') \times \int DA^t(t'') e^{\frac{i}{\hbar}\int_{t'}^{t} L(A^t;t'')dt''} \\ &= (\infty) \times \int DA^t(t'') e^{\frac{i}{\hbar}\int_{t'}^{t} L(A^t;t'')dt''}! \end{aligned} \tag{6.43b}$$

The infinity we obtained from the path integral is a mathematical artifact of using a over-complete set of $A_\mu$ fields to represent the physical electric and magnetic fields, and would be absent if we worked only with the 'physical' $A^t$ field.

To remove the unphysical $\Lambda$ field from the path integral, one way is to fix the gauge by introducing a constraint on the functional form of $(\vec{A}, \varphi)$ in the path integral. The constraint can be introduced through a $\delta$-function as discussed in Chapters 1 and 3.

For examplc, wc can replace

$$\int DA(\vec{x}''t'')e^{\frac{i}{\hbar}\int_{t'}^{t} L(t'')\mathrm{d}t''} \rightarrow \int D\lambda(\vec{x}''t'')\,DA(\vec{x}''t'')e^{\frac{i}{\hbar}\int_{t'}^{t} L(A^t;t'')\mathrm{d}t''\mathrm{d}\vec{x}''}\,e^{i\int(x''t'')\,F(A_\mu)\,\mathrm{d}t''\,\mathrm{d}\vec{x}''}, \tag{6.44a}$$

$$\text{where } F(A_\mu) = \nabla.\vec{A} + \frac{1}{c}\frac{\partial\varphi}{\partial t} \text{ (Lorentz gauge)} \tag{6.44b}$$

$$\text{or } F(A_\mu) = \nabla\cdot\vec{A} \text{ (Coulomb gauge).} \tag{6.44c}$$

Alternatively, one may use a different mathematical trick where we replace the path integral by

$$\begin{aligned}\int DA(t'')e^{\frac{i}{\hbar}\int_{t'}^{t} L(t'')\mathrm{d}t''} &\rightarrow \int D\Lambda(t'')\det\left[\tfrac{\delta F}{\delta\Lambda}\right]e^{-\int \mathrm{d}t''\mathrm{d}x''\frac{F^2}{2\alpha}}DA^t(t'')e^{\frac{i}{\hbar}\int_{t'}^{t} L(A^t;t'')\mathrm{d}t''}\\ &= \int DA(t'')\det\left[\tfrac{\delta F}{\delta\Lambda}\right]e^{-\int \mathrm{d}t''\mathrm{d}x''\frac{F^2}{2\alpha}}e^{\frac{i}{\hbar}\int_{t'}^{t} L(A^t;t'')\mathrm{d}t''}\end{aligned} \tag{6.45}$$

and take the limit $\alpha \rightarrow 0$ at the end when computing physical observables. In this limit, all contributions from 'pure gauges' $\Lambda$ vanish. This is called the Faddeev–Popov technique. We note that in practice the gauge can be fixed quite easily in U(1) gauge theory and these sophisticated techniques are often not necessary. These techniques become important in treating more complicated non-Abelian gauge theories.

**Homework Q6**

Compute the free energy of the EM field in the path integral formulation and compare the result obtained from canonical quantization. You can fix the gauge by any method you like.

The problem associated with gauge invariance in quantizing gauge theories demonstrates clearly that gauge theories are theories with redundant variables. We have discussed in this section mathematical techniques invented to 'remove' these redundancies. Because of the redundancy, the real mathematical structure of the physical field variables becomes 'hidden' and unclear. We shall address this question again in Chapter 12, where the possibility of representing some simple gauge theories in terms of 'string-like' variables which represent more directly the physical degrees of freedom in the gauge theories will be introduced and discussed.

## References


1 Sakurai, J.J. (1994) *Modern Quantum Mechanics*, Addison-Wesley, USA.

2 Wilczek, F. (1990) *Fractional Statistics and Anyon Superconductivity*, World Scientific, Singapore, New Jersey, London, Hong Kong.

3 Wu, T.T. and Yang, C.N. (1977) *Phys. Rev. D* **16**, 1018.

4 Nagaosa, N. (1995) *Quantum Field Theory in Condensed Matter Physics*, Springer, Berlin, Heidelberg, New York, Barcelona, Hong Kong, London, Milan, Paris, Singapore, Tokyo.

# 7
# Introduction to Effective Field Theory, Phases, and Phase Transitions

## 7.1
## Introduction to Effective Field Theory: Boltzmann Equation and Fluid Mechanics

As we know, Newtonian mechanics is based on the picture of particles. However, the identity of particles is lost when we consider theories like the Boltzmann equation or fluid mechanics. In these theories, although the microscopic constituents of the system are particles (atoms and molecules), the system is described by continuous field variables. We shall study in this section how the Boltzmann and Navier–Stokes equations are derived. The derivations provide us with examples of how an 'effective' field theory can be constructed from an underlying microscopic theory of rather different nature and the regime of validity of these 'effective' field theories.

First, we review the derivation of the Boltzmann equation.

In real fluids there are typically $10^{22}$–$10^{23}$ atoms/cm$^3$. A complete description of the behavior of fluids requires us to keep track of the motion of these $10^{22}$ or so atoms in Newtonian mechanics.

A trick physicists found to keep track of the particles is to introduce the mathematical object/concept of a distribution function $f(\vec{r},\vec{k},t)$, which is a useful tool when a large number of particles is present.

The distribution function $f(\vec{r},\vec{k},t)>0$ is defined as the *probability density* of finding a particle with momentum $\vec{k}$ appearing at position $\vec{r}$ at time $t$ – notice the introduction of *probability* here, that is, we shall not keep track of the detailed motion of all particles, but only the probability of when and where they may travel. Note that the distribution function defined in this way is 'classical'. The same function cannot be defined rigorously in quantum mechanics because of the uncertainty principle $\Delta k \Delta x \geq \hbar$. We hope that the mathematics can be simplified by working with $f(\vec{r},\vec{k},t)$ but not the positions of individual particles.

Formally, $f(\vec{r},\vec{k},t)$ can be defined as

$$f(\vec{r},\vec{k},t) = \left\langle \sum_i \delta(\vec{r}-\vec{r}_i(t))\delta(\vec{k}-\vec{k}_i(t)) \right\rangle, \tag{7.1a}$$

where $\vec{r}_i(t)$ and $\vec{k}_i(t)$ are the position and momentum of the $i$th particle at time $t$. $\langle\ldots\rangle$ is an average, which is usually taken to be the ensemble average over all possible

*Introduction to Classical and Quantum Field Theory.* Tai-Kai Ng

ISBN: 978-3-527-40726-2

initial positions and momenta of particles consistent with the imposed boundary conditions.

**Question**
What is the advantage of introducing an ensemble average? Is it necessary? Can one work with a given unique initial condition of particles?

Note that a large number of measurable physical quantities can be computed with knowledge of $f(\vec{r}, \vec{k}, t)$. For example,

$$\text{density of particles } n(\vec{r}, t) = \int \frac{d^3k}{(2\pi)^3} f(\vec{r}, \vec{k}, t) \tag{7.1b}$$

$$\text{and current density } \vec{j}(\vec{r}, t) = \int \frac{d^3k}{(2\pi)^3} \vec{v}(\vec{k}) f(\vec{r}, \vec{k}, t). \tag{7.1c}$$

To determine $f(\vec{r}, \vec{k}, t)$, we start with Newton's law for a particle with momentum $\vec{k}_i$ at position $\vec{r}_i$,

$$\dot{\vec{r}}_i = \vec{v}(\vec{r}_i), \qquad \dot{\vec{k}}_i = \vec{F}_i(t), \tag{7.2a}$$

where

$$\vec{F}_i = \vec{F}^{\text{ext}}(\vec{r}_i) + \sum_{j \neq i} \vec{F}^{\text{int}}(\vec{r}_i - \vec{r}_j) \tag{7.2b}$$

is the total force acting on the $i$th particle and $\vec{F}^{\text{int}}$ describes the interaction between particles.

Therefore, if a particle with momentum $\vec{k}_i$ is located at $\vec{r}_i$ at time $t$, then, at time $t + dt$, it is located at $\vec{r}_i + \delta\vec{r}_i = \vec{r}_i + \vec{v}(\vec{r}_i)dt$, with momentum $\vec{k}_i + \delta\vec{k}_i = \vec{k}_i + \vec{F}_i(t)dt$.

Since this motion is deterministic, we must have as a result

$$f(\vec{r} + \delta\vec{r}, \vec{k} + \delta\vec{k}, t + \delta t) = f(\vec{r}, \vec{k}, t)$$

$$\Rightarrow \frac{\partial f}{\partial t}\delta t + \vec{v}(\vec{k})\delta t \cdot \nabla_{\vec{r}} f + \vec{F}\delta t \cdot \nabla_{\vec{k}} f = 0 \tag{7.3a}$$

$$\text{or } \frac{\partial f}{\partial t} + \vec{v} \cdot \nabla_{\vec{r}} f + \vec{F} \cdot \nabla_{\vec{k}} f = \frac{df}{dt} = 0.$$

The equation seems to look fine so far, except that the force term is not totally correct because $\sum_{j \neq i} \vec{F}^{\text{int}}(\vec{r}_i - \vec{r}_j)$ actually depends on the positions of other particles. In a more rigorous derivation, the force term should be replaced by

$$\vec{F}.\nabla_{\vec{k}} f \to \vec{F}^{\text{ext}}.\nabla_{\vec{k}} f + \int d^d r' \vec{F}(\vec{r} - \vec{r}\,')\nabla_{\vec{k}} \sum_{\vec{k}'} f^{(2)}(\vec{r}\,', \vec{k}'; \vec{r}, \vec{k}; t), \tag{7.3b}$$

where $f^{(2)}$ represents the probability that two particles with momenta $\vec{k}$ and $\vec{k}'$ appear at positions $\vec{r}$ and $\vec{r}\,'$ at time $t$, respectively.

Because of introduction of $f^{(2)}$, the equation of motion for $f$ does not close on itself when interaction between atoms exists. To determine $f$ we need to know the two-particle distribution function $f^{(2)}$. If you write down the equation of motion for $f^{(2)}$ you will find that it depends on the three-particle distribution function $f^{(3)}$. The series continues and represents a (infinite) hierarchy of equations (see Ref. [1] for example). Ultimately you need to know $f^{(N)}$, the distribution function of all $N$ particles in the system, to solve the equations of motion rigorously. Merely introducing the concept of probability does not simplify the problem.

To simplify the situation physicists have to make approximations. This is where introduction of a probability distribution function helps. The simplest plausible approximation one can make is to assume that

$$f^{(2)}(\vec{r}\,',\vec{k}\,';\vec{r},\vec{k};t) \sim f(\vec{r}\,',\vec{k}\,';t)f(\vec{r},\vec{k};t). \qquad (7.3c)$$

After making this approximation, the equation of motion for $f$ becomes a non-linear equation which closes on itself and can be solved. We note that this approximation is different from the approximation used in deriving the original Boltzmann equation [1]. We make this approximation here for illustration. The important point is that we obtain an (approximate) equation of motion for $f$ that closes on itself.

For an interacting two-particle system the motion of one particle is strongly influenced by the other and the approximation (7.3c), which neglects two-particle correlations, is certainly not valid. However, for a system with a large number of identical particles which are moving randomly on average, the approximation (7.3c) may not be bad if a particle only feels the 'average' force exerted by other particles. The fluctuations in particle correlations will be smoothed out further with the introduction of the probability distribution function where we work only with averages, justifying the approximation (7.3c) further.

Note that in this description all the physics of fluids is described by the *continuous* field variable $f(\vec{r},\vec{k},t)$. The identities of particles are lost when we make the approximation (7.3c) (or truncate the hierarchy of equations)! A natural question is: *under what situation is this continuum description valid?*

The answer to this question can be seen most simply by considering the equilibrium situation where $f(\vec{r},\vec{k},t) = f(\vec{k}) \sim e^{-\beta\varepsilon_{\vec{k}}}$ (Boltzmann distribution), and ask the question of how valid is this description? Microscopically, the atoms are moving around rapidly and if we look at a small region of space around any position $\vec{r}$, we will observe that particles are appearing and disappearing rapidly (see figure below for illustration). The description where the distribution function of particles is independent of position and time can be valid only after ensemble averages.

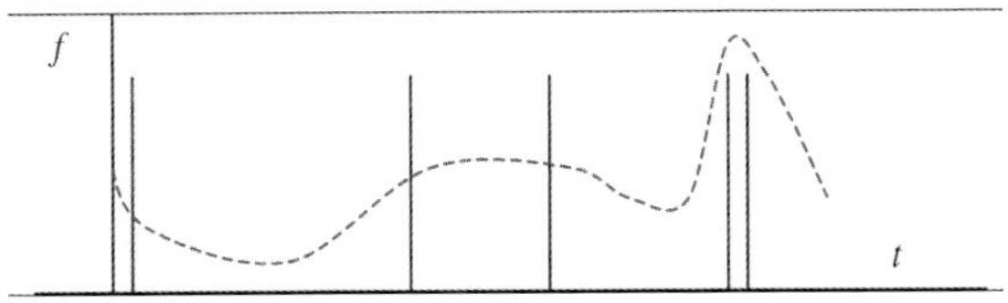


$f(r, k, t)$ (without ensemble average) at a fixed space point $r$ as a function of time before and after averaging. The dotted line represents ensemble average.

A similar picture would be obtained for fixed time if we look at $f(r, k, t)$ as a function of position around a finite region in space.

Here comes the next natural question: how reliable is the ensemble description of reality? Note that in any real experiments we are measuring only one set of particles but not the average of many sets (ensembles). However, in thermodynamics and statistical mechanics we always compare experimental results with ensemble averages.

The replacement of a single real physical system by an ensemble average is believed to be valid only when we consider macroscopic behavior of systems, that is if we consider measurements that intrinsically average over size of order $L \gg$ (average distance between particles) and time of order $T \gg$ (distance between particles)/(average speed of particles). The system is said to be 'self averaged' at macroscopic scale $\gg$ atomic dimensions when this description is valid. To see how this happens, one may imagine physically a region in space of size $L^3$ with $N$ particles on average. There are particles randomly moving in and out of this region all the time. However, when $L$ and $N$ are large enough, the fluctuation in particle number becomes negligible if we look at the particle number averaged over a time scale $\sim T$.

It should be emphasized that in general systems do not become 'self averaged' automatically when their sizes become much larger than atomic dimensions. An example is mesoscopic systems at low temperature where new physics emerges because of the absence of self averaging in the system combined with quantum mechanics (see for example Ref. [2]).

The replacement of an exact theory (Newton's equation for atoms) by an effective theory (Boltzmann equation) that is valid only at dimensions $\gg$ microscopic (atomic) dimensions is a very important tool in physics, which has been applied to numerous physical situations. When we deal with a (field) theory, we have to keep in mind that there is always a possibility that the theory we are working with is an 'effective' theory and there is a 'hidden' microscopic theory behind it which becomes important at some microscopic length and time scales. We should be careful not to apply the (field) theory to length and time scales beyond its range of validity.

### 7.1.1 Fluid Mechanics

With the realization that microscopic details often become unimportant in 'self-averaged' macroscopic systems physicists invented a new strategy to understand behaviors of macroscopic systems. An important example of this approach is the Navier–Stokes equation for fluids.

Unlike the Boltzmann equation which started from the microscopic theory of Newton's equation, the new approach ignores microscopic details and derives the equation of motion for macroscopic systems under very general principles (mainly conservation laws). The microscopic nature of the system (atoms or molecules) manifests itself only through a few macroscopic parameters appearing in the equation. To have a feeling of how the approach works, we derive the

Navier–Stokes equation in the following (see also Ref. [3]). The variables in the equation are (mass) density and current of the fluid.

The first equation that we use is the continuity equation, which expresses conservation of matter,

$$\frac{\partial \rho}{\partial t} + \nabla.(\rho \vec{v}) = 0. \tag{7.4}$$

Here, $\rho(\vec{x}, t)$ is the local (mass) density defined at macroscopic length and time scales, $\vec{v}(\vec{x}, t)$ is the corresponding velocity field, and $\rho\vec{v}$ is the current.

To proceed further, we need an equation for $\vec{v}(\vec{x}, t)$. Let us propose an equation of the form

$$\rho \frac{d\vec{v}}{dt} = \vec{f} = -\nabla p + \vec{f}', \tag{7.5}$$

which is essentially Newton's law applying to a small fluid element in the fluid. Here, $\vec{f}$ is the force felt by this small fluid element, of which one contribution is from the pressure gradient or the net force coming from the surrounding atoms acting on the fluid element. $\vec{f}'$ represents all other plausible contributions, which we shall discuss later. An important contribution to $\vec{f}'$ is the gravitational force, $\vec{f}' = -g\rho\,\hat{z}$. Note that Newton's law is applied to a moving fluid element. This is why a total derivative is presented in Equation (7.5). To go back to the laboratory frame, we have to make a Galilean transformation $F(\vec{x}, t) \to F(\vec{x} + \vec{v}t, t)$ for any space–time-dependent function $F$ with, correspondingly, $\frac{d}{dt} \to \frac{\partial}{\partial t} + \vec{v}.\nabla$. Therefore, we obtain in the laboratory frame

$$\frac{\partial \vec{v}}{\partial t} + (\vec{v}.\nabla)\vec{v} = -\frac{1}{\rho}(\nabla p - \vec{f}'), \tag{7.6a}$$

which is Euler's equation if $\vec{f}' = 0$.

**Homework Q1a**

Assume that, microscopically, the motions of atoms are described by Newton's law. Can you justify Equation (7.5), which is Newton's law applied to a collection of atoms at a volume $\gg$ (distance between atoms)$^3$? What could be missing from the equation?

The Euler equation is usually written as an equation for the current $\vec{j} = \rho\vec{v}$. In component form, we obtain

$$\frac{\partial j_i}{\partial t} = -v_i \sum_k \frac{\partial(\rho v_k)}{\partial x_k} - \rho \sum_k v_k \frac{\partial}{\partial x_k} v_i - \frac{\partial}{\partial x_i} p = -\sum_k \frac{\partial}{\partial x_k}(\rho v_k v_i + p\delta_{ik}).$$

We have used the continuity equation (7.4) to derive the above result.

Note that the equation is in the form of a continuity equation itself,

$$\frac{\partial(\rho v_i)}{\partial t} = -\sum_k \frac{\partial}{\partial x_k} \pi_{ik}, \quad \text{where} \quad \pi_{ik} = \rho v_k v_i + p\delta_{ik}, \tag{7.6b}$$

meaning that the equation is associated with a conservation law in the system.

**Homework Q1b**

By comparing with the Noether theorem, determine what is the conserved quantity here and what is the symmetry associated with this conservation law? Can you find another conservation law and continuity equation associated with liquids? Note that this extra continuity equation actually leads to another Navier–Stokes equation. You can find it in many textbooks on fluid mechanics.

#### 7.1.1.1 Friction and Viscosity

We have not discussed $\vec{f}'$ so far. Since Equation (7.5) is constructed for a small but *macroscopic* volume of particles, we expect that a plausible force that may contribute is the frictional force $\vec{f}' \sim -\eta\vec{v}$.

It should be cautioned that in usual mechanics $\vec{f}' \sim -\eta\vec{v}$ is applicable only for a particle moving in a medium at rest, that is $\vec{v}$ is the relative velocity between the particle and the surrounding medium. In a liquid where this equation is applied to the fluid element, we expect that

$$\vec{f}' \to -\eta\nabla\nabla\vec{v}, (?)$$

where $\nabla\nabla\vec{v}$ measures the difference in velocity between the fluid element and the surrounding fluid. The question mark reflects the fact that we are not sure of the correct way to introduce the gradient here.

**Question**

Why should we have $\vec{f}' \sim -\eta\nabla\nabla\vec{v}$ but not $\vec{f}' \sim -\eta\nabla\vec{v}$? (Hint: assuming a fluid element moving in a medium where $\nabla\vec{v} = \text{constant}$, how would the velocity of the fluid element changed by the friction term?)

Physicists attacked this problem by writing down the most general form of 'frictional' term that can exist. First, we realize that Equation (7.6b) is a general consequence of a conservation law (Homework Q1b) and cannot be affected by internal forces between atoms in the fluid. Therefore, the only place where internal forces between atoms can enter is through modifying $\pi_{ik}$. Friction force can only make an additional contribution $\sigma'_{ik}$ such that $\pi_{ik} = \rho v_k v_i + p\delta_{ik} + \sigma'_{ik}$. $\sigma'_{ik}$ is called the viscoelasticity tensor.

The general form for $\sigma'_{ik}$ can be determined by the following arguments: first, it must be a symmetric tensor if the fluid is isotropic. Second, it must be zero when $\partial_i v_j$ is zero. Therefore, for small $\partial_i v_j$, we expect that $\sigma'_{ik}$ can be expanded in powers of the $\partial_i v_j$. To lowest order, the most general symmetric tensor one can construct is of the form

$$\sigma'_{ik} = \eta(\partial_i v_k + \partial_k v_i) + \left(\xi - \frac{2}{3}\eta\right)\delta_{ik}(\nabla.\vec{v}), \tag{7.7}$$

where $\eta$, $\xi$ are parameters determined by the microscopic dynamics of the fluid. It can be shown that, to describe dissipation, we must have $\eta > 0$, $\xi > 0$. For incompressible fluids where $\rho = \text{constant}$ and $\nabla.\vec{v} = 0$, we obtain the simplest version of the Navier–Stokes equation

$$\frac{\partial \vec{v}}{\partial t} + (\vec{v}.\nabla)\vec{v} = -\frac{1}{\rho}\nabla p + \frac{\eta}{\rho}\nabla^2\vec{v}. \tag{7.8}$$

Note that the microscopic properties of the fluid enter only through the parameters $\eta$, $\xi$. They can be computed from, for example, the Boltzmann equation or from linear response theory (see for example Ref. [1]).

The fact that the Navier–Stokes equation is almost universally applicable to all fluids is an example of so-called (universal) emergent phenomena where the effective theories of different microscopic systems with different microscopic constituents and Hamiltonians all appear the same at length scales much larger than the atomic scale. The reason why this happens in this case is that the macroscopic behavior of fluids is almost entirely governed by conservation laws that are insensitive to microscopic details of the system. This is in fact, a phenomenon that occurs quite often in physics. We shall see more examples later.

#### 7.1.1.2 Limitation of Hydrodynamics: An Example

In hydrodynamics, one can show (after some approximations) that, for a fluid at rest with average density $n_0$, the energy functional that enters the partition function can be written as

$$H[n] = E[n_0] + \int d^d x[a(\nabla n(\vec{x}\,))^2 + b(n(\vec{x}\,) - n_0)^2],$$

where $a, b > 0$. The functional is valid when the fluctuation in particle density $|n - n_0|$ in the system is small.

Now, let us evaluate the free-energy density for the fluid. Using our result in Chapter 1, we find that

$$f = f[n_0] - \frac{k_B T}{2}\frac{1}{(2\pi)^d}\int d^d k \ln\left(\frac{k_B T}{ak^2 + b}\right). \tag{7.9a}$$

The integral diverges in the large-$k$ limit and $f \to \infty$! What is wrong?

The mistake we made is that the energy functional is only valid at length scales $l \gg l_0$, the atomic dimension. Correspondingly, it is only valid for momentum scales $k \ll l_0^{-1}$. Therefore, the free energy should be

$$f = f[n_0] - \frac{k_B T}{2}\frac{1}{(2\pi)^d}\int_{|k| < l_0^{-1}} d^d k \ln\left(\frac{k_B T}{ak^2 + b}\right) + f_{\text{atomic}}, \tag{7.9b}$$

where the large-$k$ divergence is cut off by $k_c \sim l_0^{-1}$. $f_{\text{atomic}}$ is the free-energy contribution from short distances $l < l_0$ where the hydrodynamic description is not valid and a microscopic theory is needed to evaluate $f_{\text{atomic}}$.

## 7.2
## Landau Theory of Phases and Phase Transitions

The most successful example of effective field theory is perhaps the Landau theory of phases and phase transitions. We shall describe this theory here. In macroscopic systems different phases are often observed for the same material. For example, different phases of water, magnetic and non-magnetic states of magnetic materials, superconducting and normal states of metals, and so on. Different states of these systems exist under different temperatures, different pressures, or other different external factors. The macroscopic properties of different phases are different, although the microscopic constituents remain identical. A system can migrate between different phases (phase transition) when certain external parameters change. For example, water can go from solid to liquid to gas phase as pressure decreases or as temperature increases and a superconductor can go from superconducting to normal phase as magnetic field or temperature increases.

Phases and phase transitions are observed in systems with very diverse microscopic constituents. In fact, it is hard to find materials we encounter in physics which do not undergo any phase transition. It seems that phases and phase transitions are general properties of very large classes of Hamiltonians, and are governed by general rules which are quite independent of the microscopic details of the systems. The question is then: what are these rules? And what is the relation between these rules and the microscopic Hamiltonian?

Experimentally, it was realized very early that free energies develop a non-analytic behavior across phase transitions and can be used to classify different families of phase transitions.

Let $\tau$ be the parameter which controls the phase transition such that the system is in one phase if $\tau < \tau_c$, and in another phase if $\tau > \tau_c$. It was discovered that although the free energy (per unit volume) $f(\tau)$ itself is always a continuous function across phase transitions, the derivatives of the free energy against $\tau$ can be discontinuous across phase transitions. A phase transition is called *first order* if $\partial f/\partial\tau$ is discontinuous at $\tau = \tau_c$, and is called *second order* if $\partial f/\partial\tau$ is continuous but $\partial^2 f/\partial\tau^2$ is discontinuous. Higher-order phase transitions can be similarly defined.

### 7.2.1
### Order Parameters

The concept of an order parameter was put forth by Lev Landau. He started by asking a question: what (physically) characterizes and classifies phase transitions in general? And what is the corresponding mathematical description?

Physically, a phase transition corresponds to a drastic change in the qualitative properties of a system and Landau started by asking the question of what are the possible macroscopic parameters that can characterize these *qualitative* changes. He arrived at the conclusion that a very large class of phase transitions is characterized by change (or reduction) in symmetry properties (= broken symmetry) of the system. We shall explain what he meant by using a few examples. You are highly recommended to read his book [4] to have a better understanding.

#### 7.2.1.1 Paramagnetic ↔ Ferromagnetic Transition

The system is characterized by the local magnetization $\vec{M}(\vec{r})$. At temperatures $T > T_c$, the system is in the paramagnetic state where $\vec{M}(\vec{r})$ is pointing in random directions and the magnetizations at different space points are random and uncorrelated (or with only short-distance correlation). On the other hand, $\vec{M}(\vec{r}) \sim \vec{M}_0$ at $T < T_c$ and the magnetizations are pointing in the same direction everywhere in space. The system has higher symmetry at $T > T_c$ (magnetizations at different space points are allowed to point in any direction) but the symmetry is reduced, or broken, at $T < T_c$ (magnetizations point in the same direction everywhere).

This change in symmetry can be described more precisely using group theory. At $T > T_c$, the macroscopic behavior of the system is invariant under a group transformation corresponding to rotation in space (random magnetization orientation → random magnetization orientation), whereas, at $T < T_c$, the system changes (global direction of magnetization changes) upon rotation. The (rotational) symmetry is said to be broken when the temperature goes below $T_c$.

#### 7.2.1.2 Liquid ↔ Solid Transition

The system is characterized by the density profile of atoms $\rho(\vec{r})$. At $T > T_c$ the system is in the liquid state where $\rho(\vec{r}) \sim \rho_0$ and the atoms are uniformly distributed, whereas $\rho(\vec{r}) \to \rho_0 + \sum_{\vec{Q}} \rho_{\vec{Q}} \cos(\vec{Q}.\vec{r} + \delta)$ at $T < T_c$, where the $\vec{Q}$ are reciprocal lattice vectors and the atoms form a crystalline solid. The system has higher symmetry at $T > T_c$ because the atoms can move freely to anywhere in space whereas the symmetry is broken at $T < T_c$ because the locations of the atoms are restricted. The reduction in symmetry can be characterized by the group transformation corresponding to translation. The system is invariant under translation by an arbitrary displacement vector $\vec{r}$ at $T > T_c$, whereas the system is invariant only if the displacement vector is an integer multiple of the Bravais lattice vectors below $T_c$. The (translational) symmetry is reduced (broken).

The order parameters in the above examples are $\vec{M}(\vec{r})$ and $\rho(\vec{r})$, and the phase transitions can be characterized by the reduction in symmetry of the system upon a symmetry transformation (rotation and translation, respectively) across $T_c$. Landau's description of phase transitions using symmetry operations is found to be successful in describing a lot of phase transitions but exceptions exist. For example, the gas–liquid transition and liquid–glass transition are transitions not characterized by a change in symmetry property of the system. So, what characterizes these transitions? More generally, we want to identify and classify all plausible types of phase transitions in our world.

#### 7.2.1.3 Phase Transitions and Broken Ergodicity

From the point of view of statistical mechanics, equilibrium properties of macroscopic systems are characterized by the partition function according to the so-called *ergodic assumption* which states that macroscopic systems migrate through all possible configurations in the allowed phase space of the system with probability given by the Boltzmann statistical weight $e^{-\beta E}$, irrespective of its dynamics.

The occurrence of phase transitions associated with broken symmetry shows that the ergodic assumption is not always correct in realistic systems. To see how this occurs, let us consider the ferromagnetic system at temperatures $T < T_c$ and write the partition function as

$$Z = \int d\vec{M}\, e^{-\beta H(\vec{M})}, \tag{7.10a}$$

where

$$e^{-\beta H(\vec{M})} = \mathrm{Tr}\left[e^{-\beta H}\delta(\vec{M} - \frac{1}{V}\int d^d r \vec{M}(\vec{r}))\right] \tag{7.10b}$$

represents the probability density that the average magnetization of the system is $\vec{M}$ and

$$\int d\vec{M} = \int M dM d\Omega\, (\vec{M} = M\hat{n}),\ \Omega = \text{solid angle}. \tag{7.10c}$$

If the Hamiltonian function is invariant under rotation, then $H(\vec{M}) = H(M)$ should be independent of the direction of $\vec{M}$, meaning that magnetizations pointing in different directions in space should contribute equally to $Z$. However, in real magnets we observe that the magnetization is pointing only in one direction (although any direction is equally likely) at $T < T_c$, implying that the ergodic assumption is broken. The breakdown of the ergodic assumption has been studied since it was proposed and the typical answer to the question is that, for macroscopic systems, it may take a very long time (millions of years for example) for the system to truly equilibrate and the ergodic assumption may not be applicable in our everyday time scale. This is essentially the case for systems with broken symmetry.

Note that the entropy of the system is lowered in a phase with broken symmetry (ordered phase) because of the self restriction of the system to move in only a part of the phase space. Therefore, phase transitions of this kind always proceed from a higher-temperature disordered phase (higher entropy) to a lower-temperature ordered phase with broken symmetry (lower entropy).

More generally, phase transitions are reflections of broken ergodicity. In a disorder-to-order phase transition, the system is trapped in one part of phase space and cannot migrate to the other parts within the time scale of the experiment. In the case of phase transitions associated with broken symmetry, the different parts of phase space which broke up at low temperature are characterized by different orientations/configurations of the order parameter. *To look for other types of phase transitions, we are actually asking whether there exist other ways of breaking up the phase space into separated pieces where the system cannot travel from one part to the other within the time scale of the experiment.*

The gas–liquid transition and liquid–glass transition are examples of phase transitions where no symmetry is broken but the behavior of the system changes qualitatively. In the case of the gas–liquid transition, the size of the system shrinks drastically when going from the gas to the liquid phase, and the available phase space for atomic motion shrinks in this transition. Physically, the atoms are freely moving in the gas phase but are 'forced' to stay close to one another in the liquid phase. In the liquid–glass transition, the atomic motion 'freezes' when going from the liquid to the glass phase and the atoms are trapped in 'fixed' (in the time scale of the experiment) but random positions in the glass. The available phase space of the motion shrinks again, although there is no broken symmetry associated with the transition. These are termed phase transitions with 'broken ergodicity' nowadays, and are different from phase transitions associated with broken symmetry proposed by Landau. More recently, a new kind of phase and phase transition called topological phase and topological phase transition was proposed. In this case the phase space is broken up into pieces with different topological indices and are not classified simply by an order parameter. A topological phase transition occurs if the system is trapped into one of these pieces of phase space. You can look at Ref. [5] for some examples.

### 7.2.2 Landau's Phenomenological Theory of Phase Transitions

Landau proposed a general mathematical description of phases and phase transitions based on his idea of broken symmetry. We shall introduce his theory here. Phase transitions associated with broken symmetries are usually second order or weakly first order transitions. We shall again consider the paramagnetic–ferromagnetic and liquid–solid transitions as examples.

To formulate a theory of phase transitions, first we have to identify the proper mathematical variables (order parameters) that characterize phases. In the two examples we discussed above, the proper variables are obviously the magnetization field $\langle \vec{M}(\vec{r})\rangle$ and the density field $\langle \rho(\vec{r})\rangle$. The free energy $F$ of the system is a functional of these fields.

First we consider the paramagnetic–ferromagnetic transition. In the paramagnetic phase $\langle \vec{M}(\vec{r})\rangle = 0$ and magnetization develops gradually as the system goes from the paramagnetic to the ferromagnetic state. Around the phase-transition region $\langle \vec{M}(\vec{r})\rangle$ is expected to be small and Landau argued that one can expand the free energy in a power series in $\langle \vec{M}(\vec{r})\rangle$ in this region, that is

$$F = \int \mathrm{d}^d r f[\vec{M}(\vec{r})] \sim \int \mathrm{d}^d r \begin{pmatrix} f_0(T) + \vec{\sigma}_1(T).\vec{M}(\vec{r}) + a(T)|\vec{M}(\vec{r})|^2 \\ +\vec{\sigma}_2(T).\vec{M}(\vec{r})|\vec{M}(\vec{r})|^2 + b(T)|\nabla\vec{M}(\vec{r})|^2 + \ldots \end{pmatrix} \tag{7.11a}$$

We have neglected the average $\langle \ldots \rangle$ notation for brevity. Note a non-trivial extension of equilibrium thermodynamics here. We have assumed in Equation (7.11a) that the free energy per unit volume $f[\vec{M}(\vec{r})]$ can fluctuate from one point of space to another because $\vec{M}(\vec{r})$ may differ at different regions of space in a real system. While this is correct physically, we shall see later that the spatial variation of $\vec{M}(\vec{r})$ should be treated more carefully in the correct statistical mechanics analysis. Note that gradient terms also exist in the free-energy expansion because of spatial variation of the order parameter $\vec{M}(\vec{r})$.

Next, we observe that if the Hamiltonian is rotationally invariant in space, then the $\vec{\sigma}_{1(2)}$ terms should be zero since $\vec{\sigma}_{1(2)}$ specifies a preferred direction in space. Therefore, we obtain, to fourth order,

$$f[\vec{M}(\vec{r})] \sim f_0(T) + a(T)|\vec{M}(\vec{r})|^2 + c(T)|\vec{M}(\vec{r})|^2|\vec{M}(\vec{r})|^2 + b(T)|\nabla\vec{M}(\vec{r})|^2. \tag{7.11b}$$

Note that we keep only the gradient term to second order for reasons which we shall not discuss in detail here. This is justified by the so-called renormalization group (RG) analysis which you can find in many other textbooks (see for example Ref. [6]).

The thermodynamic state of the system is obtained by minimizing the free energy with respect to $\vec{M}(\vec{r})$. Note that the Taylor expansion makes sense only if $\vec{M}(\vec{r})$ is small. Therefore, we must have $c(T) > 0$ and $b(T) > 0$ so that the free energy does not diverge to $-\infty$ for large $\vec{M}(\vec{r})$ or $\nabla\vec{M}(\vec{r})$. In case these parameters are found to be negative, we have to keep higher-order terms in the expansion to guarantee smallness of $\vec{M}(\vec{r})$ after minimizing the energy. Note also that, provided that $c(T) > 0$ and $b(T) > 0$, the sign of $a(T)$ can be arbitrary and the absolute minimum of the free energy occurs at $\nabla\vec{M}(\vec{r}) = 0$.

Minimizing the energy, we obtain $2a(T) + 4c(T).|\vec{M}(\vec{r})|^2 = 0$.

Therefore,

$$|\vec{M}(\vec{r})|^2 = -\frac{a}{2c} \neq 0 \qquad \text{when } a(T) < 0. \tag{7.12a}$$

Note that only the magnitude of $\vec{M}(\vec{r})$ is determined by the equation and there exists in principle infinitely many possible solutions with the same free energy, corresponding to magnetizations pointing in different directions in space. The system picks up one solution randomly as the physical state (broken symmetry) in reality.

Landau observed that a phase transition from a state with $\vec{M}(\vec{r}) = 0$ to $\vec{M}(\vec{r}) \neq 0$ can occur if we assume that $a(T)$ changes sign at a temperature $T_c$. For small $\Delta T = T - T_c$, $a(T)$ can be expanded in a power series of $\Delta T$. Keeping only the lowest-order term, he obtained $a(T) = a(T - T_c)$.

The corresponding free-energy density is, after minimization, of the form

$$f[T] \sim \begin{cases} f_0(T), & T > T_c, \\ f_0(T) - \dfrac{a(T_c - T)^2}{4c(T)}, & T < T_c. \end{cases} \tag{7.12b}$$

Note that $\partial f/\partial T$ is continuous across $T_c$ but $\partial^2 f/\partial T^2$ is discontinuous, if we assume that $f_0(T)$ and $c(T)$ are non-zero and are smooth functions across $T_c$. Thus, Landau's theory describes general second-order phase transitions.

**Homework Q2**
Write down a free energy such that the transition is a third-order phase transition.

### 7.2.2.1 Weakly First-Order Phase Transition

Landau's theory can be generalized to describe weakly first-order phase transitions. We consider the liquid–solid transition as an example here.

We proceed as before to write down a phenomenological free energy in terms of the order parameter field $\rho(\vec{r})$. Note that in the ordered (solid) state $\rho(\vec{r})$ is of the form $\rho(\vec{r}) = \rho_0 + \sum_{\vec{Q}} \rho_{\vec{Q}} \cos\vec{Q}.(\vec{r}+\vec{\delta})$, where the $\vec{Q}$ are reciprocal lattice vectors. Therefore, it is more convenient to take the Fourier-transformed fields $\rho_{\vec{Q}}$ as order parameter fields and write $F = \int d^d r f[\rho_{\vec{Q}}]$. Note that $F$ is independent of $\vec{\delta}$ because of translational invariance.

For small $\rho_{\vec{Q}}$, we may expand $f$ in a Taylor series, obtaining

$$f[\rho_{\vec{Q}}] \sim f_0(\bar{\rho},T) + \sum_{\vec{Q}} V_{\vec{Q}} \rho_{-\vec{Q}} + \sum_{\vec{Q}} A(\vec{Q},T))\rho_{\vec{Q}}\rho_{-\vec{Q}} + \sum_{\vec{Q}_1,\vec{Q}_2} B(\vec{Q}_1,\vec{Q}_2,T)\rho_{\vec{Q}_1}\rho_{\vec{Q}_2}\rho_{-\vec{Q}_1-\vec{Q}_2} + \sum_{\vec{Q}_1,\vec{Q}_2,\vec{Q}_3} C(\vec{Q}_1,\vec{Q}_2,\vec{Q}_3,T)\rho_{\vec{Q}_1}\rho_{\vec{Q}_2}\rho_{\vec{Q}_3}\rho_{-\vec{Q}_1-\vec{Q}_2-\vec{Q}_3} + \cdots, \tag{7.13}$$

where $V_{\vec{Q}}$ corresponds to an external potential that couples to density fluctuations with wave vector $\vec{Q}$, and $V_{\vec{Q}} = 0$ for translationally invariant systems. Note that unlike the paramagnetic–ferromagnetic transition third-order ($\rho^3$) terms are not forbidden by symmetry. Furthermore, the stability requirement implies that $C(\vec{Q}_1,\vec{Q}_2.\vec{Q}_3,T) > 0$ as in the case of ferromagnets.

We first consider the simpler situation $B(\vec{Q}_1,\vec{Q}_2,T) = 0$. In this case, minimizing the free energy leads to

$$A(\vec{Q},T))\rho_{\vec{Q}} + \sum_{\vec{Q}_1,\vec{Q}_2} C(\vec{Q}_1,\vec{Q}_2,\vec{Q}-\vec{Q}_1-\vec{Q}_2,T)\rho_{\vec{Q}_1}\rho_{\vec{Q}_2}\rho_{\vec{Q}-\vec{Q}_1-\vec{Q}_2} = 0. \tag{7.14}$$

Since $C(\vec{Q}_1,\vec{Q}_2.\vec{Q}_3,T) > 0$, $\rho_{\vec{Q}} \neq 0$ only if $A(\vec{Q},T) < 0$ for some $\vec{Q}$. In particular, a solid–liquid transition occurs if the function $A(\vec{Q},T)$ is positive at $T > T_c$ with a minimum at $|\vec{Q}| = Q_0$, $A(Q_0\hat{n},T) \to 0$ as $T \to T_c$, and becomes negative at $T < T_c$.

Next, we consider the more general situation with $B(\vec{Q}_1,\vec{Q}_2,T)$ non-zero. Just below the transition, a solid state can be thought of as a superposition of density waves with wavelength $\lambda = 2\pi/Q_0$ and identical amplitudes $\bar{\rho}$ but traveling in different directions, that is

$$\rho(\vec{r}) = \rho_0 + \bar{\rho} \sum_{|\vec{G}|=Q_0} \cos\vec{G}.(\vec{r}+\vec{\delta}), \tag{7.15a}$$

where the $\vec{G}$ are primary reciprocal lattice vectors of the underlying Bravais lattice with $|\vec{G}| = Q_0$. In this case, the free energy takes the form

$$f[\rho_{\vec{Q}}] \sim f_0(T) + \sum_{\vec{G}} A(\vec{G},T))\bar{\rho}^2 + \sum_{\vec{G}_1,\vec{G}_2} B(\vec{G}_1,\vec{G}_2,T)\bar{\rho}^3 + \sum_{\vec{G}_1,\vec{G}_2,\vec{G}_3} C(\vec{G}_1,\vec{G}_2,\vec{G}_3,T)\bar{\rho}^4$$

$$= f_0(T) + \bar{A}(T)\bar{\rho}^2 + \bar{B}(T)\bar{\rho}^3 + \bar{C}(T)\bar{\rho}^4 f[\rho_{\vec{Q}}] \quad (7.51b)$$

where $\bar{A}(T) = A_0(T - T_c)$.

For $\bar{B}(T) = 0$, the free energy is of the same form as in the case of the paramagnetic–ferromagnetic transition and Landau's theory describes a second-order phase transition. However, the situation becomes different if $\bar{B}(T) \neq 0$. In this case the mean-field equation $2\bar{A}(T)\bar{\rho} + 3\bar{B}(T)\bar{\rho}^2 + 4\bar{C}(T)\bar{\rho}^3 = 0$ has three solutions at

$$\bar{\rho}(T) = (0, \frac{-3\bar{B}(T) \pm \sqrt{9\bar{B}(T)^2 - 32\bar{A}(T)\bar{C}(T)}}{8\bar{C}(T)}). \quad (7.16)$$

The corresponding free energy has two minima (and one maximum) as a function of $\bar{\rho}(T)$ (assuming that $\bar{B}(T) > 0$) if $9\bar{B}(T)^2 > 32\bar{A}(T)\bar{C}(T)$. (see figure below)

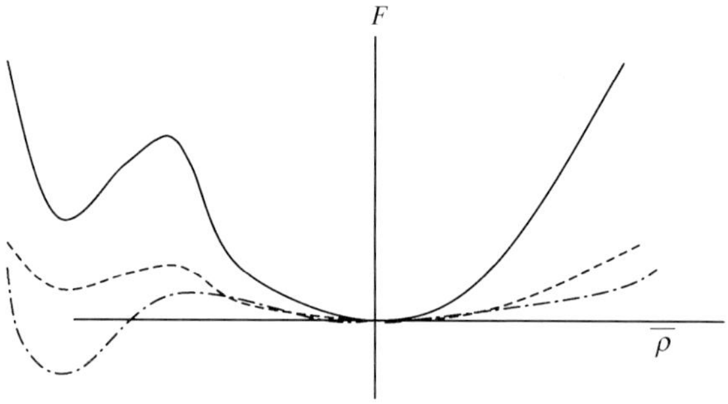


Temperature dependence of free energy: solid line – highest temperature $T_1$, dashed line – intermediate temperature $T_2$, and dash-dotted line – lowest temperature $T_3$. A first-order transition occurs at some temperature $T_2 > T > T_3$.

Note that the absolute energy minimum is located at $\bar{\rho}(T) = 0$ when $\bar{A}(T)$ is large (solid line).

Let us now examine what would happen when $\bar{A}(T)$ decreases. The free-energy value $F[\bar{\rho}_1]$ at $\bar{\rho}_1(T) = \frac{-3\bar{B}(T) - \sqrt{9\bar{B}(T)^2 - 32\bar{A}(T)\bar{C}(T)}}{8\bar{C}(T)}$ will decrease when $\bar{A}(T)$ decreases, until $\bar{A}(T) \to \frac{\bar{B}(T)^2}{4\bar{C}(T)}$, at which $F[\bar{\rho}_1] \to 0$.

What is interesting is that, for $\bar{A}(T) < \frac{\bar{B}(T)^2}{4\bar{C}(T)}$, the absolute free energy minimum of the system is shifted from $\bar{\rho}(T) = 0$ to $\bar{\rho}(T) = \bar{\rho}_1(T)$ (dash-dotted line), and there is a discontinuous change in the system where a state with density modulation suddenly appears, corresponding to a first-order phase transition at $\bar{A}(T) = \frac{\bar{B}(T)^2}{4\bar{C}(T)}$.

**Homework Q3**

Compute the free energy of the system and show that the above situation describes a first-order phase transition.

Note that a second-order phase transition will appear at $\bar{A}(T) = 0$ if $\bar{B}(T) = 0$. The replacement of the second-order phase transition by a (weakly) first-order phase transition is a general property of Landau's phenomenological theory if a third-order ($\rho^3$) term exists in the Landau free-energy functional [4].

Landau's phenomenological theory of phase transitions is found to be successful in describing general properties of phase transitions associated with symmetry breaking, especially for second-order phase transitions. The theory does not usually describe accurately the critical behaviors at the transition points because of subtle reasons which we shall not discuss in this book (see Ref. [6] for example). Like hydrodynamics, Landau's theory provides an example of (universal) *emergent phenomena*, where a large variety of microscopic systems develop similar macroscopic behaviors near phase transitions.

#### 7.2.2.2 **Effective Free Energies in Landau Theory**

Landau's theory of phase transitions, if interpreted correctly, is an effective (classical) field theory describing phase transitions. The emergence of *universal* macroscopic behavior in different systems during phase transitions implies that to understand the macroscopic behavior of phase transitions it is not necessary to work with the precise microscopic Hamiltonians. We just have to identify correctly the 'effective' theory that describes the macroscopic behaviors of phase transitions. In the following, we shall clarify the role of Landau theory as an effective field theory for thermodynamics.

We start by clarifying the theoretical foundation of Landau's theory. Recall that in Landau's theory a non-trivial generalization of usual thermodynamics is made. We assumed that the free energy per unit volume, $f[\phi(\vec{r})]$, can fluctuate from one point of space to another because the order parameter, which we denote by $\phi(\vec{r})$ here, can differ at different regions of space.

For the definition of the local free energy $f[\phi(\vec{r})]$ to be valid, we must be able to define $\phi(\vec{r})$ in a region of space which is large enough for thermodynamics to make sense and the function $\phi(\vec{r})$ has to vary slowly enough within this region so that it can be considered as essentially constant. To see how this can work, let us take magnetization $\vec{M}(\vec{r})$ as an example and divide space into boxes of size $L^3$.

First, let us consider the $i$th box and ignore the others. We can define the local free energy $F_i[\vec{M}_i]$ with order parameter $\vec{M}_i$ as before, that is

$$e^{-\beta F_i(\vec{M}_i)} = \mathrm{Tr}_{(i)}\left[e^{-\beta H_i(\vec{M})}\delta(\vec{M}_i - \frac{1}{L^3}\int \mathrm{d}^d r_i \vec{M}(\vec{r}))\right], \tag{7.17a}$$

where the trace is taken over the phase space associated with the $i$th box. For a general order parameter $\phi_i$, we may define the local free energy similarly, with

$$e^{-\beta F_i(\phi_i)} = \mathrm{Tr}_{(i)}\left[e^{-\beta H_i(\psi)}\delta(\phi_i - \frac{1}{L^3}\int \mathrm{d}^d r_i \phi[\psi(\vec{r})])\right], \tag{7.17b}$$

where $\phi[\psi(\vec{r})]$ is in general a functional of the microscopic field variables $\psi(\vec{r})$ which define the (microscopic) Hamiltonian. Examples of $\phi[\psi(\vec{r})]$ include the magnetization $\vec{M}(\vec{r})$ which defines the average spin direction of electrons in the box, or the superconductor order parameter which represents the amplitude of the electron-pair wavefunction. In both cases the $\psi(\vec{r})$ represent physical electron operators.

Next, we consider the interaction between different boxes characterized by $\phi_i$ ($i=1, \ldots, N$). We can formally define the free energy $F[\phi_1, \phi_2, \phi_3, \ldots]$ as

$$e^{-\beta F[\phi]} = \mathrm{Tr}\left[e^{-\beta H(\psi)} \prod_i \left(\delta(\phi_i - \frac{1}{L^3}\int d^d r_i \phi[\psi(\vec{r})])\right)\right], \tag{7.18a}$$

where the trace is taken now over the whole system. Note that the partition function and free energy of the whole system are, by definition,

$$Z = e^{-\beta F} = \prod_i \int d\phi_i e^{-\beta F[\phi]}, \tag{7.18b}$$

The free energy $F$ is obtained after we trace out all possible configurations.

The approach where the original microscopic Hamiltonian $H[\psi]$ with trace over the microscopic variables $\psi$ is replaced by a different effective Hamiltonian $F[\phi]$ with trace over the order parameter field $\phi$ in evaluating the free energy is an example of effective (classical) field theory. Note that, by construction, the same partition function is obtained after tracing out the respective fields. Therefore, the two 'Hamiltonians' are 'equivalent' as far as thermodynamic quantities are concerned, that is

$$Z = \mathrm{Tr}(e^{-\beta H(\psi)}) = \int D\phi e^{-\beta F[\phi]}.$$

**Homework Q4 (Gaussian approximation)**

Evaluate the partition function and free energy for

$$F[\phi] = \int d^d r \left[a(T-T_c)\phi^2 + c(\nabla\phi)^2 + b\phi^4\right]$$

at both $T >$ (<) $T_c$ by expanding $F[\phi] = F[\phi_0] + F'[\phi - \phi_0]$ to second order in $\delta\phi = \phi - \phi_0$, where $\phi_0$ is the Landau order parameter that minimizes $F[\phi]$ (Gaussian approximation). What happens to the free energy when $T \to T_c$ in the Landau approximation and in the Gaussian approximation?

(Note that the free energy can be evaluated exactly for quadratic energy functionals (see Chapter 1).

In this regard, Landau has made a very drastic approximation when he replaced the trace over the order parameter field $\phi$ by minimizing $F[\phi]$ with respect to $\phi$ to obtain the free energy of the system.

The approximation made by Landau effectively assumed that the partition function is dominated by the contribution from a single configuration $\phi = \phi_0$ which minimizes $F[\phi]$, and the sum over all possible $\phi$ configurations in the free energy is replaced by the contribution from the $\phi = \phi_0$ configuration alone. This is an example of mean-field approximation which implicitly assumes broken ergodicity or broken symmetry. By definition, the contribution to the free energy is indeed strongest from the configuration that minimizes $F[\phi]$. Whether its contribution is so strong that all other configurations can be neglected (or whether the system actually goes into a broken-symmetry state) is not so obvious. We shall see examples where Landau's theory breaks down in the next few chapters.

#### 7.2.2.3 Continuum Limit and Landau Theory

How space is divided into boxes in the above example is quite arbitrary. In principle the boxes can be of arbitrary shape, and the resulting energy functional $F[\phi]$ that comes out would depend on how the boxes are constructed. This arbitrariness introduces uncertainty in actual calculations. It would be much better if we can find a way of defining the order parameter field $\phi$ without referring to particular way(s) of dividing up spaces. This is possible if we define the order parameter in *Fourier space.*

Assume that we can define a local order parameter $\phi(\vec{r}\,) = \phi[\psi(\vec{r}\,)]$ in the configuration space where the microscopic Hamiltonian is defined. We can then define corresponding Fourier transforms,

$$\phi(\vec{q}\,) = \int d^d r e^{i\vec{q}\cdot\vec{r}}\,\phi(\vec{r}\,), \tag{7.19a}$$

and introduce the corresponding free energy $F_Q[\phi]$ in Fourier space through

$$e^{-\beta F_Q[\phi]} = \mathrm{Tr}\left[e^{-\beta H(\psi)}\prod_{\vec{q}}(\delta(\phi(\vec{q}\,) - \int d^d r e^{i\vec{q}\cdot\vec{r}}\,\phi[\psi(\vec{r}\,)])\right]. \tag{7.19b}$$

It is more interesting to introduce a momentum-cutoff-dependent free energy $F_\Lambda[\phi]$ defined by

$$e^{-\beta F_\Lambda[\phi]} = \mathrm{Tr}\left[e^{-\beta H[\psi]}\prod_{|\vec{q}\,|<\Lambda}(\delta(\phi(\vec{q}\,) - \int d^d r e^{i\vec{q}\cdot\vec{r}}\,\phi[\psi(\vec{r}\,)])\right], \tag{7.20a}$$

where only $\phi(\vec{q}\,)$ with $|\vec{q}\,| < \Lambda$ enters $F_\Lambda[\phi]$ and $\phi(\vec{q}\,)$ with $|\vec{q}\,| > \Lambda$ are traced over. Therefore,

$$Z = e^{-\beta F} = \int \prod_{|\vec{q}\,|<\Lambda} d\phi(\vec{q}\,)e^{-\beta F_\Lambda[\phi]} = \int^\Lambda D\phi(\vec{q}\,)e^{-\beta F_\Lambda[\phi]}. \tag{7.20b}$$

We can also introduce a corresponding inverse Fourier transformed field

$$\bar{\phi}(\vec{r}\,) = \int_{|\vec{q}\,|<\Lambda} d^d q e^{i\vec{q}\cdot\vec{r}}\,\phi(\vec{q}\,) \tag{7.20c}$$

$$\text{and write } Z = e^{-\beta F} = \int D\bar{\phi}(\vec{r}\,)e^{-\beta F_\Lambda[\bar{\phi}]}. \tag{7.20d}$$

Now, let us ask the question of how $F_\Lambda[\phi]$ would depend on $\vec{q}$ if $\Lambda$ is very small (meaning much less than any microscopic momentum scale (or inverse length scale) in the system) and $\phi$ is also small, meaning that the system is at the disordered phase or close to the critical point of a second-order phase transition. In this case, we may expand $F_\Lambda[\phi]$ in power series of both $\phi$ and $|\vec{q}\,|^2$. Assuming space isotropy, we obtain

$$F_\Lambda[\phi] = \sum_{|\vec{q}|<\Lambda} \begin{pmatrix} a(\vec{q}\,;T)\phi(\vec{q}\,)\phi(-\vec{q}\,) + b(\vec{q}_1,\vec{q}_2;T)\phi(\vec{q}_1)\phi(\vec{q}_2)\phi(-\vec{q}_1-\vec{q}_2) \\ +c(\vec{q}_1,\vec{q}_2,\vec{q}_3;T)\phi(\vec{q}_1)\cdots\phi(-\vec{q}_1-\vec{q}_2-\vec{q}_3)+\ldots \end{pmatrix}$$

$$\text{and } a(\vec{q}\,;T) = a_0(T) + a_1|\vec{q}\,|^2 + \ldots, \text{etc.} \tag{7.21a}$$

In particular, if we keep only the $|\vec{q}\,|^2$ term in $a(\vec{q}\,;T)$, we obtain

$$F_\Lambda[\phi] = \sum_{|\vec{q}\,|<\Lambda} ([a_0(T) + a_1(T)|\vec{q}\,|^2]\phi(\vec{q}\,)^2 + b(T)\phi(\vec{q}_1)\phi(\vec{q}_2) \times\phi(-\vec{q}_1-\vec{q}_2) + c(T)\phi^4 + \ldots) \tag{7.21b}$$

and

$$F_\Lambda[\bar{\phi}] = \int d^d r(a_0(T)\bar{\phi}^2 + a_1(T)(\nabla\bar{\phi})^2 + b(T)\bar{\phi}^3 + c(T)\bar{\phi}^4 + \ldots), \tag{7.21c}$$

which is precisely Landau's free energy. Note that, similar to the case of the Navier–Stokes equation, the form of the energy functional is very general. Only the coefficients $a$, $b$, $c$ differ for different microscopic systems.

#### 7.2.2.4 **Quantum Phase Transitions**

The phenomenon/concept of phase transitions, order parameters, and effective Hamiltonians can be introduced in quantum systems as in classical thermodynamics. We shall make a brief introduction to the topic here.

Imagine a quantum Hamiltonian $H(g)$ for a system of large volume $V \to \infty$, characterized by the parameter $g$. For example, the Heisenberg model on a cubic lattice,

$$H(g) = \sum_{\langle i,j\rangle}(S_{ix}S_{jx} + S_{iy}S_{jy}) + g\sum_{\langle i,j\rangle} S_{iz}S_{jz}, \tag{7.22}$$

where $\langle i,j\rangle$ = set of nearest-neighbor sites, and $\vec{S}$ are spins with magnitude $^1/_2$, or a model of interacting electrons,

$$H(g) = \sum_{\vec{k}\sigma} c^+_{\vec{k}\sigma}c_{\vec{k}\sigma} + g_s\sum_{\vec{q}}\rho_{\vec{q}}\rho_{-\vec{q}} + g_a\sum_{\vec{q}}\sigma_{\vec{q}}\sigma_{-\vec{q}},$$
$$\text{where } \rho_{\vec{q}} = \sum_{\vec{k}\sigma} c^+_{\vec{k}+\vec{q}\sigma}c_{\vec{k}\sigma} \text{ and } \sigma_{\vec{q}} = \sum_{\vec{k}\sigma}\sigma c^+_{\vec{k}+\vec{q}\sigma}c_{\vec{k}\sigma}. \tag{7.23}$$

A quantum phase transition occurs if the qualitative properties of the *ground-state wavefunction* change at a certain critical point $g = g_c$. For example, $g_c = 0$ is a point separating the ferromagnetic and anti-ferromagnetic phases of the Heisenberg model in Equation (7.22). As in the case of classical phase transitions, it is hard to have a general definition of what is meant by a change in 'qualitative properties of the system', and one criterion of a phase transition is to look at the ground-state energy of the system as a function of g.

Let us consider the ground-state energy of the system per unit volume $E(g)$ as a function of g. Following the example of classical phase transitions, a quantum phase transition is said to occur at $g = g_c$ if $E(g)$ becomes non-analytic at the point, that is if some power of derivatives $d^n E(g)/dg^n$ becomes discontinuous at $g = g_c$. Note that this definition merges smoothly with the definition of phase transitions in thermodynamics since the free energy becomes the ground-state energy at temperature $T \to 0$.

In analogy with Landau's phenomenological theory of phase transitions, we may define an order parameter $\phi$ characterizing a phase transition if the phase transition is characterized by broken symmetry. We can also construct an effective action for $\phi$ describing the ground-state and low-energy properties of the system close to the (quantum) critical point $g_c$. With the effective action we may study the dynamics of the order parameter field $\phi$. The effective actions can be derived by integrating out the microscopic degrees of freedom in the path integral approach, similar to the derivation of effective free energy from the classical Hamiltonian we outlined in the previous section, and the ground state of the system can be obtained by minimizing the energy with respect to $\phi$ as in Landau theory. The (low) energy spectrum of the system usually changes qualitatively across the transition. An example of a quantum phase transition is a superconductor–insulator transition. We shall study this in Chapter 11. You can also look at Ref. [7] for a more thorough discussion of quantum phase transitions.

## 7.3 Other Examples of Effective Classical and Quantum Field Theories

The use of effective theory goes beyond application to phase-transition phenomena or hydrodynamics and has a much deeper physical ground. It provides us with guidance in tackling two fundamental problems: (1) how can we construct the microscopic laws of physics from what we observe in experiments and (2) how can we understand the huge complexity of the macroscopic world starting from a few simple laws in physics?

Physicists realize nowadays that because of the limitation of experimental techniques what we observe in physics experiments today reflects just a small part of what is actually going on in our microscopic world and the existing laws of physics are only effective theories built on the variables we are able to observe and, to some extent, control. So, the question is, given what we observe, to what extent can we determine the microscopic laws of physics?

The converse question also exists: given that most of the macroscopic phenomena we observe in our daily life are based microscopically on the laws of gravity, electromagnetism, and quantum mechanics, how can we understand the apparent existence of complicated and diversified rules that give the phenomena (for example, evolution of life) in our macroscopic world? In other words, we want to understand how effective laws of nature can evolve from microscopic laws of physics under different external constraints.

Let us try to obtain some insight to this question by assuming that the basic laws of physics have the form of a Schrödinger equation,

$$i\hbar\frac{\partial}{\partial t}|\psi(a,b,\ldots)> = H|\psi(a,b,\ldots)> \tag{7.24}$$

where $a$, $b$, and so on are variables characterizing the wavefunction.

Suppose that we can only monitor a subset of these variables $\{a\}$ in experiments, whereas the complementary set $\{b\}$ is not observable to us. What is the effective equation of motion for $\{a\}$? Is it still a Schrödinger equation? Or will a totally different equation of motion come out? This is the type of question we want to answer when studying general effective field theory.

### 7.3.1 Projective Hilbert Space (Mori) Approach and Correlation Functions

Let us go a little bit further in the above example. Suppose that we have a microscopic Hamiltonian characterized by an $N \times N$ matrix, and the set $\{a\}$ corresponds to the first $M$ components of vectors characterizing the Hilbert space, and $\{b\}$ the remaining $N - M$ components. The Schrödinger equation can be rewritten in the form

$$i\hbar\frac{\partial}{\partial t}\begin{bmatrix} a(t) \\ b(t) \end{bmatrix} = \begin{bmatrix} H_{aa} & H_{ab} \\ H_{ba} & H_{bb} \end{bmatrix}\begin{bmatrix} a(t) \\ b(t) \end{bmatrix}, \tag{7.25a}$$

where $H_{\alpha\beta}$ is an $N_\alpha \times N_\beta$ matrix, $N_a = M$, and $N_b = N - M$. $a(t)$ and $b(t)$ are $M$-component and $(N - M)$-component vectors, respectively.

Since the Schrödinger equation is linear, an effective equation of motion for $a(t)$ can be obtained by eliminating $b(t)$ from the equation of motion. We obtain

$$b(t) = \left(i\hbar\frac{\partial}{\partial t} - H_{bb}\right)^{-1} H_{ba}a(t) \tag{7.25b}$$

and

$$\begin{aligned} i\hbar\frac{\partial}{\partial t}a(t) &= H_{aa}a(t) + H_{ab}b(t) = \left[H_{aa} + H_{ab}\frac{1}{\left(i\hbar\frac{\partial}{\partial t} - H_{bb}\right)}H_{ba}\right]a(t) \\ &= H_{\text{eff}}\left[i\hbar\tfrac{\partial}{\partial t}\right]a(t), \end{aligned} \tag{7.25c}$$

that is we obtain a modified Schrödinger equation with an effective time-dependent Hamiltonian for $a(t)$. An approach of this kind where some degrees of freedom in the system are eliminated to obtain an effective theory for the remaining variables forms the theoretical basis of most effective field theory. A similar approach has been used in constructing effective free energies in classical field theory (see previous section).

A practical application of this approach is the calculation of correlation functions of the form $\langle \hat{\phi}(t)\hat{\phi}^{+}(t')\rangle$ that we discussed in Chapter 5. To illustrate this, we consider temperature $T=0$, where the correlation function becomes $\langle \hat{\phi}(t)\hat{\phi}^{+}(t')\rangle = \langle G|\hat{\phi}\, U(t-t')\hat{\phi}^{+}|G\rangle$.

We note that the correlation function is of the form $\langle \alpha|U(t-t')|\alpha\rangle$, where $|\alpha\rangle = \hat{\phi}^{+}|G\rangle$ is a quantum state which is in general not an eigenstate of the system. Now, imagine dividing the total Hilbert space into two parts; one part consists of the state $|\alpha\rangle$ only and the other $\beta = \{|m\rangle\}$ is the rest. We now apply the projective dynamic approach (Equation (7.25a-c) to the system. Let $|\alpha(t)\rangle = a(t)|\alpha\rangle$. We obtain from Equation (7.25b)

$$\begin{aligned} i\hbar\frac{\partial}{\partial t}a(t) &= \left[\varepsilon_\alpha + H_{\alpha\beta}\frac{1}{\left(i\hbar\frac{\partial}{\partial t}-H_{\beta\beta}\right)}H_{\beta\alpha}\right]a(t) \\ &= \left[\varepsilon_\alpha + \sum_{m,m'} H_{\alpha m'}\frac{1}{\left(i\hbar\frac{\partial}{\partial t}-H_{\beta\beta}\right)_{m'm}}H_{m\alpha}\right]a(t) \\ &= \int dt'\left[\varepsilon_n\delta(t-t') + \sum_{m,m'} H_{nm'}G_{m'm}(t-t')H_{mn}\right]a(t') \\ &= \int dt'[\varepsilon_n\delta(t-t') + \Sigma_n(t-t')]a(t'), \end{aligned} \tag{7.26a}$$

where

$$\begin{aligned} &\varepsilon_n = \langle n|H|n\rangle, \quad H_{nm} = \langle n|H|m\rangle, \quad \text{and} \\ &G_{m'm}(t-t') = \langle m'|\left(i\hbar\frac{\partial}{\partial t}-H_{\beta\beta}\right)^{-1}|m\rangle. \end{aligned} \tag{7.26b}$$

The correlation function $\langle \hat{\phi}(t)\hat{\phi}^{+}(t')\rangle$ is a solution of the above integral equation with boundary condition $a(0) = \langle|\phi|^2\rangle$ (see Chapter 5).

Introducing the Fourier transform $G_\phi(\omega) = \int\langle \hat{\phi}(t)\hat{\phi}^{+}(t')\rangle e^{i\omega t}dt$, we obtain

$$G_\phi(\omega+i\delta) = \frac{\langle|\phi|^2\rangle}{\omega+i\delta-\varepsilon_c-\Sigma_n(\omega+i\delta)}, \tag{7.26c}$$

where $\Sigma_n$ is the self energy for the state $|n\rangle$. This is called the Mori formalism [7] to evaluate correlation functions.

**Homework Q5a**

Using the above approach, show that the Fourier transform of the fermionic retarded correlation function $G_R(k;t-t') = i\langle c_k(t)c_k^{+}(t') + c_k^{+}(t')c_k(t)\rangle\theta(t-t')$ can be written as

$$G_R(k;\omega) = \frac{1}{\omega+i\delta-\varepsilon_k-\Sigma_k(\omega+i\delta)}, \tag{7.27}$$

where $c_k(c_k^{+})$ is a fermion annihilation (creation) operator.

**Homework Q5b**

Show that in general $a(t)$ in Equation (7.25c) does not satisfy $\frac{\partial}{\partial t}(a^+(t)a(t)) = 0$, that is probability is generally speaking not conserved after projection into the subspace $|\alpha\rangle$.

Note that, in this approach, the self energy is nothing but an effective Hamiltonian for the system when all other states except the state $|\alpha\rangle = \widehat{\phi}^+|G\rangle$ are eliminated. The self energy defined in Equations (7.26a) and (7.26b) can be calculated in perturbation theory if we treat the off-diagonal matrix elements $H_{nm}$ as perturbation. An example of evaluating the self energy to lowest (second) order is given in Appendix A.

The systematic elimination of part of the degrees of freedom in a system to construct an effective theory for the remaining degrees of freedom is often difficult to carry out rigorously in realistic calculations and many other methods for obtaining effective field theory are developed. One powerful method starts from 'guessing' the general form of equation of motion or free energy from symmetry considerations. This is what we have shown in the derivation of the Navier–Stokes equation and the Landau effective free energy. Depending on the physical problem of interest, physicists have constructed many different approaches and we cannot enumerate them all in this book. We shall discuss two more examples in the following. The first example is for quantum systems and the second for classical systems. First, we consider quantum systems.

## 7.3.2
## Quantum Principle of Least Action and Applications

One can introduce a 'quantum' action in quantum mechanics (this is not the classical Lagrangian and action before quantization) defined by

$$S = \int_a^b dt < \psi(t)| \left[ i\hbar\frac{\partial}{\partial t} - H \right] |\psi(t)>. \tag{7.28}$$

Minimization of the action ($\delta S = 0$) with respect to an arbitrary wavefunction $<\psi(t)|$ leads to the Schrödinger equation $i\hbar\frac{\partial}{\partial t}|\psi(t)> = H|\psi(t)>$.

Now, imagine that we are interested only in a subset of parameters $\{q_i\}$ that characterize (partially) the wavefunction, that is $|\psi(t)> = |q_i(t); Q_j(t)>$, where $\{Q_j\}$ is the rest of the parameters characterizing the wavefunction which we do not observe or control. In the projective dynamics approach, we derive the effective dynamics for $\{q_i\}$ by eliminating the variables $\{Q_j\}$ exactly. The principle of least action provides us with an alternative, approximate approach where we derive the equations of motion for the variables $\{q_i\}$ through $\frac{\delta S}{\delta q_i} = 0$, but neglecting the dynamical effect of $\{Q_j\}$. The method can be used to derive an effective semiclassical equation of motion or an effective Schrödinger equation for the variables $\{q_i\}$.

As an example, let us consider a system with two identical particles with Hamiltonian

$$H = -\frac{\hbar^2}{2m}\nabla_1^2 - \frac{\hbar^2}{2m}\nabla_2^2 + U(\vec{x}_1 - \vec{x}_2) + V(\vec{x}_1, t) + V(\vec{x}_2, t). \tag{7.29}$$

Here, $U(\vec{x})$ is an attractive interaction between the particles and $V(\vec{x},t)$ is the time-dependent potential the particles see. First, let us parameterize the Schrödinger wavefunction $|\psi(t)> = \psi(\vec{x}_1, \vec{x}_2, t)$ by

$$\psi(\vec{x}_1, \vec{x}_2, t) = \psi_0(\vec{x}_1 - \vec{R}(t), \vec{x}_2 - \vec{R}(t)) e^{i\vec{k}(t).((\vec{x}_1+\vec{x}_2)/2 - \vec{R}(t))}, \tag{7.30}$$

where $\psi_0(\vec{x}_1 - \vec{R}, \vec{x}_2 - \vec{R}) \sim \sqrt{\rho(\vec{x}_1 - \vec{R}, \vec{x}_2 - \vec{R})} e^{i\vartheta(\vec{x}_1 - \vec{x}_2)}$ is a two-body wavefunction centered around $\vec{R}(t) = \langle \psi(t) | (\frac{\vec{x}_1 + \vec{x}_2}{2}) | \psi(t) \rangle$, the center of mass. The effect of the time-dependent potential is to generate a center of mass motion in this approximation. Let us consider the situation where we are not interested in the relative motion between the particles described by $\psi_0$, but are interested in the center of mass motion of the whole 'molecule' characterized by $\vec{R}(t), \vec{k}(t)$, where $\hbar\vec{k}(t)$ is the total momentum carried by the wavefunction. Therefore, we neglect the dynamics of $\psi_0$ and treat it as a fixed, time-independent wavefunction (except for the implicit dependence on time through $\vec{R}(t)$). The corresponding approximate quantum action is

$$\begin{aligned} S &= \int_a^b \mathrm{d}t \int \mathrm{d}^d x_1 \int \mathrm{d}^d x_2 \left[ \begin{array}{l} \psi_0^*(i\hbar) \left( -\dot{\vec{R}}.(i\vec{k} + (\nabla_1 + \nabla_2)) + i\dot{\vec{k}}.(\frac{(\vec{x}_1 + \vec{x}_2)}{2} - \vec{R}) \right) \psi_0 \\ -\frac{\hbar^2}{2M} \vec{k}^2 \rho - \left[ \sum_{i=1,2} \left[ \frac{\hbar^2}{2m} |\nabla_i \psi_0|^2 + V(\vec{x}_i, t)\rho \right] - U(\vec{x}_1 - \vec{x}_2)\rho \right] \end{array} \right] \\ &= \int_a^b \mathrm{d}t \left( \dot{\vec{R}}.\hbar\vec{k} - \frac{(\hbar\vec{k})^2}{2M} - V_{\text{eff}}(\vec{R}, t) \right), \end{aligned} \tag{7.31}$$

where $M = 2m$ and

$$V_{\text{eff}}(\vec{R}, t) = \int \mathrm{d}^d x_1 \int \mathrm{d}^d x_2 \left[ \sum_{i=1,2} \left[ \frac{\hbar^2}{2m} |\nabla_i \psi_0|^2 + V(\vec{x}_i, t)\rho \right] + U(\vec{x}_1 - \vec{x}_2)\rho \right]. \tag{7.32}$$

The other terms vanish after evaluating the integral.

Minimizing the action with respect to $\vec{R}(t)$ and $\vec{p}(t) = \hbar\vec{k}(t)$, we obtain

$$\vec{p}(t) = M\dot{\vec{R}}(t) \text{ and } \dot{\vec{p}}(t) = -\nabla V_{\text{eff}}(\vec{R}; t), \tag{7.33}$$

which is Newton's equation of motion for a classical particle with mass $M$ moving under an effective potential energy $V_{\text{eff}}(\vec{R}, t)$ which includes a 'quantum correction' from the wavefunction $\psi_0$, that is we have derived an effective semiclassical dynamics for the molecule. A corresponding quantum theory for the molecule can be obtained by quantizing the action (7.31) again to obtain an effective Schrödinger equation for the center of mass motion of the molecule.

Alternatively, we can also construct an effective Schrödinger equation directly by introducing the center of mass and relative coordinates $\vec{R} = (\vec{x}_1 + \vec{x}_2)/2$

and $\vec{r} = (\vec{x}_1 - \vec{x}_2)$ of the two-body wavefunction. We let

$$\psi(\vec{x}_1, \vec{x}_2, t) = \psi(\vec{R}, \vec{r}, t) = \psi_{CM}(\vec{R}, t)\psi_0(\vec{r}), \tag{7.34a}$$

where $\psi_0(\vec{r})$ is a known trial wavefunction describing the relative motion of the two particles and $\psi_{CM}(\vec{R}, t)$ is a wavefunction to be determined from the quantum principle of least action. In this case, it is straightforward to show that the action in terms of $\psi_{CM}$ becomes

$$S = \int_a^b dt \langle \psi_{CM}(\vec{R}, t) | \left[ i\hbar \frac{\partial}{\partial t} - H_{eff} \right] | \psi_{CM}(\vec{R}, t) \rangle, \tag{7.34b}$$

where

$$H_{eff} = -\frac{\hbar^2}{2M} \nabla_R^2 + V_{eff}(\vec{R}, t). \tag{7.34c}$$

**Homework Q6**
Derive $V_{eff}(\vec{R}, t)$. What is the relation between this effective Schrödinger equation and the one derived by quantizing the effective classical action we derived previously?

#### 7.3.2.1 Quasi-particles

In the above example, $\vec{R}(t)$ corresponds to the center of mass of a two-particle wavefunction. The approach can be generalized to introduce the important concept of quasi-particles with the following construction. Let us imagine a microscopic system with $N$ degrees of freedom $\vec{X} = (\vec{x}_1, \vec{x}_2, \ldots, \vec{x}_N)$ described by a wavefunction $\psi_R(\vec{x}_1, \vec{x}_2, \ldots, \vec{x}_N; \vec{r}_1, \vec{r}_2, \ldots, \vec{r}_M)$ characterized by $M$ special points in the wavefunction located at $\vec{R} = (\vec{r}_1, \vec{r}_2, \ldots, \vec{r}_M)$. These special points are a generalization of $\vec{R}(t)$ and may or may not belong to the set of coordinate points in $\vec{X}$. We may imagine a situation where the low-energy dynamics of the system can be characterized by the points $\vec{R}$ which can be treated as coordinates of particle-like excitations. The statistics of these particles (called quasi-particles) is characterized by the Berry phase when we exchange two such points in the wavefunction adiabatically.

An example of this construction is the generalization of the above two-particle system to a $2N$-particle system which forms $N$ molecules. In this case the molecules are 'quasi-particles' of the original $2N$-particle system and $\vec{R}(t)$ denote the positions of the molecules. The statistics of the molecules is determined by the Berry phase under exchange of two molecules and is in general different from the statistics of the underlying particles. For example, a molecule formed by two fermions is a boson, and so on.

Another less trivial example where this description applies is the boson system(s) we discussed in Chapters 3 and 4, where we showed that a classical theory described by canonical variables

$$(q, p) = (q_1, q_2, \ldots, q_N; p_1, p_2, \ldots, p_N)$$

can be described, after quantization, as a system of bosons representing collective excitations of the original classical system. Wavefunctions of these bosons are, in general, complicated functionals of the $(q, p)$ coordinates which can be represented in the following way:

$$|\psi(t)\rangle = \int D\vec{r}\, \phi(\vec{r}_1, \vec{r}_2, \ldots, \vec{r}_M; t)\psi_R(\{p, q\}; \vec{r}_1, \vec{r}_2, \ldots, \vec{r}_M), \tag{7.35}$$

where $\psi_R(\{p, q\}; \vec{r}_1, \vec{r}_2, \ldots, \vec{r}_M)$ represents a product of harmonic oscillator wavefunctions at lattice sites $\vec{X} = (\vec{x}_1, \vec{x}_2, \ldots, \vec{x}_n)$ where the canonical variables $(q, p)$ are defined. The harmonic oscillators are at their ground state except for the lattice points $\vec{r} = (\vec{r}_1, \vec{r}_2, \ldots, \vec{r}_M)$ where the wavefunctions are at their first excited states, representing one boson at these locations. (We assume here that only one boson exists at these sites for simplicity.) The wavefunction $\phi(\vec{r}_1, \vec{r}_2, \ldots, \vec{r}_M; t)$ represents the wavefunction of the $M$-boson system which satisfies a Schrödinger equation for bosons originating from the boson representation of the Hamiltonian.

Perhaps the most interesting example of this kind of construction is quasi-particles in the fractional quantum Hall effect (FQHE). The FQHE occurs when we have a two-dimensional electron liquid moving under a constant magnetic field perpendicular to the plane. At special fillings it is now well established that the ground state of the system can be described by Laughlin-type wavefunctions (see for example, Ref. [5] & [8]. Excitations of the system are described precisely by wavefunctions of the form (7.35) where the special points $\vec{R}$ are 'zeros' of the wavefunction created by magnetic flux. 'Zeros' means that the wavefunction is zero if the positions of particles coincide with these special points. The quasi-particles in FQHE are very special because they carry fractional charges and fractional statistics. The statistics of these particles can be determined by a Berry phase calculation. We shall not go into this subject in this book. You can find out more details in, for example, Ref. [8]. A brief introduction to fermionic quasi-particles in Fermi liquid theory is available in Chapter 10 of this book.

### 7.3.3 (Generalized) Langevin and Fokker–Planck Equations

The generalized Langevin and Fokker–Planck equations are classical equations of motion describing dynamics of the order parameter or hydrodynamic fields in macroscopic systems [6, 9]. They are natural extensions of the idea of effective free energies $F_{\text{eff}}[\phi]$ in thermodynamics to dynamics of macroscopic systems. The approach is exact in the linear response regime around equilibrium. First, let us discuss the Langevin equation.

We shall assume as in Chapter 5 that small fluctuations of the order parameter $\phi$ or hydrodynamic fields around the equilibrium configuration $\langle\phi\rangle = \bar{\phi} = 0$ can be described by an equation of motion of the form

$$\widehat{L}\left(\frac{\partial}{\partial t}, \nabla\right)\phi(\vec{x}, t) = F(\vec{x}, t), \tag{7.36}$$

where $F(\vec{x}, t)$ is an external force acting on the $\phi$ field. The response function of the $\phi$ field to the external force is given by

$$\phi(\vec{x}, t) = \hat{L}^{-1}F(\vec{x}, t) = \int d^d x' \int_{-\infty}^{t} \chi(\vec{x} - \vec{x}', t - t')F(\vec{x}', t')dt'. \tag{7.37}$$

As we have discussed in Chapter 5, the variable $\phi$ has statistical fluctuations in time and $\langle \phi(\vec{x}, t)\phi(\vec{x}', t')\rangle = C(\vec{x} - \vec{x}', t - t') \neq 0$ in general. Langevin proposed that these fluctuations can be understood as arising physically from a fluctuating force $F_f(\vec{x}, t)$ acting on $\phi$ with average $\langle F_f(\vec{x}, t)\rangle = 0$ and

$$\langle F_f(\vec{x}, t)F_f(\vec{x}', t')\rangle = \gamma(\vec{x} - \vec{x}', t - t') \neq 0. \tag{7.38}$$

The force–force correlation function $\gamma(\vec{x}, t)$ can be determined from linear response theory by noting that with Equation (7.37) we can write

$$\begin{aligned}\langle \phi(\vec{x}, t)\phi(\vec{y}, \tau)\rangle = \int d^d x' \int d^d y' \int_{-\infty}^{t} dt' \int_{-\infty}^{\tau} d\tau' \chi(\vec{x} - \vec{x}', t - t')\chi(\vec{y} - \vec{y}', \tau - \tau') \\ \times \langle F_f(\vec{x}', t')F_f(\vec{y}', \tau')\rangle,\end{aligned} \tag{7.38a}$$

from which, after Fourier transformation (assuming translational invariance in time and space), we obtain

$$C(\vec{q}, \omega) = |\chi(\vec{q}, \omega)|^2 \gamma(\vec{q}, \omega). \tag{7.38b}$$

**Exercise: derive Eq. (7.38b)**

Recall that $C(\vec{q}, \omega) = \frac{2k_B T}{\omega} \mathrm{Im}\chi(\vec{q}; \omega)$ from the fluctuation–dissipation theorem in the hydrodynamic regime (Chapter 5) and therefore $\gamma(\vec{q}, \omega)$ is related to $\chi(\vec{q}; \omega)$. The relation between $\gamma(\vec{q}, \omega)$ and $\chi(\vec{q}; \omega)$ can be seen more clearly if we introduce $\Gamma(\vec{q}, \omega) = \chi(\vec{q}, \omega)^{-1}$. Therefore,

$$C(\vec{q}, \omega) = -\frac{2k_B T}{\omega} \frac{\mathrm{Im}\Gamma}{(\mathrm{Re}\Gamma)^2 + (\mathrm{Im}\Gamma)^2} = -\frac{2k_B T}{\omega} |\chi|^2 \mathrm{Im}\Gamma$$

and

$$\gamma(\vec{q}, \omega) = -\frac{2k_B T}{\omega} \mathrm{Im}\Gamma(\vec{q}, \omega), \tag{7.39}$$

which is also called the second fluctuation–dissipation theorem [9].

The physical meaning of this result can be seen more clearly if we go to the small-frequency limit and assume that

$$\begin{aligned}\chi(\vec{q}, \omega) &\sim \chi(\vec{q}, 0) + \omega^2 \frac{d\mathrm{Re}\chi(\vec{q}, \omega)}{d(\omega^2)}\bigg|_{\omega=0} + i\omega \frac{d\mathrm{Im}\chi(\vec{q}, \omega)}{d(\omega)}\bigg|_{\omega=0} \\ &= a(q) + b(q)\omega^2 + ic(q)\omega,\end{aligned} \tag{7.40a}$$

in accordance with the general allowable form of $\chi$ (Homework Q6, Chapter 5). Therefore,

$$\begin{aligned}\chi^{-1} \sim L \sim a(q) - i\frac{c(q)}{a(q)^2}\omega - \frac{1}{a(q)^2}\left(b(q) + \frac{c(q)^2}{a(q)}\right)\omega^2 \\ \sim a(q) + \eta(q)\frac{\partial}{\partial t} + M(q)\frac{\partial^2}{\partial t^2}\end{aligned} \tag{7.40b}$$

after Fourier transforming back in time and $\phi$ satisfies an equation of motion of the form

$$M(q)\frac{\partial^2}{\partial t^2}\phi(\vec{q}, t) = -a(q)\phi(\vec{q}, t) - \eta(q)\frac{\partial}{\partial t}\phi(\vec{q}, t) + F(\vec{q}, t), \tag{7.41}$$

which is Newton's equation of motion for a particle in a harmonic potential $\frac{a}{2}\phi^2$ subjected to frictional force $-\eta\dot{\phi}$ and external force $F$. Note that this form of equation of motion is obtained just by assuming that $\chi(\vec{q}\,; \omega)$ has the general form $\chi(\vec{q}, \omega) \sim a(q) + b(q)\omega^2 + ic(q)\omega$ at small $\omega$.

Therefore, the second fluctuation–dissipation theorem says that in the small-frequency limit the force–force correlation is related to the coefficient of friction of the $\phi$ field by

$$\gamma(\vec{q}, \omega \to 0) = 2k_B T\eta(\vec{q}\,). \tag{7.42}$$

The presence of the friction term in the equation of motion guarantees that the $\phi$ field decays to the equilibrium value $\langle\phi\rangle = \bar{\phi}$ in the absence of external force, and ensures the stability of the equilibrium state. The harmonic potential term is related to the effective free energy $F_{\text{eff}}[\phi]$ through the coefficient $a(q)$, where

$$a(q) = \chi(q, 0) = \left.\frac{\delta^2 F_{\text{eff}}}{\delta\phi(\vec{q}\,)\delta\phi(-\vec{q}\,)}\right|_{\phi=\bar{\phi}} > 0. \tag{7.43}$$

Note that the linear order term $\left.\frac{\delta F_{\text{eff}}}{\delta\phi}\right|_{\phi=\bar{\phi}}$ vanishes if $\bar{\phi}$ minimizes the free energy $F_{\text{eff}}[\phi]$.

The Langevin equation of motion is often generalized phenomenologically to non-equilibrium situations by assuming an equation of motion of the form [6, 9]

$$0 = -\frac{\delta F_{\text{eff}}[\phi]}{\delta\phi(\vec{r}\,)} - \eta\frac{\partial}{\partial t}\phi(\vec{r}, t) + F(\vec{r}, t), \tag{7.44a}$$

where it is assumed that the motion is slow enough so that it is enough to keep only the first-order time-derivative term in the equation of motion. $\eta = \eta(\vec{q} \to 0)$ and

$$\langle F_f(\vec{x}, t)F_f(\vec{x}\,', t')\rangle \to 2\eta k_B T\delta(t-t')\delta(\vec{x} - \vec{x}\,'). \tag{7.44b}$$

In this approximation $-\frac{\delta F_{\text{eff}}}{\delta\phi}$ represents a generalized force acting on the system at arbitrary field configuration $\phi$.

#### 7.3.3.1 Fokker–Planck Equation

The Fokker–Planck equation assumes that the fluctuating force in the Langevin equation is 'real' and, as a result, we can define a classical probability distribution functional for the $\phi$ field, $P[\phi]$, that changes in time under the fluctuating and external forces. The Fokker–Planck equation is an equation governing the time evolution of $P[\phi]$. We shall derive the corresponding Fokker–Planck equation for the generalized Langevin equation (7.44a, b) in the following as example.

The Fokker–Planck equation can be derived by assuming that the time evolution of $\langle\phi(t)\rangle = \int D\phi\phi P[\phi;t]$ satisfies the Langevin equation, where $P[\phi;t]$ is the probability distribution of the $\phi$ field at time $t$ [5], that is

$$\frac{\partial}{\partial t}\langle\phi(t)\rangle = \frac{1}{\eta}\left[-\left\langle\frac{\delta F_{\text{eff}}[\phi]}{\delta\phi}\right\rangle + \langle F_{\text{f}}\rangle + \langle F_{\text{ext}}\rangle\right] \tag{7.45a}$$

or

$$\begin{aligned}\int D\phi\phi\frac{\partial}{\partial t}P[\phi;t] &= \frac{1}{\eta}\int D\phi\left[-\frac{\delta F_{\text{eff}}[\phi]}{\delta\phi} + F_{\text{f}} + F_{\text{ext}}\right]P[\phi;t]\\ &= -\frac{1}{\eta}\int\phi D\left(\left[-\frac{\delta F_{\text{eff}}[\phi]}{\delta\phi} + F_{\text{f}} + F_{\text{ext}}\right]P[\phi;t]\right)\\ &= -\frac{1}{\eta}\int\phi\frac{\delta}{\delta\phi}\left(\left[-\frac{\delta F_{\text{eff}}[\phi]}{\delta\phi} + F_{\text{f}} + F_{\text{ext}}\right]P[\phi;t]\right)D\phi.\end{aligned} \tag{7.45b}$$

Comparing both sides of the equation, we obtain

$$\frac{\partial}{\partial t}P[\phi;t] \sim -\frac{1}{\eta}\frac{\delta}{\delta\phi}\left(\left[-\frac{\delta F_{\text{eff}}[\phi]}{\delta\phi} + F_{\text{f}} + F_{\text{ext}}\right]P[\phi;t]\right). \tag{7.45c}$$

The effect of the fluctuating force has to be treated carefully because of its singular behavior (Equation (7.44b)). We start by rewriting Equation (7.45c) in an integral form

$$P[\phi;t+\Delta t] - P[\phi;t] = -\frac{1}{\eta}\int_t^{t+\Delta t}\frac{\delta}{\delta\phi}\left(\left[-\frac{\delta F_{\text{eff}}[\phi]}{\delta\phi} + F_{\text{f}} + F_{\text{ext}}\right]P[\phi;t']\right)\text{d}t', \tag{7.46a}$$

which can be iterated to second order to obtain, for small $\Delta t$,

$$\begin{aligned}P[\phi;t+\Delta t] - P[\phi;t] \sim &-\frac{1}{\eta}\int_t^{t+\Delta t}\frac{\delta}{\delta\phi}\left(\left[-\frac{\delta F_{\text{eff}}[\phi]}{\delta\phi} + F_{\text{f}} + F_{\text{ext}}\right]P[\phi;t]\right)\text{d}t'\\ &+\frac{1}{\eta^2}\int_t^{t+\Delta t}\frac{\delta}{\delta\phi}\left(\left[-\frac{\delta F_{\text{eff}}[\phi]}{\delta\phi} + F_{\text{f}} + F_{\text{ext}}\right]\right.\\ &\left.\times\int_t^{t'}\frac{\delta}{\delta\phi}\left(\left[-\frac{\delta F_{\text{eff}}[\phi]}{\delta\phi} + F_{\text{f}} + F_{\text{ext}}\right]P[\phi;t]\right)\text{d}t''\right)\text{d}t'.\end{aligned} \tag{7.46b}$$

The second term is in general of order $(\Delta t)^2$ and can be neglected in the limit $\Delta t \to 0$ except for the fluctuation force, which after averaging gives a finite contribution

$$\frac{1}{\eta^2}\int_t^{t+\Delta t}\left(\int_t^{t'}\langle F_f(t')F_f(t'')\rangle dt''\right)dt'\frac{\delta^2}{\delta\phi^2}P[\phi;t] = \frac{k_B T}{\eta}\Delta t\frac{\delta^2}{\delta\phi^2}P[\phi;t]. \tag{7.46c}$$

Therefore, after averaging we obtain the Fokker–Planck equation

$$\eta\frac{\partial P[\phi;t]}{\partial t} - \frac{\delta}{\delta\phi}\left(\frac{\delta F_{\text{eff}}}{\delta\phi}P[\phi;t]\right) + F_{\text{ext}}\frac{\delta P[\phi;t]}{\delta\phi} - k_B T\frac{\delta^2 P[\phi;t]}{\delta\phi^2} = 0. \tag{7.47}$$

To check the correctness of the equation, we consider the equilibrium state described by $\partial_t P = 0$, $F_{\text{ext}} = 0$. In this case, Equation (7.47) becomes

$$\frac{\delta}{\delta\phi}\left(\frac{\delta F_{\text{eff}}}{\delta\phi}P_0[\phi] + k_B T\frac{\delta P_0[\phi]}{\delta\phi}\right) = 0 \tag{7.48}$$

with solution $P_0[\phi] \sim e^{-\beta F_{\text{eff}}[\phi]}$, in agreement with thermodynamics [6, 9].

### Fokker–Planck Equation and Imaginary-Time Quantum Mechanics

In the absence of external force $F_{\text{ext}}$, the above Fokker–Planck equation can be transformed into a Schrödinger-like equation by writing $P[\phi;t] = P_s[\phi;t]e^{-\frac{\beta}{2}F_{\text{eff}}[\phi]}$.

It is easy to obtain by direct substitution

$$\eta\frac{\partial P_s}{\partial t} = \frac{1}{2}\left(\frac{\delta^2 F_{\text{eff}}}{\delta\phi^2} - \frac{1}{2k_B T}\left(\frac{\delta F_{\text{eff}}}{\delta\phi}\right)^2\right)P_s + k_B T\frac{\delta^2}{\delta\phi^2}P_s, \tag{7.49}$$

which is a Schrödinger equation at imaginary time if we identify $P_s[\phi;\, t]$ as a 'wavefunction' for the field variable $\phi$. The advantage of this representation is that we can now borrow techniques we learnt from quantum field theory to tackle this equation [5].

#### Homework Q7a

Derive the above equation.

#### Homework Q7b

Using results in Chapter 3, write down the (imaginary-time) action and Hamiltonian for the 'quantum field' that satisfies the above Schrödinger equation.

## References


1 Reichl, L.E. (1980) *A Modern Course on Statistical Physics*, University of Texas Press.

2 Lee, P.A. and Ramakrishnan, T.V. (1985) *Rev. Mod. Phys.* **57**, 287.

3 Landau L.D. and Lifshitz, E.M. (2000) *Fluid Mechanics*, Butterworth-Heinmann.

4 Landau L.D. and Lifshitz E.M. (2000) *Statistical Physies (Part I)*, Butterworth-Heinmann.

5 Wen X.G. (2004) *Theory of Quantum Many-Body Systems,* Oxford University Press.
6 Parisi G. (1988) *Statistical Field Theory,* Addison-Wesley.
7 Sachdev S. (1999) *Quantum Phase Transitions,* Cambridge University Press.
8 Nagaosa N. (1995) *Quantum Field Theory in Condensed Matter Physics,* Springer.
9 Forster D. (1975) *Hydrodynamic Fluctuations, Broken Symmetry and Correlation Functions,* Addison-Wesley.

# 8
# Solitons, Instantons, and Topology in QFT

We introduce in this chapter some important examples of non-linear differential equations with (non-perturbative) solutions called solitons and/or instantons. The important and unique role of these solutions as particle-like solutions in non-linear field equations, as representing quantum tunneling processes in the path integral approach and as non-perturbative objects in restoring broken symmetry in partition functions, will be discussed. The role of topology in stabilizing these solutions will be introduced.

## 8.1
## Introduction to Solitons

Roughly speaking, solitons and instantons are special solutions of non-linear (differential) field equations that are not adiabatically connected to any solutions of the corresponding linearized equations, and are particularly stable. Solitons are solutions of a time-dependent field equation, and instantons are the corresponding time-independent solutions, or solutions of a corresponding equation at imaginary time. *The fact that they are not adiabatically connected to solutions of the corresponding linearized equations means that they cannot be obtained via perturbation theory and special techniques have to be derived to study them.*

The word soliton was used to describe the 'particle-like' behavior of these special solutions. You can imagine these solutions as describing wave packets which do not disperse as time goes on and maintain more or less their original shape when they travel (see e.g. Ref. [1]).

To understand what the above means, we shall study several examples of non-linear differential equations. First, we start with linear differential equations, and ask the question of under what situation can we have non-dispersive wave packets?

An example of a field equation where wave packets do not disperse is the one-dimensional field equation

$$\left(\frac{1}{c^2}\frac{\partial^2}{\partial t^2} - \frac{\partial^2}{\partial x^2}\right)\phi(x,t) = 0. \tag{8.1}$$

*Introduction to Classical and Quantum Field Theory.* Tai-Kai Ng

ISBN: 978-3-527-40726-2

It is easy to see that any well-behaved function of the form $\phi(x, t) = f(x - ct)$ is a solution of the field equation. In this case, the solution describes a wave packet that travels with a constant speed $c$ in space. The shape of the wave packet does not change when it travels.

The situation becomes totally different if we add a mass term in the equation,

$$\left(\frac{1}{c^2}\frac{\partial^2}{\partial t^2} - \frac{\partial^2}{\partial x^2} + m^2c^2\right)\phi(x,t) = 0. \tag{8.2}$$

It is easy to show by direct substitution that $f(x - ct)$ is generally not a solution of the equation. To see why this is the case physically, we Fourier transform the equation and write $\phi(x,t) = \int \mathrm{d}\, k a(k) e^{ikx - \omega_k t}$.

It is straightforward to see that

$$\omega_k = \sqrt{c^2k^2 + m^2c^4}, \tag{8.3a}$$

with corresponding group velocity

$$v_g(k) = \frac{\mathrm{d}\omega_k}{\mathrm{d}k} = \frac{ck}{\sqrt{k^2 + m^2c^2}}. \tag{8.3b}$$

Note that the group velocity is different for different values of $k$ in general, and is identical for all $k$ only when $m = 0$. Only in this particular limit do all $k$ components of the wave packet travel with the same speed and the wave packet does not disperse.

Now, we turn to non-linear field equations. We assume that the Lagrangian formulation is applicable and the classical field equations are obtained by minimizing an action $S[\phi]$. As a starting point, we consider the 1D $\phi^4$ model

$$L[\phi] = \int_{-\infty}^{\infty} \mathrm{d}x \left[\frac{1}{2c^2}\left(\frac{\partial\phi}{\partial t}\right)^2 - \frac{1}{2}\left(\frac{\partial\phi}{\partial x}\right)^2 - \frac{1}{4}(\phi^2 - 1)^2\right], \tag{8.4a}$$

with corresponding energy (Hamiltonian) function

$$H[\phi] = \int_{-\infty}^{\infty} \mathrm{d}x \left[\frac{1}{2c^2}\left(\frac{\partial\phi}{\partial t}\right)^2 + \frac{1}{2}\left(\frac{\partial\phi}{\partial x}\right)^2 + \frac{1}{4}(\phi^2 - 1)^2\right]. \tag{8.4b}$$

The corresponding equation of motion is

$$\frac{1}{c^2}\frac{\partial^2\phi}{\partial t^2} - \frac{\partial^2\phi}{\partial x^2} - \phi + \phi^3 = 0. \tag{8.4c}$$

First, we consider the solution which minimizes energy. Note that the energy is always minimized when $\phi$ is time and space independent. It is easy to see that there are two solutions with (absolute) minimum energy $H[\phi] = 0$,

$$\phi = \phi_0 = \pm 1, \tag{8.4d}$$

or two degenerate 'ground states'. To obtain low-energy excitations, we usually write $\phi = \phi_0 + \phi'$. The equation of motion for $\phi'$ for $\phi_0 = \pm 1$ is

$$\frac{1}{c^2}\frac{\partial^2\phi'}{\partial t^2} - \frac{\partial^2\phi'}{\partial x^2} + 2\phi' \pm 3\phi'^2 + \phi'^3 = 0. \tag{8.5a}$$

Assuming that $\phi'$ is small and neglecting $\phi'^2$ and $\phi'^3$ terms, we obtain the linearized equation of motion

$$\frac{1}{c^2}\frac{\partial^2\phi'}{\partial t^2}-\frac{\partial^2\phi'}{\partial x^2}+2\phi'=0. \tag{8.5b}$$

The solutions of this linearized equation are wave solutions $\phi' \sim e^{i(kx-\omega_k t)}$ with $\omega_k = c^2k^2 + 2c^2$. The non-linear terms can be treated as perturbative corrections to the (approximate) linear solutions.

Now, let us ask whether there exist solutions that are fundamentally different from the wave solutions described above. A good way to start is to first look for static solutions of the field equation different from the $\phi = \phi_0 = \pm 1$ solutions, that is solutions satisfying

$$\frac{\partial^2\phi}{\partial x^2} = -\phi + \phi^3. \tag{8.6}$$

To see what kinds of solutions we may have, let us compare the equation with Newton's equation of motion

$$m\frac{\mathrm{d}^2x}{\mathrm{d}t^2} = -x + x^3 = -\frac{\mathrm{d}V}{\mathrm{d}x}, \tag{8.7a}$$

where

$$V(x) = -\frac{1}{4}(1-x^2)^2. \tag{8.7b}$$

The equation describes a particle moving in an 'inverted' double well potential $V(x)$ as shown in the figure below.

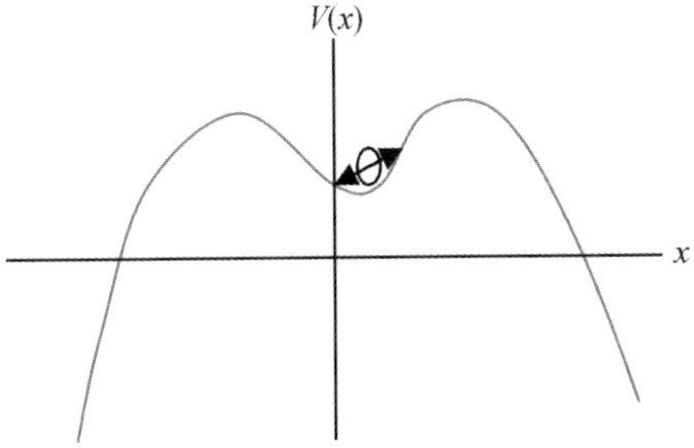


There exist two kinds of solutions to the equation: (i) solutions that oscillate between the two potential 'maxima' and (ii) solutions where the particle 'escapes' to infinity when 'time' goes to $\pm\infty$. Note that the energy of the solution of the field equation $\sim$ – the energy of the solution of the fictitious Newton problem. Therefore, the second kind of solutions with $x \to \pm\infty$ at $t \to \pm\infty$ (or $\phi(x \to \pm\infty) \to \pm\infty$ in the corresponding field theory problem) is unphysical because their (potential) energies diverge. In fact, to minimize the energy, we should have solutions that 'maximize' the energy in the fictitious one-body problem. These are the solutions with $x \to \pm 1$ at $t \to \pm\infty$. The 'ground state' of the field theory system corresponds to $\phi(x) = \pm 1$. An alternative solution we can construct is a solution that interpolates between the $\phi(x) = \pm 1$ solutions with $\phi(x \to \infty) = -\phi(x \to -\infty) = 1$. This is called the *kink solution*. Note that we can also construct an *anti-kink* solution with $\phi(x \to \infty) = -\phi(x \to -\infty) = -1$.

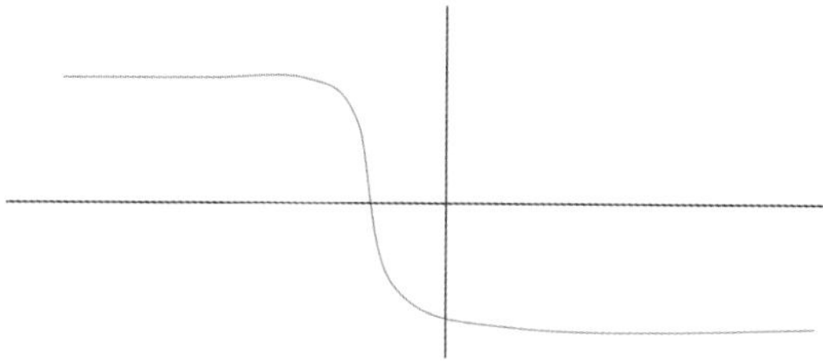


**Homework Q1a**
Borrow your knowledge in classical mechanics to solve the static kink (or anti-kink) solution. (Answer: $\phi(x) = \tanh\left(\frac{1}{\sqrt{2}}(x - x_0)\right)$.)

The time-dependent kink solution can be obtained from the static solution rather easily if you note that the equation of motion is in fact *Lorentz invariant*. Therefore, from space–time symmetry we expect that there is a time-dependent solution that corresponds to a static kink solution traveling with a constant velocity.

**Homework Q1b**
Show that the time-dependent kink solution is given by $\phi(x,t) = \tanh\left(\frac{1}{\sqrt{2}}\frac{(x-x_0)-ut}{\sqrt{1-u^2}}\right)$, where $u = v/c$, $v =$ velocity of the kink.

The energy of the kink (anti-kink) solution can be calculated easily. For a static kink, it is equal to $E_0 = \frac{2\sqrt{2}}{3}$. When the kink travels, its energy is $E(u) = \frac{E_0}{\sqrt{1-u^2}}$. Note that the energy obeys the Einstein mass–energy relation for relativistic particles because of the covariant form of the equation of motion.

**Homework Q1c**
Verify the above energy expression for the traveling kink solution.

Historically, the interesting properties of the kink solution led physicists to speculate that one may identify some soliton solutions of non-linear Lorentz-invariant field equations in three dimensions as representing 'composite particles' in a field theory. For example, it has been speculated that hadrons can be considered as soliton solutions of the quark field. To make this identification precise, the stability of these solutions in a 'quantum' environment has to be clarified. This is where topology enters.

### 8.1.1 Stability of Solitons and Topology

From our construction we know that kink (anti-kink) solutions are local minima in the energy functional $H[\phi]$, that is they are stable as far as classical mechanics is concerned. When quantum mechanics (QM) is included, the stability of these solutions depends typically on the 'barrier height' separating the kink solution from the ground state (vacuum). This is because in QM a field configuration (or wavefunction) can decay by quantum tunneling. A particle trapped in a box remains there forever only if the tunneling barrier height is infinitely high. The barrier height is the maximum potential energy the particle has to go through to travel out of the box. Similarly, the barrier height of the kink solution is the maximum energy the field configuration has to go through when $\phi$ changes continuously from the kink configuration ($\phi(x) = \tanh(\frac{1}{\sqrt{2}}(x-x_0))$) to the 'vacuum' ($\phi(x) = \phi_0 = \pm 1$) solution. The energy barrier of such a change (see figure below) is of order $\sim(L/2)E_0$, where $L$ = length of our system, and goes to infinity when $L \rightarrow \infty$. (see figure below for illustration)

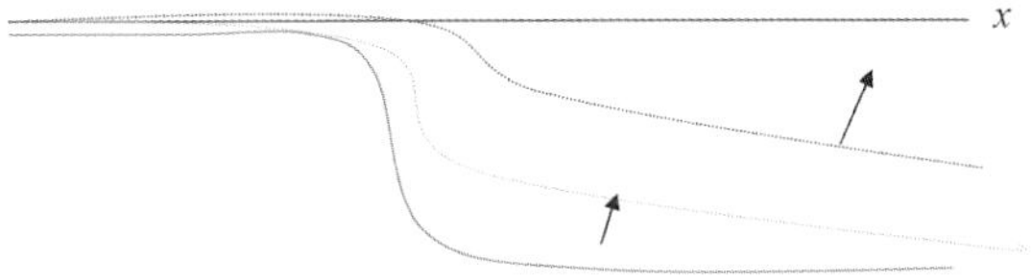


Therefore, the kink solutions are very stable. The stability arises because the system possesses two degenerate ground states and the boundary regions of the kink solutions ($x \rightarrow \pm\infty$) are in different ground states. To overturn (half of) the system from one ground state to another, we have to overcome an energy barrier of the order of the size of the system.

The above idea can be formulated rigorously with the language of topology.

Let us imagine classifying solutions of the $\phi^4$ model according to the boundary condition of the solution at $x \rightarrow \pm\infty$. To make sure that the energy of the solution is finite, the system must have $\phi(x) = \phi_0 = \pm 1$ at $x \rightarrow \pm\infty$, and there are altogether four possible boundary conditions or four different sectors of solutions. There exists an energy barrier $\sim LE_0$ between solutions in different sectors. The ground-state solutions have $\phi(x \rightarrow \infty) = \phi(x \rightarrow -\infty)$, whereas kink (anti-kink) solutions are solutions with $\phi(x \rightarrow \infty) = -\phi(x \rightarrow -\infty)$. The solutions are said to belong to different topological sectors.

These considerations can be generalized beyond the $\phi^4$ model and beyond 1 + 1 dimensions. As long as a field equation has degenerate ground states with different field values, it is plausible to construct solutions with different boundary values (corresponding to different ground states) at different boundary regions of space. The classification of plausible boundary conditions with a given degenerate set of ground states at different dimensions is the territory of topology.

**Another Example: Sine-Gordon Model**

Another example of a 1D non-linear field equation where solutions are classified by different topological sectors is the sine-Gordon model. The Lagrangian is

$$L[\phi] = \int_{-\infty}^{\infty} dx \left[ \frac{1}{2c^2}\left(\frac{\partial\phi}{\partial t}\right)^2 - \frac{1}{2}\left(\frac{\partial\phi}{\partial x}\right)^2 + (\cos\phi - 1) \right], \tag{8.8a}$$

with equation of motion

$$\frac{1}{c^2}\frac{\partial^2\phi}{\partial t^2} - \frac{\partial^2\phi}{\partial x^2} + \sin\phi = 0. \tag{8.8b}$$

The static solution of the equation has degenerate minima at $\phi = 2N\pi$ and there are infinite numbers of degenerate ground states characterized by $\phi(x \to \pm\infty) = 2N\pi$ with different $N$. Again, kink-like solutions exist joining ground states with different values of $N$. We shall not go into the mathematical details of these solutions here. Interested readers can look at Ref. [1].

#### 8.1.1.1 **Topological Index for One-Dimensional Scalar Fields**

For $\phi^4$-like models, one can define a topological charge (or index) by

$$Q = \frac{1}{2\phi_0}(\phi(x \to \infty) - \phi(x \to -\infty)) = \frac{1}{2\phi_0}\int dx \frac{\partial\phi}{\partial x}. \tag{8.9}$$

Obviously, $Q$ characterizes the different sectors of solutions in $\phi^4$-like models. For the $\phi^4$ model, $Q = 0, \pm 1$. For the sine-Gordon model, $Q = N_R - N_L$ can be any integer.

### 8.1.2
### Multi-Kink Solutions and Difference Between Solitary Wave and Soliton

If one wants to identify kink solutions as particles, it is important to ask whether multi-kink solutions exist and whether they are stable. In the $\phi^4$ model, since we have only two degenerate ground states, only solutions with *alternate kink and anti-kink configurations* can exist (see figure below). However, for the sine-Gordon model, because of the infinite number of degenerate ground states, we can have solutions with arbitrary kink and anti-kink configurations. (see figure below)

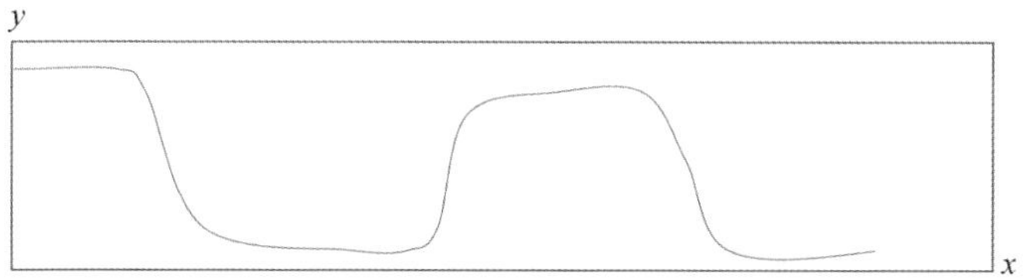


*Multi-kink–anti-kink solution in $\phi^4$ model*

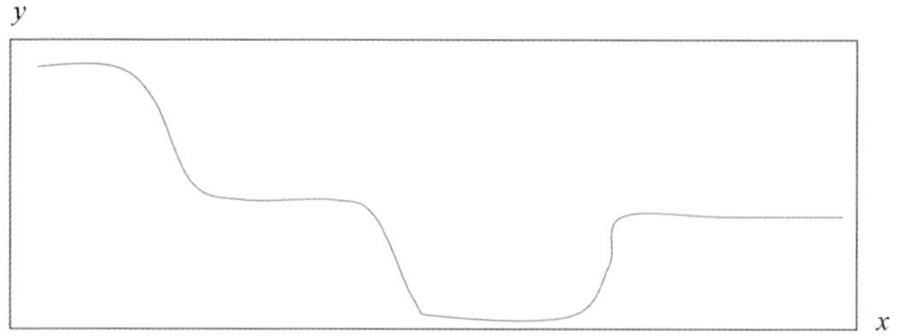


*Multi-kink–anti-kink solution in sine-Gordon model*

Note that a solution describing kink–anti-kink pair(s) is *not* protected by topology or boundary conditions. The particles in a kink–anti-kink pair may collide with each other and disappear afterward, corresponding to annihilation of a pair of particles and anti-particles. Well, such decay is not obviously possible because of energy conservation. The kink–anti-kink solution cannot simply disappear with nothing left. A more sensible question to ask is whether the *shape* of a kink–anti-kink pair will change after colliding with another kink–anti-kink pair. If the shape does not change, we may imagine that the solution corresponds to a stable particle. This kind of solution is called a soliton. If the shape of the kink changes after collision, the solution is called a solitary wave and whether it may be identified as a real particle is questionable. Mathematically, it is known that kinks in the $\phi^4$ model are solitary waves, whereas kinks in the sine-Gordon model are solitons.

#### 8.1.2.1 Quantization of Solitons/Solitary Waves

To find out whether mathematically the soliton (or solitary wave) solutions can be identified as real particles, we must see how they are affected when quantum mechanics is introduced. Mathematically, we must quantize the solutions quantum mechanically and see how they behave after quantization. Quantization of solitons is often carried out in the semiclassical approximation (expansion of the action around the classical local energy minima, a method we shall use very often in the next few chapters). The quantized kinks–anti-kinks in the $\phi^4$ model become quantum particles interacting with each other through exchange of wave-like modes of the system (representing fluctuations above the classical solution) in perturbation theory. The quantized kinks–anti-kinks in the sine-Gordon model behave similarly. We shall not discuss quantization of solitons further in this book because of the highly technical nature of the subject. You can look at Refs. [1, 2] for some examples.

## 8.2 Introduction to Instantons

Instantons appear in two different but strongly related problems. First of all, they represent solutions of the non-linear equation $\delta H[\phi]=0$, corresponding to local minima in the energy function $H[\phi]$ that enters in classical thermodynamics, that is $Z=\int D\phi e^{-\beta H[\phi]}$. These local minima contribute to partition functions non-perturbatively and are important if they are stable. They may also represent solutions of the equation of motion of a classical field theory at imaginary time – and contribute to the thermodynamics of the corresponding quantum system. We shall first illustrate the important role of instantons in classical $\phi^4$-field theory in one dimension.

### 8.2.1 Instantons in 1D Classical Theories

Let us study the thermodynamics of a one-dimensional system with energy function

$$H[\phi]=\int_{-\infty}^{\infty}\mathrm{d}x\left[\frac{1}{2}\left(\frac{\partial\phi}{\partial x}\right)^2+\frac{1}{4}(\phi^2-1)^2\right], \tag{8.10}$$

which is the $\phi^4$ model we studied before except that time does not enter in classical thermodynamics. As we have seen in the previous section, a static kink solution exists in this model, with energy $E_0 = \frac{2\sqrt{2}}{3}$. We also have solutions representing alternating kink–anti-kink solutions. If the kinks and anti-kinks are sufficiently far away, the energy of the solution with a total number of kinks + anti-kinks $= N$ is roughly $NE_0$. To see how important they are thermodynamically, let us estimate their contributions to the partition function with the assumption that they can be treated as a dilute gas of non-interacting particles.

For a single kink and a pair of kink–anti-kink solutions, their contributions to the partition function are ($Z_0 = 1$)

$$Z_1 \sim 2\int_0^L \mathrm{d}x e^{-\beta E_0} \quad \text{and} \quad Z_2 \sim 2\int_0^L \mathrm{d}x \int_0^x \mathrm{d}y e^{-\beta 2E_0}, \tag{8.11a}$$

respectively. The integrals come from the fact that the available number of states for a single-kink solution is proportional to $\int_0^L \mathrm{d}x$ and is proportional to $\int_0^L \mathrm{d}x \int_0^x \mathrm{d}y$ for a kink–anti-kink solution. The factor of two comes because we can start with either a kink or an anti-kink on the left.

We can construct $Z_3$, $Z_4$ and so on similarly. After summing up all multiple kink–anti-kink contributions, we obtain

$$Z_{\text{Kink}} \sim \sum_N Z_N \sim 2\sum_N \left[ \int_0^{x_2} \mathrm{d}x_1 \int_0^{x_3} \mathrm{d}x_2 \cdots \int_0^L \mathrm{d}x_N \right] e^{-\beta N E_0}$$

$$= 2\sum_N \frac{1}{N!} \left( L e^{-\beta E_0} \right)^N = 2e^{\Delta L}, \tag{8.11b}$$

where $L$ = length of the system and $\Delta = e^{-\beta E_0}$.

The importance of the kink–anti-kink solutions can be seen by evaluating the average number of kinks + anti-kinks present in the system at temperature $T$, which is

$$\langle N_{\text{Kink}} \rangle = \frac{\sum_N \frac{N}{(N)!} \left( L e^{-\beta E_0} \right)^N}{Z_{\text{Kink}}} = L e^{-\beta E_0}. \tag{8.11c}$$

Note that $\langle N_{\text{kink}} \rangle > 1$ as long as $k_B T > \frac{E_0}{\ln L}$, implying that kink contributions to the free energy are important at any finite temperature when $L \to \infty$.

Recall that the classical ground state of the system is two-fold degenerate and the system is expected to be in one of the two-fold states ($\phi = \pm 1$) at temperature $T = 0$. This is an example of broken symmetry discussed in the last chapter where in principle the $\phi = \pm 1$ states should contribute equally to the partition function but the physical system is 'trapped' at only one of the two possible states at low temperature.

However, the system is no longer in one of the two degenerate ground states when kinks exist. The existence of a large number of kinks in the system implies that solutions with the four different boundary conditions all contribute equally to the partition function and broken symmetry is restored at any temperature $T > 0$ when $L \rightarrow \infty$.

The absence of broken symmetry can also be checked by direct computation of the correlation function $\langle \phi(x)\phi(y) \rangle$. At $T = 0$ where $\phi(x) = \pm 1$, $\langle \phi(x)\phi(y) \rangle = \langle \pm 1 . \pm 1 \rangle = 1$. At finite temperature we may compute the correlation approximately noting that $\phi(x)$ changes sign when passing through a kink (anti-kink). Performing a similar analysis as before, we obtain $\langle \phi(x)\phi(y) \rangle \sim e^{-\Delta|x-y|}$. Note that the correlation function decays exponentially at any finite temperature $T$.

**Homework Q2a**
Show that $\langle \phi(x)\phi(y) \rangle \sim e^{-\Delta|x-y|}$ in the 'non-interacting-kink' model.

The above conclusion is common to any classical system with two-fold-degenerate ground states and with finite-energy, kink-like solutions joining the two ground states. All these systems have no long-range order at temperatures $k_B T > \frac{E_0}{\ln L}$. A well-known example is the 1D Ising model where we know that the system is disordered at any finite temperature (see figure below). This is an example of the Mermin–Wagner theorem which says that one-dimensional classical systems with discrete symmetry are always disordered at any finite temperature $T > 0$.

*Kink in Ising model*

**Homework Q2b**
Evaluate the correlation function $\langle S_i S_j \rangle$ for the 1D 'Ising model' and show that it has the same form as $\langle \phi(x)\phi(y) \rangle$ discussed here.

The fact that instantons appear as classical solutions joining different degenerate ground-state configurations of a field theory system means that instantons are very effective in restoring symmetries of an *otherwise symmetry-broken ground state*. As long as the average instanton number in a system is non-zero, the system is *not* in a state with broken symmetry. There are lots of examples of instantons that join different ground-state configurations and restore broken symmetry in condensed matter physics. We shall see more examples later.

**Question**
Can plane-wave-like (perturbative) solutions upon the $\phi = \pm 1$ states restore the broken symmetry of the system? Why? If your answer is yes, can you construct an

(approximate) criterion as to when will the perturbative solutions restore broken symmetry?

#### 8.2.1.1 Instantons in Quantum Mechanics: Quantum Tunneling

Let us consider the quantum mechanics of a particle moving in a double-well potential $V(x) = \frac{1}{4}(x^2-1)^2$. The action for this system is

$$S(t_2, t_1) = \int_{t_1}^{t_2} dt \left[ \frac{1}{2c^2} \left( \frac{\partial x}{\partial t} \right)^2 - \frac{1}{4}(x^2-1)^2 \right], \tag{8.12}$$

and the Green's function measuring the probability of the particle traveling from point $x'$ (at time $t_1$) to point $x$ (at time $t_2$) is $\langle x | U(t_2, t_1) | x' \rangle = \int Dx(t) e^{\frac{i}{\hbar} S(t_2, t_1)}$.

A common approximation to evaluate the Green's function is the saddle-point approximation or mean-field approximation, where we approximate the path integral by summing over paths close to the classical paths satisfying $\delta S = 0$. (We shall see how this is done in practice in the next few chapters.)

This approach has a problem at low energy, because classically the particle is located at either $x = 1$ or $x = -1$ points in the ground state. In fact, at low energy $E < {}^1/_4$, there is no classical trajectory that allows the particle to go from one side of the potential well to another. The particle is not aware of the existence of the other potential well at all! Therefore, the particle cannot travel to the other side of the potential well in the saddle-point approximation.

However, we know that this result is wrong for a quantum particle in a double well! Because of quantum tunneling, the particles will have equal probabilities of spending their times in the two wells even in the ground state! The possibility of *quantum tunneling* does not appear in classical theory at low energy and a better approximation that goes beyond the saddle-point approximation has to be used if we want to include quantum tunneling events.

Magically, the situation changes completely if we go to imaginary time, that is if we consider the partition function

$$Z = \int Dx(\tau) e^{-S(x(\tau))},$$

where

$$S(x(\tau)) = \int_0^{\beta} d\tau \left[ \frac{1}{2c^2} \left( \frac{\partial x}{\partial \tau} \right)^2 + \frac{1}{4}(x^2-1)^2 \right]. \tag{8.13}$$

Note that the partition function of a quantum particle in a double-well potential is almost the same as the partition function for the classical $\phi^4$ model, except that $L_\phi \to \beta/\hbar$ and $\beta_\phi \to 1$. In particular, the equation of motion $\delta S = 0$ has a kink solution at imaginary time! Mathematically, $V(x) \to -V(x)$ in the equation of motion when we go to imaginary time. In particular, there is no (low-energy) classical solution for Newton's equation with $V(x) = \frac{1}{4}(x^2-1)^2$ that connects points $x = \pm 1$, but there are solutions with $V(x) = -\frac{1}{4}(x^2-1)^2$. (see figure next page)

(No) (Yes)

Using what we have learnt in the $\phi^4$ model, we can evaluate the partition function including only instanton solutions. Borrowing Equation (8.11), we obtain

$$Z_{\text{Kink}} \sim e^{\Delta L} \rightarrow Z_{\text{DW}} \sim e^{\Delta\beta}, \tag{8.14}$$

where $\Delta = e^{-E_0}$.

**Homework Q3a**
Derive the above results.

At low temperature $T \rightarrow 0$, we expect that $Z \rightarrow e^{-\beta E_G}$. This result implies that the ground-state energy of the double-well problem is $E_G \sim -e^{-E_0}$. Note that $E_G = 0$ classically. The lowering of the energy to $E_G \sim -e^{-E_0}$ is a result of quantum tunneling.

For example, for a two-level system linked by a tunneling term $t$, with

$$H = 0[|1\rangle\langle 1| + |2\rangle\langle 2|] + t[|1\rangle\langle 2| + |2\rangle\langle 1|],$$

the ground-state energy and wavefunction are $E_G = -t$ and $|G\rangle = \frac{1}{\sqrt{2}}(|1\rangle + |2\rangle)$, respectively. The classical energy (0) is lowered by quantum tunneling.

Therefore, *instantons in quantum systems represent quantum tunneling events*! The effective tunneling matrix element is $t \sim e^{-E_0}$ in the double-well problem.

**Homework Q3b**
Solve the Schrödinger equation for the double-well problem in the WKB approximation where we approximate $\psi \sim \exp\left[i \int^x k(x)dx\right]$. Show that it reproduces essentially the instanton result if we allow $k(x)$ to be an imaginary number. The semiclassical approximation corresponds to keeping only real $k(x)$.

The WKB approximation and instanton analysis give essentially identical results in the double-well problem. However, the instanton approach has the advantage that it can be applied to quantum field theories (QFTs) more readily compared with the WKB approximation. We shall see more examples later.

### 8.2.2 Winding Number

Another interesting example of instanton application in quantum mechanics is the rotor problem. In this problem, we consider a unit vector fixed at the origin and restricted to rotate only in the $x$–$y$ plane. The action for this model is

$$S = \int_{t_i}^{t_f} dt \frac{m(\dot{\vec{r}})^2}{2} = \frac{I}{2}\int_{t_i}^{t_f} dt \dot{\vartheta}^2, \tag{8.15}$$

where $\vec{r} = r\hat{r}$ and $\dot{\vec{r}} = r\dot{\vartheta}\,\hat{\vartheta}$ in plane coordinates. $I = mr^2$ is the moment of inertia. The eigenenergies of this problem are $\varepsilon_n = \frac{n^2}{2I}$ and are quantized (we set $\hbar = 1$ in the following). The question is: how can we obtain this spectrum from the path integral approach?

We shall evaluate the partition function for this problem using the imaginary-time path integral, with techniques we learnt in Chapter 4. To see the importance of winding number, first we evaluate the partition function in an 'incorrect' way.

Naïvely, the partition function is

$$Z_0 = \int D\vartheta\, e^{-\frac{I}{2}\int_0^\beta \dot{\vartheta}^2 d\tau}, \tag{8.16a}$$

with periodic boundary condition

$$\vartheta(0) = \vartheta(\beta). \tag{8.16b}$$

Can you see what is wrong? In fact, we can see that the above result is wrong without actually evaluating the path integral. If we replace $\vartheta$ by $x$, the coordinate of a particle, and $I \to m$ (mass), we see that the above action is the action for a free particle moving in one dimension. The energy spectrum is given by $\varepsilon_k = \frac{k^2}{2m}$ and is not quantized!

**Exercise**

Show that

$$Z_0 = \int \frac{dk}{2\pi} e^{-\beta\frac{k^2}{2I}} = \sqrt{\frac{I}{2\pi\beta}}.$$

For a free particle moving in one dimension, the energy is quantized if the one-dimensional world is 'closed', that is if we impose periodic boundary conditions that the two points $x = 0$ and $x = L$ are identical points in the one-dimensional universe. The same holds for the rotor model. The angles $\vartheta$ and $\vartheta + 2n\pi$ ($n$ = arbitrary integer) in fact represent the same points in the $\vartheta$ space. The question is: how can we incorporate this condition in the path integral approach?

The answer is that we should modify the boundary condition $\vartheta(\beta) = \vartheta(0)$. In fact, the more general condition that is allowed should be $\vartheta(\beta) = \vartheta(0) + 2n\pi$ for arbitrary $n$ because $\vartheta$ and $\vartheta + 2n\pi$ represent the same points in the $\vartheta$ space. The $n \neq 0$ configuration represents an instanton solution of the problem. $n$ is called the winding number because it describes how many times $\vartheta$ has wound up between $\tau = 0$ and $\tau = \beta$. To include the winding number contributions in the path integral, we can reparameterize the $\vartheta$ field as

$$\vartheta(\tau) = \vartheta'(\tau) + \frac{2\pi n}{\beta}\tau,$$

where

$$\vartheta'(\beta) = \vartheta'(0), \tag{8.17}$$

and sum over all possible values of $n$ in the path integral, that is we obtain

$$Z \to \sum_n Z_n = \sum_n \int D\vartheta' e^{-\frac{I}{2}\int_0^\beta \dot{\vartheta}'^2 d\tau - \frac{I}{2\beta}(2\pi n)^2} = Z_0 \sum_n e^{-\frac{2\pi^2 I}{\beta} n^2} \tag{8.18}$$

with periodic boundary condition $\vartheta'(\beta) = \vartheta'(0)$. The different values of $n$ represent different topological sectors (winding numbers) in the path integral.

The sum $\sum_n e^{-\alpha n^2}$ can be evaluated by using the mathematical identity $\sum_n \delta(n-x) = \sum_m e^{2\pi i m x}$, where

$$\sum_n e^{-\alpha n^2} = \sum_n \int e^{-\alpha x^2} \delta(x-n) dx = \sum_m \int d\, x e^{2\pi i m x} e^{-\alpha x^2} = \sqrt{\frac{\pi}{\alpha}} \sum_m e^{-\frac{\pi^2}{\alpha} m^2}. \tag{8.19}$$

This is an example of duality transformation; the number field $m$ is the dual field of $n$.

Putting $\alpha = \frac{2\pi^2 I}{\beta}$ into the expression for $Z$, we obtain

$$Z = \sum_m e^{-\beta \frac{m^2}{2I}} = \sum_m e^{-\beta \varepsilon_m}, \tag{8.20}$$

where $\varepsilon_m = \frac{m^2}{2I}$ are exactly the eigenenergies for the rotor model.

**A Comment: Relation between Quantum Field Theory and Statistical Mechanics**
You may be astonished why the simple $\phi^4$ model we considered can be applied to so many different problems. In particular, the instanton analysis indicates that a one-dimensional classical system is equivalent to a zero-dimensional quantum system. This correspondence between classical and quantum systems is not accidental and the equivalence between a '*d*'-dimensional quantum system at imaginary time and a '(*d* + 1)'-dimensional classical system can be proved rigorously in many models. That is why we often use the notation of (*d* + 1)-dimensional quantum field theory. '*d*' is the spatial dimension and '1' is time, which often acts as an extra dimension in a corresponding classical system. This is particularly true for Lorentz-invariant systems where space and time are symmetric. You will see many other examples when you explore quantum field theory further.

## 8.3 Vortices and Kosterlitz–Thouless Transition

So far, we used mainly an example of kink solutions in the 1D system to illustrate the use of solitons and instantons in QFT. With this simple example we already observe that the properties of soliton/instanton solutions are often strongly dimension dependent. A kink solution in the 2D $\phi^4$ model will cause an energy $\sim L =$ length of the system and costs infinite energy to form in the thermodynamic limit $L \to \infty$. As a result, kink solutions do not contribute significantly to the path integral or partition function in dimensions larger than one. (Note that, on the other hand, a local 'hill-like' solution where $\phi(\vec{x} \to \infty) = \pm 1$ but $\phi(\vec{x} \to 0) = \mp 1$ as shown in the figure below is energetically favorable but is not protected by topology and can be destroyed easily.)

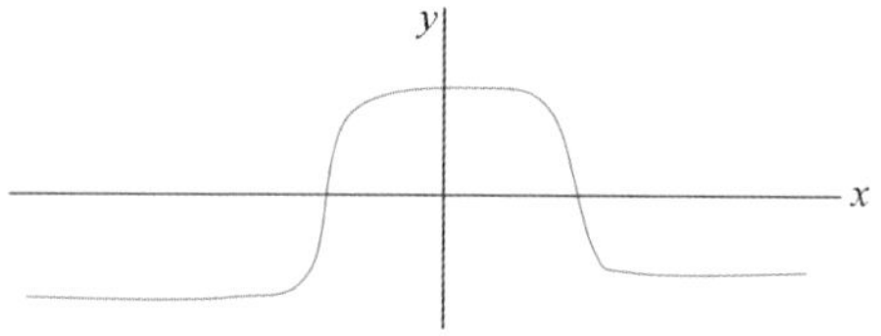


Cross-section of a "hill-like" solution in 2D $\phi^4$ -model

In the remaining part of this chapter we shall introduce two more examples of 'important' soliton or instanton solutions which live at dimensions >1. We shall first study the important topic of vortices and the Kosterlitz–Thouless transition in two dimensions in this section.

Vortices are instanton solutions of the planar classical Heisenberg model (or $X$–$Y$ model) with Hamiltonian $H = -J\sum_{\langle i,j \rangle} \vec{S}_i\vec{S}_j$, where $J>0$ and $\langle i, j\rangle$ are nearest neighbors on a lattice (which you can assume to be a square lattice for simplicity). The $\vec{S}_i$ are unit vectors located on lattice points which are restricted to lie on a plane (see figure below).

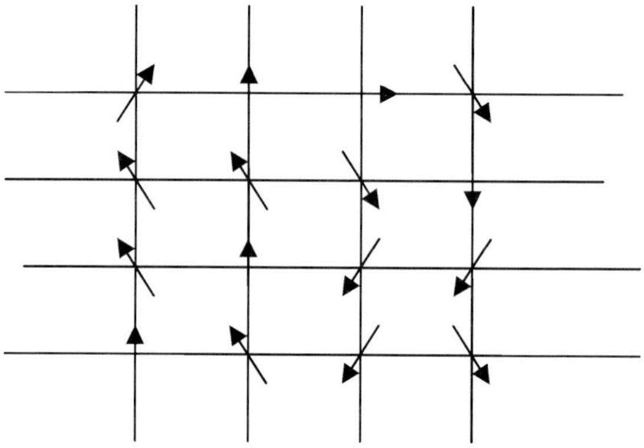

We can characterize the spins by their angle $\vartheta$ with respect to the $x$ axis, that is $\vec{S} = (\cos\vartheta, \sin\vartheta, 0)$, and the Heisenberg model becomes

$$H[\vartheta] = -J\sum_{\langle i,j \rangle} \cos(\vartheta_i - \vartheta_j). \tag{8.21a}$$

In the limit $J \gg k_B T$, we expect that the adjacent spins will be almost pointing parallel to each other and the discrete $\vartheta_i$ can be approximated by a continuous function, that is

$$\vartheta_i \sim \vartheta(\vec{x}_i) \quad \text{and} \quad \vartheta_{i+\delta} - \vartheta_i \sim \vec{\delta}.\nabla\vartheta(\vec{x}_i).$$

Making this replacement, we obtain the continuum Heisenberg model

$$H[\vartheta] \to -\frac{J}{a^2}\int d^2x\left[1-\frac{a^2}{2}(\nabla\vartheta)^2\right], \tag{8.21b}$$

where $a \sim$ lattice spacing. We assume that $\nabla\vartheta$ is small and keep only terms up to second order in $\cos(\vec{\delta}.\nabla\vartheta)$ in Equation (8.21b). We shall also neglect the constant term $\sim -\frac{J}{a^2}\int d^2x[1]$ in the following.

Minimizing $H[\vartheta]$ with respect to $\vartheta$, we obtain the equation of motion

$$\nabla.\nabla\vartheta = 0, \tag{8.22a}$$

which determines the stable classical configurations in the model.

To study the solutions to Equation (8.22a), we first examine general 2D vector functions $\vec{A} = (A_x, A_y)$ with $\nabla \times \vec{A} = 0$. For smooth functions, $\nabla \times \vec{A} = 0$ implies that $\vec{A} = \nabla\theta$. However, we have seen in Chapter 6 that in two dimensions there exists a singular function $\vec{A}(r,\theta) = \frac{1}{r}\widehat{\theta}$ which has the unusual property that $\nabla \times \vec{A} = 0$ is zero everywhere except at $r \to 0$, where

$$\nabla \times \vec{A} = 2\pi\delta^{(2)}(r)\,\widehat{z}, \tag{8.22b}$$

or $\oint \vec{A}.d\vec{l} = 2\pi$ around $r=0$. We also found that these functions can be expressed as $\vec{A} = \nabla\vartheta$, if we allowed the $\vartheta$ field to be multi-valued.

Interestingly, a multi-valued $\vartheta$ field is physically allowed in the Heisenberg model since the $\vartheta$ represent angles, and $\vartheta$ and $\vartheta + 2n\pi$ represent the same point in $\vartheta$ space. For example, we may imagine a $\vartheta$-field configuration $\vartheta(r,\theta) = \pm\theta + \theta_0$, where $\theta_0$ is any constant angle. In this case $\nabla\vartheta = \pm\frac{1}{r}\widehat{\theta}$ and $\nabla.\nabla\vartheta = 0$ except at $r=0$, where $\nabla \times \nabla\vartheta = \pm 2\pi\delta^{(2)}(r)$. This is an example of a vortex (anti-vortex) solution. The corresponding spin configuration of a vortex solution is shown in the figure below.

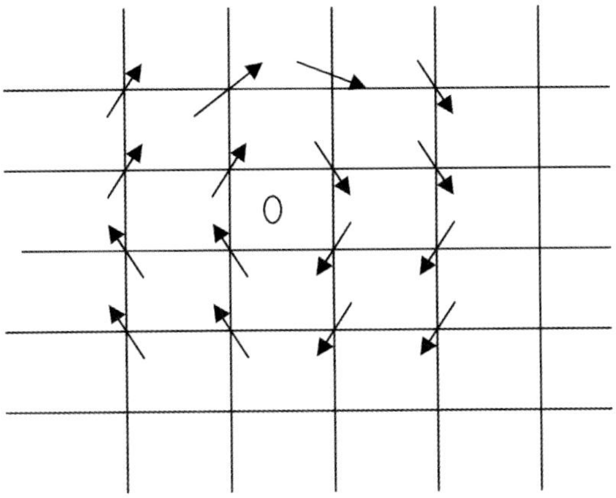


Note the existence of the singular point $r=0$ (center of vortex) where the value of $\vartheta$ is not well defined. However, this is not a problem in the lattice model where the singularity at $r \to 0$ is not occupied by any physical spin. In other words, the lattice provides a cutoff where the mathematical singularity is avoided.

The energy of a vortex is

$$E = \frac{J}{2}\int d^2x(\nabla\vartheta)^2 \sim \frac{J}{2}(2\pi)\int_a^L r\mathrm{d}r\left(\frac{1}{r}\right)^2 = \pi J \ln\left(\frac{L}{a}\right), \tag{8.23a}$$

which diverges logarithmically as the size of the system $L$. Note that the short-distance divergence is cut off by the lattice spacing $a$.

You may wonder whether vortex solutions are important at all in evaluating the partition function of the 2D Heisenberg model because of the logarithmic divergence in energy. Vortex solutions are important because of two reasons: first of all, they are topologically stable. Once a vortex is formed, it is very difficult to destroy. The topology is reflected in the winding number around a vortex, where the angle $\vartheta$ changes by $\pm 2\pi$ (winding number $= \pm 1$) when we go around a vortex (anti-vortex) (see figure below).

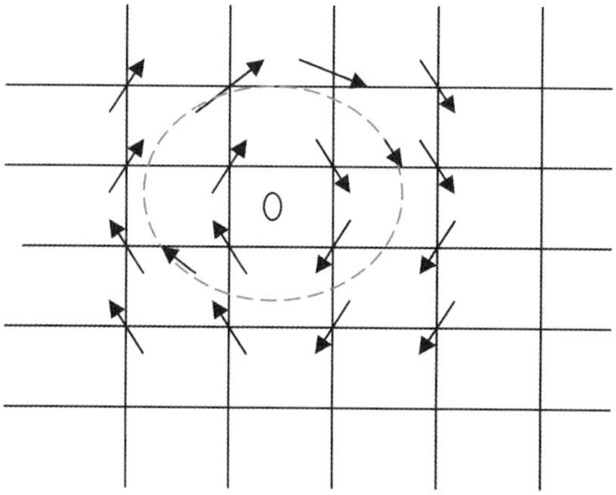


Secondly, we note that the available phase space for a vortex solution is of order $\sim(L/a)^2$. The associated entropy is therefore of order $2\ln(L/a)^2$. The free energy of a single vortex is thus of order

$$F = E - TS \sim \pi J \ln\left[\frac{L}{a}\right] - 2Tk_B \ln\left[\frac{L}{a}\right] = (\pi J - 2k_B T)\ln\left[\frac{L}{a}\right], \tag{8.23b}$$

and becomes negative when $T > \pi J/2k_B \sim T_{KT}$. Thus, we expect that there may exist a phase transition where vortex solutions are unimportant at temperatures $T < T_{KT}$ but become important at $T > T_{KT}$. Such a phase transition is found to exist and is called the Kosterlitz–Thouless (KT) transition nowadays. Before we discuss the KT transition, we first discuss the low-temperature phase where vortices are unimportant.

## 8.3.1
## Low-Temperature Spin–Spin Correlation Function

We examine the spin–spin correlation function in the Heisenberg model at $T < T_{KT}$ where vortices are unimportant. In this case, we can forget about the singular configurations in the $\vartheta$ field and assume that $\vartheta(\vec{x})$ is an everywhere smooth function.

In this case, the spin–spin correlation function is [3]

$$\langle \vec{S}(\vec{x}).\vec{S}(\vec{y})\rangle \sim \langle e^{i\vartheta(\vec{x})}e^{i\vartheta(\vec{y})}\rangle = \frac{1}{Z}\int D\vartheta e^{i\vartheta(\vec{x})}e^{i\vartheta(\vec{y})}e^{-\frac{J}{2k_B T}\int d^2x(\nabla\vartheta)^2}, \tag{8.24}$$

where $Z = \int D\vartheta e^{-\frac{J}{2k_B T}\int d^2x(\nabla\vartheta)^2}$.

Introducing $\vartheta(\vec{q}) = \int d^2 x e^{i\vec{q}.\vec{x}}\vartheta(\vec{x})$, we can write

$$\begin{aligned}\langle \vec{S}(\vec{x}).\vec{S}(\vec{y})\rangle &= \frac{1}{Z}\prod_{\vec{q}}\int d\vartheta_{\vec{q}} e^{-\frac{J}{2k_B T}\vec{q}^2\vartheta(\vec{q})\vartheta(-\vec{q})+i\vartheta(\vec{q})(e^{i\vec{q}.\vec{x}}-e^{i\vec{q}.\vec{y}})}\\ &= \prod_{\vec{q}>0}\frac{1}{Z_{\vec{q}}}\int d\vartheta_{\vec{q}}\int d\vartheta_{-q} e^{-\frac{J}{2k_B T}\vec{q}^2\vartheta(\vec{q})\vartheta(-\vec{q})+i[\vartheta(\vec{q})(e^{i\vec{q}.\vec{x}}-e^{i\vec{q}.\vec{y}})+(\vec{q}\to-\vec{q})]}\\ &= e^{-\frac{k_B T}{J}\int\frac{d^2q}{(2\pi)^2}\frac{(1-\cos[\vec{q}.(\vec{x}-\vec{y})])}{q^2}} \sim \frac{1}{|\vec{x}-\vec{y}|^{\frac{k_B T}{2\pi J}}}.\end{aligned} \tag{8.25}$$

We find that the spin–spin correlation function decays as a power law at any finite temperature $T>0$, indicating that the system is in a critical state and true long-range magnetic order exists only at $T=0$. This is an example of the Mermin–Wagner theorem in 2D, which says that a classical system with continuous symmetry cannot develop true long-range order except at temperature $T=0$.

### 8.3.2
### Kosterlitz–Thouless Transition

To describe the KT transition we first have to derive a free-energy expression for vortices. The derivation starts from the assumption that after taking into account the vortex contributions, the vector field $\nabla\vartheta$ can be written as a sum of two parts, a regular part $\nabla\vartheta'$, where $\vartheta'$ is a single-valued function, and a singular part which represents vortices (or configurations with non-zero winding number), that is

$$\nabla\vartheta = \nabla\vartheta' + \vec{A}, \tag{8.26}$$

where $\nabla.\vec{A} = 0$ and $\nabla\times\vec{A} = \rho(\vec{r})\hat{z} = 2\pi\sum_i q_i\delta^{(2)}(\vec{r}-\vec{r}_i)\hat{z}$, where $\vec{r}_i$ are the locations of the vortices and $q_i = \pm 1$ is the vorticity (winding number) of the vortices. The energy of the system is $H[\vartheta] = \frac{J}{2}\int d^2x(\nabla\vartheta)^2$. To proceed further, we introduce a fictitious $\vec{j}$ field (a duality transformation) and write [3]

$$\begin{aligned}Z[\vartheta] &= \int D\vartheta e^{-\frac{J}{2k_B T}\int d^2x(\nabla\vartheta)^2} = Z\left[\vartheta,\vec{j}\right]\\ &= \int D\vec{A}\int D\vartheta'\int D\vec{j} e^{-\frac{1}{2}\int d^2x\left[2i\vec{j}.(\nabla\vartheta'+\vec{A})+\frac{k_B T}{J}(\vec{j})^2\right]}.\end{aligned} \tag{8.27a}$$

**Homework Q4a**

Show that, up to a constant factor, $Z[\vartheta,\vec{j}] \to Z[\vartheta]$ after integrating out the $\vec{j}$ field.

The $\vartheta'$ field can be eliminated by integration by parts, where we obtain

$$\int d^2x\vec{j}.\nabla\vartheta' \to -\int d^2x\vartheta'\nabla.\vec{j}. \tag{8.27b}$$

Integration over the $\vartheta'$ field results in the constraint $\nabla.\vec{j}=0$.

In 2D, the constraint $\nabla.\vec{j}=0$ can be satisfied by letting

$$j_x = \frac{\partial a_0}{\partial y}, \quad j_y = -\frac{\partial a_0}{\partial x}, \quad \text{or} \quad \vec{j} = \nabla \times (a_0\,\widehat{z}\,). \tag{8.27c}$$

With this transformation, the partition function becomes

$$Z\left[\vartheta,\vec{j}\right] \to Z\left[\vec{A},a_0\right] = \int d\vec{A}\int da_0 e^{-\frac{1}{2}\int d^2x\left[2i\nabla\times(a_0\widehat{z}).\vec{A}) + \frac{k_BT}{J}(\nabla a_0)^2\right]}. \tag{8.27d}$$

Integrating by parts again, we obtain

$$\int d^2x\nabla\times(a_0\,\widehat{z}\,).\vec{A} \to -\int d^2xa_0\,\widehat{z}\,.\nabla\times\vec{A} = -\int d^2xa_0(\vec{r})\rho(\vec{r}). \tag{8.27e}$$

Therefore,

$$Z\left[\vec{A},a_0\right] \to Z[\rho,a_0] = \sum_{\{q_i\}}\int da_0 e^{-\frac{1}{2}\int d^2x\left[-2ia_0\rho + \frac{k_BT}{J}(\nabla a_0)^2\right]}. \tag{8.27f}$$

The $a_0$ field can be integrated out, to obtain

$$Z[\{q_i\}] = \sum_{\{q_i\}} e^{-\frac{\pi J}{2k_BT}\sum_{i,j}q_iq_j\ln\left|\frac{\vec{r}_i-\vec{r}}{aj}\right|}, \tag{8.28}$$

where we have used $\rho(\vec{r}) = 2\pi\sum_i q_i\delta^{(2)}(\vec{r}-\vec{r}_i)$. Equation (8.28) is the partition function of a collection of charges $\{q_i\}$ interacting with each other through the (2D) Coulomb potential $\ln\left(\frac{r}{a}\right)$.

**Homework Q4b**

Derive Equation (8.28) from Equation (8.27f).

There is a subtlety here. The derived energy function $\sum_{i,j} q_iq_j\ln\left(\frac{|\vec{r}_i-\vec{r}_j|}{a}\right)$ includes contributions from $i=j$, which are infinity formally. This divergence is an unphysical result coming from replacing our lattice model by an effective field theory (continuum) model. The $i=j$ term corresponds to the energy needed to create a vortex core, which is the region where the continuum approximation $\vartheta_{i+\delta}-\vartheta_i \sim \vec{\delta}.\nabla\vartheta(\vec{x}_i)$ breaks down. Physically, the energy for creating a single vortex must be finite in the original classical lattice model. To cure the problem, we simply replace

$$\frac{\pi J}{2}\sum_{i,j} q_iq_j\ln\left(\frac{|\vec{r}_i-\vec{r}_j|}{a}\right) \to \frac{\pi J}{2}\sum_{i\neq j} q_iq_j\ln\left(\frac{|\vec{r}_i-\vec{r}_j|}{a}\right) + \sum_i \mu_0 q_i^2, \tag{8.29a}$$

where $\mu_0 \sim J$ is the energy needed to create the vortex core and

$$Z[\{q_i\}] \rightarrow \sum_{\{q_i\}} e^{-\frac{\pi J}{2k_B T}\sum_{i\neq j} q_i q_j \ln|\frac{\vec{r}_i - \vec{r}}{aj}| + \frac{\mu}{k_B T}\sum_i q_i^2}, \tag{8.29b}$$

correspondingly.

To study the KT transition, we shall work with the Hamiltonian before the $a_0$ field is eliminated, that is

$$\frac{H}{k_B T} = \int d^d x \left( \frac{\mu_0}{k_B T} n(\vec{x}) - 2\pi i a_0(\vec{x})\rho(\vec{x}) + \frac{k_B T}{2J}(\nabla a_0(\vec{x}))^2 \right), \tag{8.30a}$$

where

$$\begin{aligned} n(\vec{x}) &= \sum_i \delta^{(2)}(\vec{r} - \vec{r}_i) = n_+(\vec{x}) + n_-(\vec{x}), \\ \rho(\vec{x}) &= \sum_i q_i \delta^{(2)}(\vec{r} - \vec{r}_i) = n_+(\vec{x}) - n_-(\vec{x}). \end{aligned} \tag{8.30b}$$

## 8.3.3
## Screening and KT Transition

The KT transition will be studied using a mean-field theory [4] in the following. The idea of mean-field theory is the same as Landau's theory for second-order phase transitions. We replace the exact partition function (8.30) by a mean-field free energy

$$F_{\text{eff}} = \int d^d x \left( \mu_0 \langle n(\vec{x}) \rangle - 2\pi i a_0(\vec{x}) k_B T \langle \rho(\vec{x}) \rangle + \frac{(k_B T)^2}{2J}(\nabla a_0(\vec{x}))^2 \right), \tag{8.31}$$

where $\langle n(\vec{x}) \rangle$ and $\langle \rho(\vec{x}) \rangle$ are (ensemble) averaged density and charge density of vortices, respectively. $a_0(\vec{x})$ is determined by minimizing $F_{\text{eff}}$ with respect to itself.

The average densities are determined by assuming an effective one-particle Hamiltonian

$$\mu^{\pm}_{\text{eff}}(\vec{x}) = \mu_{\text{eff}} \mp 2\pi i k_B T a_0(\vec{x}) \tag{8.32a}$$

for the vortices, where

$$\langle n_{\pm}(\vec{x}) \rangle = a^{-2} e^{-\frac{\mu^{\pm}_{\text{eff}}(\vec{x})}{k_B T}}. \tag{8.32b}$$

For a neutral system the average densities of vortices and anti-vortices are the same, that is $\langle \rho(\vec{x}) \rangle = 0$. Minimizing the free energy with respect to $a_0(\vec{x})$, we obtain $a_0(\vec{x}) = 0$ and a self-consistent solution

$$\langle n(\vec{x}) \rangle = n_0 = 2a^{-2} e^{-\frac{\mu_{\text{eff}}}{k_B T}} \tag{8.33}$$

as long as $\mu_{\text{eff}}$ is finite.

To determine $\mu_{\text{eff}}$, we consider the energy of adding a vortex with charge (vorticity) $Q = \pm 1$ to the system at $\vec{x} = 0$. In this case, the additional mean-field free energy is

$$F_{\text{eff}}(Q) = \mu_0 + \int d^d x \left( -2\pi i \delta a_0(\vec{x}) k_B T(\delta\langle\rho(\vec{x})\rangle + Q\delta(\vec{x})) + \frac{(k_B T)^2}{2J}(\nabla \delta a_0(\vec{x}))^2 \right), \tag{8.34}$$

where $\delta a_0$ and $\delta\langle\rho\rangle$ are the changes induced by the additional charge. The free-energy expression can be simplified by using Equation (8.32b), where we obtain

$$\begin{aligned} \delta\langle\rho(\vec{x})\rangle &= a^{-2} e^{-\frac{\mu^{+}_{\text{eff}}(\vec{x})}{k_B T}} - a^{-2} e^{-\frac{\mu^{-}_{\text{eff}}(\vec{x})}{k_B T}} \\ &= a^{-2} e^{-\frac{\mu_{\text{eff}}}{k_B T}} \left(e^{2\pi i \delta a_0(\vec{x})} - e^{-2\pi i \delta a_0(\vec{x})}\right) \sim 2\pi i n_0 \delta a_0(\vec{x}). \end{aligned} \tag{8.35a}$$

The last result is obtained by assuming that $\delta a_0(\vec{x})$ is small. This is justified since the presence of one additional charge (vortex) should have a very small effect on a macroscopic system with a finite density of charges.

Substituting this into Equation (8.34), we obtain

$$F_{\text{eff}}(Q) \to \mu_0 + k_B T \int d^d x \left( -2\pi i \delta a_0(\vec{x}) Q \delta(\vec{x}) + n_0 (2\pi \delta a_0(\vec{x}))^2 + \frac{k_B T}{2J}(\nabla \delta a_0(\vec{x}))^2 \right). \tag{8.35b}$$

Minimizing the energy with respect to $\delta a_0(\vec{x})$, we obtain

$$-\frac{k_B T}{J} \nabla^2 \delta a_0(\vec{x}) + 8\pi^2 n_0 \delta a_0(\vec{x}) = 2\pi i Q \delta(\vec{x}), \tag{8.35c}$$

which is the (2D) Poisson–Boltzmann equation. The equation can be solved by Fourier transforming, and we obtain

$$\delta a_0(\vec{q}) = \frac{2\pi i Q}{\frac{k_B T}{J} q^2 + 8\pi^2 n_0}. \tag{8.36}$$

Putting this back into the free energy (8.35b), we find that the free-energy cost for the extra charge is, after Fourier transforming,

$$\begin{aligned} \mu_{\text{eff}} &= F_{\text{eff}}(Q) = \mu_0 + \frac{k_B T}{2} \int d^2 q \left( \frac{1}{\left(\left(\frac{k_B T}{J}\right) q^2 + 8\pi^2 n_0\right)} \right) \\ &\sim \mu_0 + \frac{\pi J}{2} \ln\left(1 + \frac{k_B T}{8\pi^2 J a^2 n_0}\right). \end{aligned} \tag{8.37}$$

**Homework Q5**

Derive the above expressions (8.36) and (8.37).

Note that $\mu_{\text{eff}}$ is a function of the vortex (anti vortex) density $n_0$. Equation (8.37) and Equation (8.33) form a self-consistent set of equations for $n_0$.

**Homework Q6**

Show that the solution for $n_0$ at small $x = \frac{8\pi^2 Ja^2 n_0}{k_B T}$ has the form $n_0 = 0$ for $T < T_c = \frac{\pi J}{2k_B}$

and

$n_0 \sim (e^{-\frac{\mu'}{k_B T}})^{\frac{T}{T-T_c}}$ for $T > T_c$. What is $\mu'$?

Physically, vortices are bound together at low temperature and are liberated at $T = T_c$ because of gain in entropy. This is the physics described in mean-field theory.

It should be noted that a vortex–anti-vortex pair separated by a distance $d$ has excitation energy $E_{\text{pair}} = 2\mu_0 + \frac{\pi J}{2} \ln \frac{d}{a}$ and has a high probability of being excited even at $T < T_c$ if $E_{\text{pair}} < k_B T$. As temperature increases larger numbers of vortex–anti-vortex pairs will be excited. The KT transition temperature describes a point where pairs of vortices overlap and the identity of individual 'pair' is lost. At this point a vortex is surrounded by a large number of vortices and anti-vortices and can no longer identify its 'partner', that is it becomes unbounded. Mathematically, the logarithmic potential of a single vortex is screened by a finite density of vortices and anti-vortices and vortices become free-particle like. Our simple mean-field theory captures the physics of screening above $T_c$ but cannot describe the vortex-pair excitations at $T < T_c$. This reflects a limitation of mean-field theory. A more accurate description of the vortex–anti-vortex system can be obtained if we use the more powerful method of renormalization groups. Interested readers can look at, for example, Refs. [3, 4] for details.

## 8.3.4
## Vortices in Superconductors and Superfluids

Besides the lattice X–Y model, vortices also exist in superconductors and superfluids. They are classical solutions obtained from minimizing the Ginsburg– Landau (GL) free energy in two dimensions,

$$F = \int d^2x \left\{ \frac{\hbar^2}{2m^*} |(\nabla - ie^* \vec{A})\phi|^2 + a|\phi|^2 + b|\phi|^4 \right\}, \tag{8.38}$$

where $\phi = \sqrt{\rho} e^{i\vartheta}$ is a complex scalar field that describes superfluidity or superconductivity. $e^* = 0$ for neutral superfluids but is non-zero for superconductors. A vortex solution centered at the origin is a solution of Equation (8.38) with $\phi(r, \vartheta) = f(r) e^{in\vartheta}$. The phase of $\phi$ has a winding number $\oint d\vec{l}.\nabla\vartheta = \pm 2n\pi$ for any integration path surrounding the origin. The singularity at $r \to 0$ is 'avoided' by having $f(r \to 0) \to 0$. We shall discuss vortices in superfluids in more detail in Chapter 11 when we discuss superconductivity.

Vortices are very important topological excitations in superfluids. They are observable in superconductor thin films and in superfluid $He^4$. Vortex lines and vortex rings are observable as solitons in 3D, and play a vital role in determining the properties of superfluids. They are also believed to be responsible for the so-called superfluid–insulator transition in superconductor films, which we shall discuss in Chapters 11 and 12.

## 8.4
## Skyrmions and Monopoles

In previous sections we discussed examples of solitons/instantons in scalar ($\phi^4$ model) and two-component vector fields (*X–Y*model). In this section we consider solitons/instantons in three-component vector fields. The first object we discuss is a soliton solution (called a skyrmion) in a 2 + 1D system with Lagrangian

$$L = \frac{1}{g}\int d^2x \left[\frac{1}{c^2}\left(\frac{\partial \vec{n}(\vec{x},t)}{\partial t}\right)^2 - (\nabla \vec{n}(\vec{x},t))^2\right], \tag{8.39a}$$

where $\vec{n}$ = unit vector. The model is called the non-linear-sigma model and is the continuum limit of the Heisenberg model. Correspondingly, it is also an instanton solution of the 2D sigma model

$$H = \frac{1}{g}\int d^2x (\nabla \vec{n}(\vec{x}))^2. \tag{8.39b}$$

The instanton solution describes (roughly) the following configuration. Imagine that $\vec{n} = \uparrow$ at the origin ($\vec{x} = 0$) and is changing continuously in a way such that $\vec{n} \to \downarrow$ when $\vec{x} \to \infty$ in all directions. This is consistent with the boundary condition that $\vec{n} = \vec{n}_0$ is a fixed vector when $\vec{x} \to \infty$ in all directions and ensures that the energy of the configuration does not diverge at $\vec{x} \to \infty$. The ground state has $\vec{n} = \vec{n}_0$ everywhere in space. (see figure below)

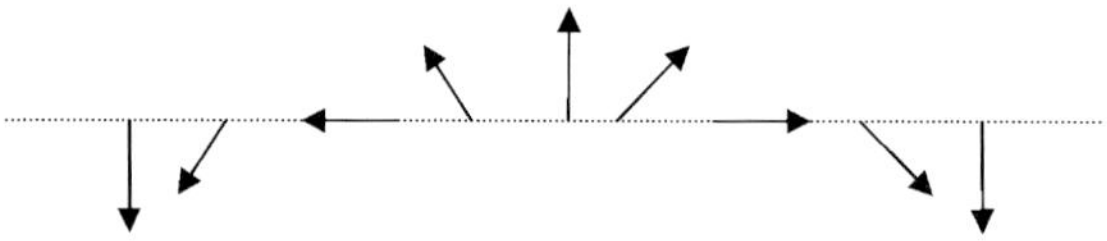

*Cross–sectional view of a skyrmion*

The existence and stability of a skyrmion solution are guaranteed by the topology of the solution. Since $\vec{n} = \vec{n}_0$ at all points at infinity, the physical coordinate plane $R_2$ can be compacted into a spherical surface, which we call $S_2$, with the origin on the two-dimensional plane mapping into the north pole of the spherical surface, whereas all points at infinity in the plane map into the south pole. Meanwhile, the phase space where the $\vec{n}$ field lives is also a spherical surface $S_2$. Thus, any continuous configuration $\vec{n}(\vec{x})$ can be represented by a mapping of $S_2$ into $S_2$. The ground-state solution $\vec{n} = \vec{n}_0$ is a trivial mapping where all points in physical space map into a single point in the $\vec{n}$-phase space. A skyrmion solution is one in which the physical space covers (or wraps around) all areas in the $\vec{n}$-phase space once. The skyrmion solution cannot be deformed continuously back to the ground-state solution without creating a discontinuity, just as we cannot destroy a kink solution without passing through an infinite-energy barrier in one dimension. In topology terms, of all plausible mappings from $S_2$ into $S_2$, skyrmion solutions belong to a topological sector different from the topological sector of the ground-state configurations.

Mappings from $S_2$ into $S_2$ can be classified by a topological number [1, 2]

$$Q = \frac{1}{8\pi} \sum_{\mu\nu=\widehat{x,y}} \int d^2x \varepsilon_{\mu\nu} \vec{n}(\vec{x}).(\partial_\mu \vec{n}(\vec{x}) \times \partial_\nu \vec{n}(\vec{x})), \tag{8.40a}$$

which measures the number of times the physical space covers (or wraps around) the $\vec{n}$ space and can only take integer values. $Q = 0$ for the ground state whereas $Q = 1$ for a single-skyrmion solution. The quantity

$$C(\vec{x}) = \varepsilon_{\mu\nu} \vec{n}(\vec{x}).(\partial_\mu \vec{n}(\vec{x}) \times \partial_\nu \vec{n}(\vec{x})) \tag{8.40b}$$

measures the infinitesimal solid angle sustained by three spins (or unit vectors) covering a small area $d^2x$ in the $x$–$y$ plane and is called the (local) chirality of the spin field [3]. (see figure below)

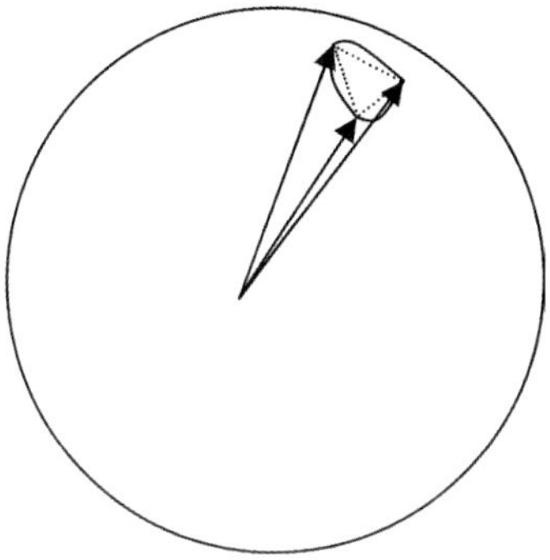

A soliton solution is obtained when the static skyrmion is changing (moving) in time according to the action (8.39a). We shall concentrate on the static instanton solution in the following.

It is not easy to solve Equation (8.39b) directly to obtain the soliton/instanton solutions because of the unit vector constraint $\vec{n}^2 = 1$. The constraint can be handled by introducing a Lagrange-multiplier term in the Hamiltonian

$$H \rightarrow \frac{1}{g} \int d^2x (\nabla \vec{n}(\vec{x}))^2 + \lambda(\vec{n}(\vec{x})^2 - 1), \tag{8.41a}$$

so that the equation of motion becomes

$$\nabla^2 \vec{n} + \lambda \vec{n} = 0. \tag{8.41b}$$

The $\lambda$ field can be eliminated by using the constraint $\lambda = \lambda \vec{n}^2 = -\vec{n}.\nabla^2 \vec{n}$. Putting it into Equation (8.41b), we obtain a non-linear equation

$$\nabla^2 \vec{n} - (\vec{n}.\nabla^2 \vec{n}) \vec{n} = 0. \tag{8.41c}$$

To obtain explicit skyrmion solutions for Equation (8.41c), we employ a trick invented by Belavin and Polyakov (see Refs. [1, 2]). They considered the identity

$$\int d^2x \left[ \partial_\mu \vec{n} \pm \sum_\nu \varepsilon_{\mu\nu} (\vec{n} \times \partial_\nu \vec{n}) \right]^2 \geq 0, \tag{8.42a}$$

which, upon expanding, becomes

$$\begin{aligned}&\int d^2x\left[(\partial_\mu\vec{n})^2+\sum_{\nu\nu'}\varepsilon_{\mu\nu}\varepsilon_{\mu\nu'}(\vec{n}\times\partial_\nu\vec{n})\cdot(\vec{n}\times\partial_{\nu'}\vec{n})\right]\\&\quad=2\int d^2x(\partial_\mu\vec{n})^2\geq\pm2\int d^2x\sum_\nu\varepsilon_{\mu\nu}\vec{n}\cdot(\partial_\mu\vec{n}\times\partial_\nu\vec{n}),\end{aligned}\tag{8.42b}$$

where we have used the identity $\vec{n}^2=1,\ \vec{n}.\partial_\nu\vec{n}=0$ and integration by parts to obtain the last result.

Thus, the energy of a solution of the system with topological number $Q$ must satisfy $E_Q\geq 4\pi|Q|$.

In particular, the energy is minimized if $E_Q=4\pi|Q|$, or when

$$\partial_\mu\vec{n}=\pm\sum_\nu\varepsilon_{\mu\nu}(\vec{n}\times\partial_\nu\vec{n}).\tag{8.42c}$$

**Homework Q7a**

Show that any configuration $\vec{n}$ that satisfies Equation (8.42c) is a solution of the field equation (8.41c).

Equation (8.42c) can be solved by introducing a complex function $w(z)$ ($z=x+iy$), which parameterizes the $\vec{n}$ field with

$$n=n_x+in_y=\frac{2w}{1+|w|^2},\qquad n_z=\frac{1-|w|^2}{1+|w|^2},\qquad\text{or}\qquad w=\frac{n}{1+n_z}.\tag{8.43a}$$

By direct substitution into Equations (8.42c) and (8.43a), it is easy to show that

$$\begin{aligned}&\partial_x n=\mp i[(\partial_y n)n_z-n\partial_y n_z],\quad\partial_y n=\pm i[(\partial_x n)n_z-n\partial_x n_z],\\&\text{and }\partial_x w=\frac{\partial_x n\mp i\partial_y n}{(1+n_z)^2}=\mp i\partial_y w,\end{aligned}\tag{8.43b}$$

which is the Cauchy–Riemann condition for $w(z)$. Hence, any $w(z)$ which is an analytic function of $z^*$ ($-$ sign) or $z$ ($+$ sign) is a solution of Equation (8.42c). One possible form of solution is

$$w(z)=\prod_{i=1}^{q}\frac{z-a_i}{\lambda}.\tag{8.43c}$$

The solution represents an instanton solution with $Q=q$, as can be seen from the following argument: note that $w(z)=\prod_{i=1}^{q}\frac{z-a_i}{\lambda}$ is a $q$th-order polynomial in $z$ and, therefore, for every spin orientation $(n, n_z)$ parameterized by $w$, there are $q$ roots for $z$ which satisfy $w=\prod_{i=1}^{q}\frac{z_w-a_i}{\lambda}$, that is there are $q$ points in the $x$–$y$ plane $z_w=x_w+iy_w$ with the same spin direction $(n, n_z)$. Therefore, the $w$ sphere must be covered $q$ times by the $z$ sphere. The energy of a solution represented by $w(z)$ can be computed using Equations (8.39) and (8.43a) and is

$$E_w=\int d^2x(\nabla\vec{n}(\vec{x}))^2=4\int d^2x\frac{|\nabla w|^2}{(1+|w|^2)^2}.\tag{8.43d}$$

**Homework Q7b**

Derive Equations (8.43b) and (8.43d). By evaluating the integral directly, show that the skyrmion solution (8.43c) has energy $E_q = 4\pi q$.

Recall that the skyrmion solutions have $E_q = 4\pi|Q|$ by construction. Therefore, the topological number is $Q = q$, in agreement with our argument.

Skyrmions exist in many physical systems. They exist as solitons in the double-layer quantum Hall effect and contribute as instantons in 1 + 1D quantum spin systems. They are responsible for the qualitative difference in properties between integer and half-integer quantum spin chains and are responsible for the existence of fractionalized spin excitations at the ends of spin chains. We shall not go into details of these topics here. Interested readers can look at, for example, Ref. [5].

## 8.4.1 Spinor ($CP_1$) Representation

The non-linear-sigma model (8.39a, b) can be represented in terms of spinors. For historical reasons this is called the $CP_1$ representation [1, 2]. The sigma model becomes a U(1) gauge theory model in this representation, and the skyrmion number $Q$ becomes the magnetic flux trapped in the system.

In the spinor representation, a unit vector $\vec{n}$ is written as

$$\vec{n}(\vec{x}) = (Z_\uparrow^*(\vec{x}) \quad Z_\downarrow^*(\vec{x}))\vec{\sigma}\begin{pmatrix} Z_\uparrow(\vec{x}) \\ Z_\downarrow(\vec{x}) \end{pmatrix} = \bar{Z}(\vec{x})\vec{\sigma}Z(\vec{x}), \tag{8.44a}$$

where $\vec{\sigma}$ is the Pauli matrix, and the $Z_\sigma$ are complex numbers satisfying $\sum_{\sigma=\uparrow\downarrow} Z_\sigma^* Z_\sigma = 1$. $Z = \begin{pmatrix} Z_\uparrow \\ Z_\downarrow \end{pmatrix}$ and $\bar{Z} = (Z_\uparrow^*, Z_\downarrow^*)$. The unit vector $\vec{n}$ can be parameterized in the spinor representation as

$$n_z = |Z_\uparrow|^2 - |Z_\downarrow|^2, \quad n_x = 2\mathrm{Re}(Z_\uparrow^* Z_\downarrow), \quad \text{and} \quad n_y = 2\mathrm{Im}(Z_\uparrow^* Z_\downarrow). \tag{8.44b}$$

Therefore,

$$(\partial_\mu \vec{n})^2 = [\partial_\mu(|Z_\uparrow|^2 - |Z_\downarrow|^2)]^2 + 4[\partial_\mu \mathrm{Re}(Z_\uparrow^* Z_\downarrow)]^2 + 4[\partial_\mu \mathrm{Im}(Z_\uparrow^* Z_\downarrow)]^2,$$

which, after some algebra, can be rewritten as

$$(\partial_\mu \vec{n})^2 = \sum_\sigma (\partial_\mu Z_\sigma^*)(\partial_\mu Z_\sigma) + \sum_{\sigma,\sigma'} (Z_\sigma^* \partial_\mu Z_\sigma)(Z_{\sigma'}^* \partial_\mu Z_{\sigma'}). \tag{8.45a}$$

We have used the constraint $\sum_{\sigma=\uparrow\downarrow} Z_\sigma^* Z_\sigma = \vec{n}.\vec{n} = 1$ in deriving Equation (8.45a), and

$$\sum_\mu (\partial_\mu \vec{n})^2 = \sum_\mu [(\partial_\mu \bar{Z}) \cdot (\partial_\mu Z) + (\bar{Z}\partial_\mu Z) \cdot (\bar{Z}\partial_\mu Z)]. \tag{8.45b}$$

**Homework Q8a**

Derive Equation (8.45a).

The Hamiltonian (8.41a) can therefore be written as

$$H = \frac{1}{g}\int d^2x \begin{pmatrix} (\partial_\mu \bar{Z}(\vec{x})).(\partial_\mu Z(\vec{x})) + (\bar{Z}(\vec{x})\partial_\mu Z(\vec{x}))\cdot(\bar{Z}(\vec{x})\partial_\mu Z(\vec{x})) \\ + i\lambda(\vec{x})[\bar{Z}(\vec{x})Z(\vec{x}) - 1] \end{pmatrix}, \tag{8.46}$$

and $Z_H = \int D\bar{Z}DZD\lambda e^{-\beta H(\bar{Z},Z,\lambda)}$ is the corresponding partition function. Note that, because of the constraint $\bar{Z}Z = 1$,

$$\bar{Z}\partial_\mu Z + (\partial_\mu \bar{Z})Z = 2\mathrm{Re}(\bar{Z}\partial\mu Z) = 0,$$

meaning that $\bar{Z}\partial_\mu Z$ must be a pure imaginary number.

The Hamiltonian can be written in a more interesting way by introducing an auxiliary field $\vec{A}$, where

$$H \to H' = \frac{1}{g}\int d^2x[((\partial_\mu \bar{Z})\cdot(\partial_\mu Z) + A_\mu^2 - 2A_\mu(i\bar{Z}\partial_\mu Z)) + i\lambda(|Z|^2 - 1)] \tag{8.47}$$

and

$$Z_H \to \int D\bar{Z}DZD\lambda DAe^{-\beta H'(\bar{Z},Z,\lambda,\vec{A})}.$$

It is easy to see that Equation (8.46) is recovered after integrating out the $A_\mu$ field.

Using again the fact that $\bar{Z}\partial_\mu Z + (\partial_\mu \bar{Z})Z = 0$, we can rewrite Equation (8.47) as

$$\begin{aligned} H' &= \frac{1}{g}\int d^2x((\partial_\mu - iA_\mu)\bar{Z})\cdot(\partial_\mu + iA_\mu)Z + i\lambda(|Z|^2 - 1) \\ &= \frac{1}{g}\int d^2x(|\partial_\mu + iA_\mu)Z|^2 + i\lambda(|Z|^2 - 1), \end{aligned} \tag{8.48}$$

which is the $CP_1$ representation of the non-linear-sigma model.

Note that the Hamiltonian now describes a U(1) gauge field ($A_\mu$) coupled to a two-component complex scalar field ($Z$). Unlike usual gauge theory there are no energy terms that involve only the $A_\mu$ field. This is because $A_\mu$ does not represent 'real' extra degrees of freedom in the system. In fact, the classical equation of motion $\frac{\delta H'}{\delta A_\mu} = 0$ leads to $A_\mu = i\bar{Z}\partial_\mu Z$.

### 8.4.2
### Meaning of Gauge Field

It is interesting to ask what is the meaning of the gauge field and where is gauge invariance coming from in the $CP_1$ representation?

Note that there is no 'phase' in the original unit vector field $\vec{n}$. However, in the spinor representation, $\vec{n} = \bar{Z}\vec{\sigma}Z$, $\vec{n}$ remains unchanged when $Z \to e^{i\vartheta}Z = e^{i\vartheta}\begin{pmatrix} Z_\uparrow \\ Z_\downarrow \end{pmatrix}$ $(\bar{Z} \to \bar{Z}e^{-i\vartheta} = e^{-i\vartheta}(Z_\uparrow^*, Z_\downarrow^*))$. The phase $e^{i\vartheta}$ is an 'unphysical' variable when a real vector is described in terms of spinors and is the origin of gauge invariance in the $CP_1$ representation.

Next, we examine the meaning of 'magnetic flux' $\nabla \times \vec{A}$ in the model. It is interesting to compare the spinor representation with the $w(z)$ representation (Equation (8.43a)). It is easy to see that we can write

$$Z_\uparrow = \frac{e^{i\vartheta}}{\sqrt{1+|w|^2}}, \quad Z_\downarrow = \frac{e^{i\vartheta} w}{\sqrt{1+|w|^2}}, \tag{8.49a}$$

and

$$\nabla \times \vec{A} = \nabla \times (i\bar{Z}\partial_\mu Z) = \frac{i}{(1+|w|^2)^2}\left[(\partial_y w)(\partial_x w^*) - (\partial_x w)(\partial_y w^*)\right]. \tag{8.49b}$$

**Homework Q8b**

Derive Equation (8.49b).

Note that for analytic functions $\partial_x w = \pm i\partial_y w$ and

$$\nabla \times \vec{A} = \frac{\pm 1}{(1+|w|^2)^2}\left[(\partial_x w)(\partial_x w^*) + (\partial_y w)(\partial_y w^*)\right]. \tag{8.50a}$$

In particular,

$$\frac{1}{2\pi}\int d^2x(\nabla \times \vec{A}) = Q \tag{8.50b}$$

from Equation (8.43d), meaning that the integral measures the skyrmion number $Q$ and the 'magnetic field' $\vec{B} = \nabla \times \vec{A}$ measures the local chirality $C(\vec{x})$ of the spin field.

The $CP_1$ representation provides a clear example of how a gauge theory can be derived from a spin model. The gauge theory description of spin models is an important tool in describing so-called spin-liquid states of spin models nowadays and is also believed to be important in describing the low-energy dynamics of strongly correlated systems. Interested readers can look at Ref. [2, 3] for more details.

### 8.4.3 Magnetic Monopoles

Lastly, we briefly discuss magnetic monopoles.

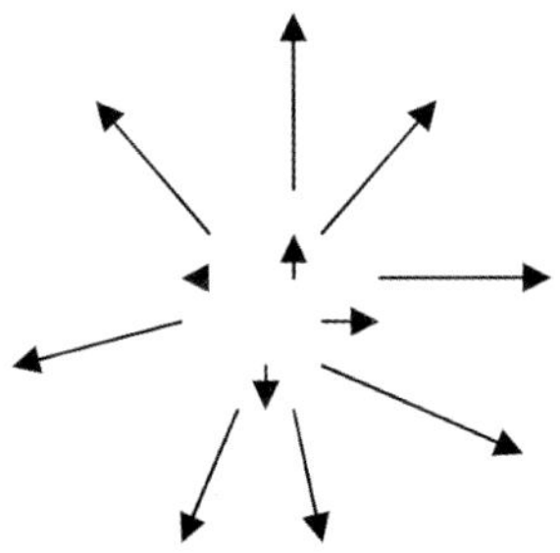

We have described magnetic monopoles in Chapter 6. You can view magnetic monopoles as a generalization of two-dimensional vortices to three dimensions where a magnetic field is coming out from a 'singular' point in space with, for example,

$$\vec{B}(\vec{x}) = \frac{1}{2}\frac{q\vec{x}}{|\vec{x}|^3} - 2\pi q \widehat{z}\, \delta(x)\delta(y)\theta(-z),$$

which represents a magnetic monopole located at the origin $\vec{x} = 0$ see figure page 185. The magnetic field of the monopole is supplied by a flux tube (solenoid) that connects the monopole to (0, 0, $-\infty$), so that $\nabla \cdot \vec{B} = 0$ is guaranteed.

Although magnetic monopoles do not exist as soliton solutions in the continuum limit of U(1) gauge theory, they exist as solitons and/or instantons in lattice U(1) gauge theory and also exist in the continuum SU(2) gauge theory coupled to a three-component scalar field (see Ref. [1]). The existence of magnetic monopoles has the interesting consequence of quantization of (magnetic charge) $\times$ (electric charge) as discussed in Chapter 6, and leads to $S = 1/2$ angular momentum quantization. The existence of magnetic monopoles as instantons in (2 + 1)-dimensional U(1) lattice gauge theory is responsible for the absence of plasma phase in the theory. We shall see this in more detail in Chapter 12.

## References


1 Rajaraman, R. (1982) *Solitons and Instantons*, North-Holland (Amsterdan, N.Y., Oxford).

2 Polyakov, A.M. (1987) *Gauge Fields and Strings*, Harwood Academic Publishers, (Chur, London, Paris, N.Y., Melbourne).

3 Nagaosa N. (1999) *Quantum Field Theory in Condensed Matter Physics*, Springer-Verlag, Berlin, Heidelberg, N.Y.

4 Minnhagen, P. (1987) *Rev. Mod. Phys.* **59**, 1001.

5 T.K. Ng, (1994) *Phys. Rev.*, **B50**, 555.

# Part Three
# A Few Examples

*Introduction to Classical and Quantum Field Theory.* Tai-Kai Ng

ISBN: 978-3-527-40726-2

# 9
# Simple Boson Liquids: Introduction to Superfluidity

## 9.1
## Saddle-Point Approximation: Semiclassical Theory for Interacting Bosons

We consider a model of a boson liquid with action

$$S = \int d\vec{r} \int_0^\beta d\tau \left\{ \bar{\psi}(\vec{r}) \left[ \frac{\partial}{\partial \tau} - \frac{\hbar^2}{2m} \nabla^2 - \mu \right] \psi(\vec{r}) + \frac{1}{2} g [\bar{\psi}(\vec{r}) \psi(\vec{r})]^2 \right\}, \tag{9.1}$$

where $\psi, \bar{\psi}$ are a coherent state representation for bosons. $g > 0$ is a repulsive interaction between bosons. The number density of bosons in the system is a constant fixed by the chemical potential $\mu$.

> Note: The model is a simplified model for cold (bosonic) atoms or $He^4$. The realistic interaction potential between atoms is the Leonard–Jones potential $V(\vec{r} - \vec{r}') = V_0 \left[ (\sigma / |\vec{r} - \vec{r}'|)^{12} - (\sigma / |\vec{r} - \vec{r}'|)^6 \right]$. The interaction potential is simplified to a contact potential in our model. It is generally believed that the physics is not altered qualitatively when we replace $V(\vec{r} - r')$ by $g\delta(\vec{r} - \vec{r}')$ $(g > 0)$. This is an example of universality.

### 9.1.1
### Semiclassical Approximation (using One-Particle QM as Example)

A powerful tool for solving the problem (approximately) is the semiclassical approximation. The idea of the approximation can be illustrated in the following simple one-body problem.

*Introduction to Classical and Quantum Field Theory.* Tai-Kai Ng

ISBN: 978-3-527-40726-2

Consider a particle moving in a potential as shown below:

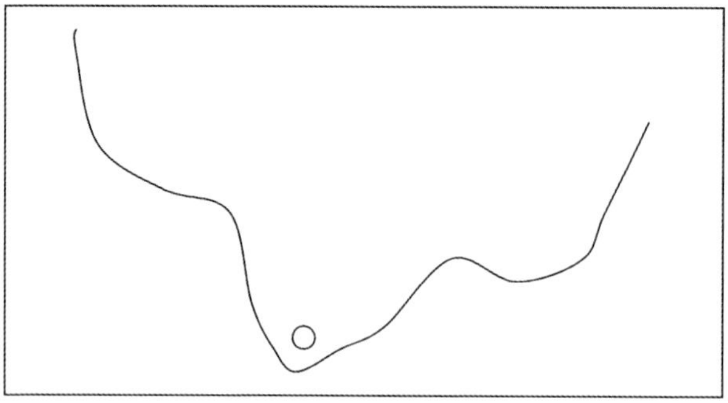

Let us first consider the ground state of the classical problem.

The classical ground state can be obtained by minimizing the kinetic energy and potential energy of the particle separately $\Rightarrow p$ (momentum) $= 0$ and the particle is localized at $x_0 =$ position of minimum energy.

To obtain the low-energy properties, we expand the potential around the minimum point:

$$V(\vec{x}) \sim V(\vec{x}_0) + \frac{1}{2}k(\vec{x}-\vec{x}_0)^2. \tag{9.2a}$$

The corresponding (approximate) Hamiltonian is

$$H = \frac{p^2}{2m} + V(\vec{x}_0) + \frac{1}{2}k(\vec{x}-\vec{x}_0)^2 \tag{9.2b}$$

$\equiv$ harmonic oscillator problem, which is exactly solvable. The classical problem can be quantized by letting $\vec{p} \to -i\hbar\nabla$.

Note that this approximation is probably valid only for properties near the ground state.

Now, we return to the boson problem.

In the coherent state representation, the classical Hamiltonian is

$$H = \int d\vec{r}\left[\frac{1}{2m}|\nabla\psi(\vec{r})|^2 - \mu\bar{\psi}(\vec{r})\psi(\vec{r}) + \frac{1}{2}g[\bar{\psi}(\vec{r})\psi(\vec{r})]^2\right]. \tag{9.3}$$

Following the above example, we first look for a classical solution $\psi(\vec{r},\tau) = \psi_0(\vec{r})$ that minimizes the energy.

It is obvious that the energy is minimized when $\nabla\psi_0(\vec{r}) = 0$, that is

$$\psi(\vec{r},\tau) = \psi_0, \quad \bar{\psi}(\vec{r},\tau) = \psi_0^*, \tag{9.4a}$$

where $\psi_0$ and $\psi_0^*$ are complex numbers chosen such that

$$E_0 = V\left[-\mu|\psi_0|^2 + \frac{1}{2}g|\psi_0|^4\right] \tag{9.4b}$$

is minimized.

The energy is minimized at $|\psi_0|^2 = N_0$, where $N_0 = \sqrt{\frac{\mu}{g}}$ is the density of 'bosons' in classical theory. This solution has the important property of broken gauge symmetry. This is because there is an infinite degeneracy of classical solutions in this problem: all possible $\psi_0 = \sqrt{N_0}e^{i\theta_0}$ with arbitrary $\theta_0$ have the same energy and should contribute equally when we evaluate the partition function. When we choose a particular $\theta_0$ as our starting classical solution, this symmetry is broken.

Next, we assume that $\psi(\vec{r}) = \psi_0 + \delta\psi(\vec{r})$ and $\delta\psi(\vec{r})$ is small, and expand the action to second order in $\delta\psi(\vec{r})$, in analogy to the one-particle problem.

(Recall that $\hat{x} \to \hat{\psi}(\vec{r})$; therefore, expansion to second order in $\delta\psi(\vec{r})$ corresponds to expansion to order $(x - x_0)^2$ in the one-particle problem.)

**Homework Q1**

Show that, to quadratic order, the Hamiltonian is

$$H' = \int d\vec{r}\left[\frac{1}{2m}|\nabla\delta\psi(\vec{r})|^2 - \mu\delta\bar{\psi}(\vec{r})\delta\psi(\vec{r}) + \frac{1}{2}g|\psi_0|^2[\delta\bar{\psi}(\vec{r}) + \delta\psi(\vec{r})]^2\right] + E_0 . \tag{9.5}$$

As in the single-particle problem, now we can replace the classical number fields by the corresponding operators, $\psi(\vec{r}) \to \hat{\psi}(\vec{r})$, $\bar{\psi}(\vec{r}) \to \hat{\bar{\psi}}(\vec{r})$ (with $[\psi(\vec{r}), \bar{\psi}(\vec{r}')] = \delta(\vec{r} - \vec{r}')$, etc.), to obtain the quantum problem in the semiclassical approximation.

Alternatively, we can write

$$\begin{aligned} \psi(\vec{r},\tau) &= \sqrt{N(\vec{r},\tau)}e^{i\theta(\vec{r},\tau)}, \\ \bar{\psi}(\vec{r},\tau) &= \sqrt{N(\vec{r},\tau)}e^{-i\theta(\vec{r},\tau)} \end{aligned} \tag{9.6a}$$

and expand the action in terms of $n$ and $\theta$, where $n = N - N_0$, $\theta_0 = 0$.

> Note: we are just expanding the classical action around the classical minimum; *no quantum mechanics (QM) is introduced yet.*

To second order in $n$ and $\theta$, we obtain $S = S_{cl} + S_2$, where $S_{cl} = \beta E_0$,

$$S_2 = \int d\vec{r}\int_0^\beta d\tau\left\{i\cdot n\partial_\tau\theta + \frac{1}{2}g\cdot n^2 + \frac{N_0}{2m}(\nabla\theta)^2 + \frac{(\nabla n)^2}{8N_0 m}\right\}. \tag{9.6b}$$

**Homework Q2a**

Derive $S_2$.

The corresponding quantum partition function is

$$Z^{(2)} = \int Dn\int D\theta e^{-S_2}, \tag{9.6c}$$

where we have used $\int d\psi \int d\bar{\psi} = \frac{1}{2}\int dn \int d\theta$.

Note that $S_2$ is quadratic in $n$ and $\theta$; thus, the path integral can be performed exactly. As before, we introduce Fourier-transformed fields

$$\vartheta(\vec{q}, i\omega) = \int d^d r \int_0^\beta d\tau e^{i(\omega\tau + \vec{q}.\vec{r})} \vartheta(\vec{r}, \tau) \tag{9.7a}$$

and

$$n(\vec{q}, i\omega) = \int d^d r \int_0^\beta d\tau e^{i(\omega\tau + \vec{q}.\vec{r})} n(\vec{r}, \tau). \tag{9.7b}$$

In terms of $n(\vec{q}, i\omega)$ and $\theta(\vec{q}, i\omega)$, we have

$$S_2 = \sum_{q, i\omega_n} \left[ \begin{array}{l} i\omega_n n(\vec{q}, i\omega_n)\theta(-\vec{q}, -i\omega_n) + \frac{1}{2}\left(g + \frac{q^2}{4N_0 m}\right) n(\vec{q}, i\omega_n) n(-\vec{q}, -i\omega_n) \\ + \frac{N_0 \vec{q}^2}{2m} \theta(\vec{q}, i\omega_n)\theta(-\vec{q}, -i\omega_n). \end{array} \right] \tag{9.7c}$$

Integrating over $n$ and $\theta$ as in previous chapters, we obtain

$$F_b = \frac{1}{V} \sum_k \left( \frac{\omega_q}{2} + \frac{1}{\beta} \ln(1 - e^{-\beta\omega_q}) \right), \tag{9.8a}$$

where

$$\omega_q = \left( \sqrt{\frac{N_0 g}{m} + \frac{q^2}{4m^2}} \right) q \sim \sqrt{\frac{N_0 g}{m}} q \tag{9.8b}$$

at small $q$.

**Homework Q2b**

Derive the above expression of the free energy.

If you compare this free energy with results in Chapter 4 for phonons, it suggests that the interacting boson system we described can be diagonalized approximately in the semiclassical approximation with excitations behaving as free bosons, that is the Hamiltonian after diagonalization can be written as

$$H = \sum_q \left( \widehat{\gamma}_q^+ \widehat{\gamma}_q + \frac{1}{2} \right) \omega_q, \tag{9.8c}$$

where $\widehat{\gamma}(\widehat{\gamma}^+)$ are canonical boson operators. To understand the system better, we need to understand the nature of the excitations described by these operators. One way to achieve this is to go back to the Hamiltonian formulation and diagonalize the quadratic Hamiltonian (9.5).

**Homework Q2c**

After Fourier transformation, the quadratic Hamiltonian $H'$ can be diagonalized by introducing the Bogoliubov transformation

$$\widehat{b}_{\vec{q}} = u(\vec{q})\widehat{\gamma}_{\vec{q}} + v(\vec{q})\,\widehat{\gamma}^{+}_{\vec{q}}\,, \quad \widehat{b}^{+}_{\vec{q}} = v(\vec{q})\widehat{\gamma}_{\vec{q}} + u(\vec{q})\,\widehat{\gamma}^{+}_{\vec{q}}\,, \tag{9.9}$$

where $\widehat{b}_{\vec{q}}(\widehat{b}^{+}_{\vec{q}}) = \int d^d r e^{+(-)i\vec{q}.\vec{r}}\delta\,\widehat{\psi}(\vec{r})(\delta\widehat{\bar{\psi}}(\vec{r}))$ and requiring that the new operators $\widehat{\gamma}, \widehat{\gamma}^{+}$ satisfy the usual boson commutation relations $[\widehat{\gamma}, \widehat{\gamma}^{+}] = \delta$, and so on.

Show that the $u$ and $v$ must satisfy $|u(\vec{q})|^2 - |v(\vec{q})|^2 = 1$. Determine $u(\vec{q}),\ v(\vec{q})$, and the excitation energy $\omega_{\vec{q}}$ by requiring that $H'$ can be written in the form $H' = \sum_q \omega_q \gamma_q^{+}\gamma_q + E_G$.

Another way to understand the nature of these excitations it is to start from the path integral (Lagrangian) formulation in terms of $n$ and $\theta$ fields and then derive the corresponding quantum problem using a canonical quantization scheme. We shall study this approach in the following.

It is customary (but not necessary) to neglect the term $\frac{(\nabla n)^2}{8N_0 m}$ in Equation (9.6b), integrate out the $n$ field, and obtain a Lagrangian in terms of the $\theta$ field only in real time. After some algebra, we obtain $S_2 \to S(\theta) = \int_0^t dt L_\theta$, where

$$L_\theta = \int d\vec{r}\left\{\frac{1}{2g}\left[\frac{\partial}{\partial t}\theta(\vec{r})\right]^2 - \frac{N_0}{2m}[\nabla\theta(\vec{r})]^2\right\}. \tag{9.10}$$

This is called the phase action for bosons and is one of the most important actions in (quantum) field theory. Note that in the absence of the $(\partial_t\theta)^2$ term this is just the continuum theory of the (classical) $X$–$Y$ model. The action (9.10) is often called the *quantum X–Y model* because of the presence of a time-derivative term (dynamics) in the action.

Let us treat $L_\theta$ as the Lagrangian of a classical system with classical field $\theta(\vec{r}, t)$ and 'quantize' the system using the canonical quantization scheme.

The conjugate momentum field to $\theta(\vec{r})$ is

$$\pi(\vec{r}) = \frac{\delta L_\theta}{\delta\dot{\theta}(\vec{r})} = \frac{1}{g}\dot{\theta}(\vec{r}) \tag{9.11a}$$

and

$$H = \int d\vec{r}\pi(\vec{r})\dot{\theta}(\vec{r}) - L_\theta = \int d\vec{r}\left[\frac{g}{2}\pi(\vec{r})^2 + \frac{c^2}{2g}(\nabla\theta)^2\right], \tag{9.11b}$$

where $c^2 = \frac{N_0}{m}g$.

Now, we can 'quantize' this classical system in the canonical quantization, that is we replace $\pi$ and $\theta$ by operators $\widehat{\pi}$ and $\hat{\theta}$ with the commutation relation

$$[\hat{\theta}(\vec{r}), \widehat{\pi}(\vec{r}')] = i\hbar\delta(\vec{r} - \vec{r}'). \tag{9.11c}$$

It is interesting to compare this with the boson commutation rule

$$[\widehat{\psi}(\vec{r}), \bar{\psi}(\vec{r}')] = \delta(\vec{r} - \vec{r}').$$

Representing $\psi, \bar{\psi}$ in terms of $N$, $\theta$, that is

$$\widehat{\psi}(\vec{r}) = e^{i\widehat{\theta}(\vec{r})}\sqrt{\widehat{N}(\vec{r})},\ \bar{\psi}(\vec{r}) = \sqrt{\widehat{N}(\vec{r})}e^{-i\widehat{\theta}(\vec{r})},$$

it can be shown that the phase and density of a boson system satisfy the commutation relation

$$[\hat{\theta}(\vec{r}), \hat{N}(\vec{r}')] = -i\delta(\vec{r} - \vec{r}'). \tag{9.11d}$$

**Homework Q2d**

Derive the above result.

When compared with the canonical commutation rule (9.11c) we derived approximately, we obtain

$$\widehat{\pi}(\vec{r}) \rightarrow -\hat{N}(\vec{r}), \tag{9.11e}$$

which is the boson density operator!

To write down the Schrödinger wavefunction of the system, we employ the 'momentum' representation ($\hat{x} = i\hbar\frac{\partial}{\partial p}$, so that $[\hat{x}, \hat{p}] = i\hbar$)

$$\begin{aligned} \hat{\theta}(\vec{r}) &= i\hbar\frac{\delta}{\delta\pi(\vec{r})} \quad \left(\hat{x} = i\hbar\frac{\partial}{\partial p}\right), \\ \widehat{\pi}(\vec{r}) &= \pi(\vec{r}) \quad\quad (\hat{p} = p). \end{aligned} \tag{9.12a}$$

In this representation the 'Schrödinger' equation of the many-particle boson system becomes

$$\hat{H}|\psi\rangle = E|\psi\rangle, \quad |\psi\rangle = |\psi[\pi(\vec{r})]\rangle,$$

where

$$\hat{H} = \int d\vec{r}\left\{\frac{g}{2}\pi(\vec{r})^2 - \frac{\hbar^2c^2}{2g}\left(\nabla\frac{\delta}{\delta\pi(\vec{r})}\right)^2\right\}. \tag{9.12b}$$

To diagonalize $H$, we introduce Fourier transforms

$$\widehat{\theta}(\vec{q}) = \int d^dr e^{+(-)i\vec{q}.\vec{r}}\,\widehat{\theta}(\vec{r}),\ \widehat{\pi}(\vec{q}) = \int d^dr e^{+(-)i\vec{q}.\vec{r}}\,\widehat{\pi}(\vec{r}).$$

Therefore, setting $\hbar = 1$,

$$\hat{H} = \sum_{\vec{q}}\left[\frac{g}{2}\widehat{\pi}(\vec{q})\,\widehat{\pi}(-\vec{q}) + \frac{\omega(\vec{q})^2}{2g}\hat{\theta}(\vec{q})\hat{\theta}(-\vec{q})\right]$$

$$= \sum_{\vec{q}}\left[\frac{g}{2}\pi(\vec{q})\pi(-\vec{q}) + \frac{\omega(\vec{q})^2}{2g}\frac{\delta}{\delta\pi(\vec{q})}\frac{\delta}{\delta\pi(-\vec{q})}\right], \tag{9.12c}$$

which describes a sum of independent harmonic oscillators.

Correspondingly, the ground-state wavefunction is

$$\Psi_0 = \prod_{\vec{q}} \exp\left[-\frac{g}{2\omega(q)}\pi(\vec{q})\pi(-\vec{q})\right] \tag{9.13a}$$

or

$$\Psi_0(\{N_{\vec{q}}\}) = \exp\left[-\frac{1}{2}\sum_{\vec{q}}\frac{g}{\omega(q)}N(\vec{q})N(-\vec{q})\right]. \tag{9.13b}$$

Introducing the inverse Fourier transform, $N(\vec{q}) \to \int \mathrm{d}^d r e^{i\vec{q}\cdot\vec{r}} n(\vec{r})$, we obtain

$$\Psi_0[n(\vec{r})] = \exp\left[-\frac{1}{2}\int \mathrm{d}^d r' D(\vec{r}' - \vec{r}) n(\vec{r}) n(\vec{r}')\right], \tag{9.13c}$$

where $D(\vec{r}) = \int \mathrm{d}^d q e^{-i\vec{q}.\vec{r}} \frac{g}{\omega(\vec{q})}$ or, in first quantized form,

$$\Psi_0[\vec{r}_1, \vec{r}_2, \ldots, \vec{r}_N] = \exp\left[-\frac{1}{2}\sum_{i\neq j} D(\vec{r}_i - \vec{r}_j')\right]. \tag{9.13d}$$

This form of wavefunction is called the Jastrow wavefunction and was first suggested by Feynman as a trial ground-state wavefunction for helium 4, where $D(\vec{r})$ is a trial function. To minimize the energy, $D(\vec{r})$ should be a function which is large and positive as $\vec{r} \to 0$, but goes to zero as $\vec{r} \to \infty$ because of strong, short-range, repulsive interaction between atoms which forbids atoms to come close. The $D(\vec{r})$ obtained in the semiclassical approximation does not have the right feature at $\vec{r} \to 0$ and does not give good energy. However, it captures the correct long-distance behavior.

The excitations are described by

$$E(\{N_{\vec{q}}\}) = E_{\mathrm{cl}} + \sum_{\vec{q}} \omega(\vec{q})\left[n_{\vec{q}} + \frac{1}{2}\right],$$

as obtained before. The nature of the excitations can be understood from the wavefunction (Feynman).

*First excited state wavefunction*

$$\Psi_{\vec{k}} = C'_{\vec{k}} N_{\vec{k}} \Psi_0 \quad (C'_{\vec{k}} = \text{normalization constant}) \tag{9.14a}$$

$$(\Psi_1 \sim p\Psi_0(p) \text{ or } x\Psi_0(x) \text{ for harmonic oscillators})$$
$$\text{or } \Psi_{\vec{k}}(\vec{r}_1, \vec{r}_2, \ldots, \vec{r}_N) = C'_{\vec{k}}\Big(\sum_i e^{i\vec{k}\cdot\vec{r}_i}\Big)\Psi_0(\vec{r}_1, \vec{r}_2, \ldots, \vec{r}_N). \tag{9.14b}$$

This form of excited state was first constructed by Feynman and corresponds to superimposing a density wave $N_{\vec{k}} \sim \sum_i e^{i\vec{k}.\vec{r}_i}$ on the ground-state wavefunction. You are encouraged to look at Ref. [1] to understand why this represents a reasonable excited state for boson systems.

Before proceeding further, we note that we have ignored the term $\frac{(\nabla n)^2}{8N_0 m}$ in our action $S_2$. It can be shown that, if we keep the term, we obtain a different excitation energy spectrum,

$$\omega_q = \sqrt{\frac{q^2}{2m}\left(\frac{q^2}{2m} + 2N_0 g\right)} \tag{9.15}$$

with everything else remaining essentially the same.

### 9.1.2
### Density–Density Response Function and One-Particle Green's Function

We shall evaluate the density–density response function and one-particle Green's function of the boson system in the semiclassical approximation here. First, we evaluate the imaginary-time density–density response function

$$\chi(\vec{q}, i\omega) = \int \langle T_\tau n(\vec{r}, \tau) n(0,0)\rangle e^{i(\vec{q}.\vec{r} + \omega\tau)} \mathrm{d}^d q \mathrm{d}\tau. \tag{9.16}$$

To evaluate $\chi$, we start from Equation (9.6b), and insert a source term in the action,

$$\tilde{S}_2 = \int \mathrm{d}\vec{r} \int_0^\beta \mathrm{d}\tau \left\{ i \cdot n\partial_\tau \theta + \frac{1}{2} g \cdot n^2 + \frac{N_0}{2m} (\nabla\theta)^2 + \frac{(\nabla n)^2}{8N_0 m} + n\xi \right\} \tag{9.17a}$$
$$\text{and } \tilde{Z} = \int DnD\theta e^{-\tilde{S}_2}.$$

The density–density response function is then

$$\langle T_\tau n(\vec{r}, \tau) n(0,0)\rangle = \frac{\int DnD\theta n(\vec{r}, \tau) n(0,0) e^{-S_2}}{Z} = \frac{\delta^2}{\delta\xi(\vec{r}, \tau)\delta\xi(0,0)} \ln \tilde{Z}\Big|_{\xi=0}.$$

Since $\tilde{S}_2$ is quadratic in $n$ and $\theta$, they can be integrated out to obtain

$$\ln \tilde{Z} = \sum_{\vec{q}, i\omega} \left( \frac{\frac{N_0 q^2}{2m}}{(i\omega)^2 - \frac{N_0 q^2}{m}\left(g + \frac{q^2}{4N_0 m}\right)} \right) |\xi(\vec{q}, \omega)|^2 \tag{9.17b}$$

and

$$\chi(\vec{q}, i\omega) = \frac{N_0 \varepsilon(q)}{((i\omega)^2 - E(q)^2)} \tag{9.17c}$$

is the density–density response function in the semiclassical approximation at imaginary time, where $\varepsilon(q) = \frac{q^2}{2m}$ is the non-interacting boson dispersion and $E(q)^2 = \varepsilon(q)(2N_0 g + \varepsilon(q))$. The corresponding retarded density–density response function is $\chi(\vec{q}, \omega) = \frac{N_0 \varepsilon(q)}{(\omega + i\delta)^2 - E(q)^2}$.

The one-particle Green's function can also be evaluated in the semiclassical approximation. We start with the amplitude–phase representation

$$\psi(\vec{r}) = e^{i\hat{\theta}(\vec{r})} \sqrt{N(\vec{r})} \sim \sqrt{N_0} + \frac{n(\vec{r})}{2\sqrt{N_0}} + i\sqrt{N_0}\theta(\vec{r}), \tag{9.18a}$$

for small fluctuations $n(\vec{r})$ and $\theta(\vec{r})$. Therefore,

$$\langle T_\tau \psi(\vec{r},\tau)\bar{\psi}(0,0)\rangle \sim N_0 + \frac{1}{4N_0}\langle T_\tau n(\vec{r},\tau)n(0,0)\rangle + N_0\langle T_\tau \theta(\vec{r},\tau)\theta(0,0)\rangle + \frac{i}{2}\langle T_\tau \theta(\vec{r},\tau)n(0,0)\rangle - \frac{i}{2}\langle T_\tau n(\vec{r},\tau)\theta(0,0)\rangle. \tag{9.18b}$$

Extending the method in computing the density–density response function, we obtain for the one-particle Green's function at imaginary time

$$G(\vec{q},i\omega) = \frac{i\omega + \varepsilon(q)}{(i\omega)^2 - E(q)^2}. \tag{9.19}$$

Note that the non-interacting Green's function $G^{(0)}(\vec{q},i\omega) = \frac{1}{i\omega - \varepsilon(q)}$ is recovered in the limit $g \to 0$.

**Homework Q3**
Derive Equation (9.19).

It is interesting to note that the structure of the Green's function in the presence of interaction is qualitatively different from the case of non-interacting bosons. When $g=0$ only one peak at $\omega = \varepsilon(q)$ exists in the spectral function. However, when $g \neq 0$, two peaks at $\omega = \pm E(q)$ exist! The spectral weight of the peak at $\omega = -E(q)$ vanishes only at $g \to 0$. This is an example of breaking down of adiabaticity and indicates that the nature of the excitation spectrum becomes qualitatively different from those of non-interacting bosons when interaction exists.

#### 9.1.2.1 A Little Bit Beyond the Semiclassical Approximation

As we have pointed out, the semiclassical approximation we employed corresponds to approximating $V(\vec{x}) \sim V(\vec{x}_0) + \frac{1}{2}k(\vec{x} - \vec{x}_0)^2$ in the vicinity of the classical potential minimum in the one-particle problem. We can improve this approximation by 'promoting' this approximation to the trial Hamiltonian approach.

In the one-particle problem, we may consider the trial Hamiltonian

$$H_{\text{trial}} = \frac{p^2}{2m} + \bar{V}(\vec{x}_0) + \frac{1}{2}\bar{k}(\vec{x} - \vec{x}_0)^2,$$

where $\bar{V}, \bar{k}, \vec{x}_0$ are no longer obtained from expanding the potential energy around the minimum at $\vec{x}_0$, but are trial parameters to be determined by minimizing the free energy (see Chapter 5). You can show with this approach that superfluidity cannot exist in 1D. The simplest version of the trial Hamiltonian we can use is, in the boson representation,

$$H' = \int d\vec{r}\left[\frac{1}{2m}|\nabla\delta\psi(\vec{r})|^2 - \mu\delta\bar{\psi}(\vec{r})\delta\psi(\vec{r}) + \frac{1}{2}g|\psi_0|^2[\delta\bar{\psi}(\vec{r}) + \delta\psi(\vec{r})]^2\right] + S_0, \tag{9.20a}$$

with the trial parameters $\mu$ and $\psi_0$. The two trial parameters are determined by minimization of free energy and the requirement that the boson density is $N$, that is

$$N = |\psi_0|^2 + <\delta\bar{\psi}(\vec{r})\delta\psi(\vec{r})>, \tag{9.20b}$$

where $\langle\ldots\rangle$ denotes thermodynamic average.

**Homework Q4**

Work out the above variational problem and show that $\mu<0$ and $\psi_. \to 0$ at any temperature in 1D. Discuss what will happen in 2D and 3D.

## 9.2 Superfluidity

### 9.2.1 Bose Condensation

Let us consider a system of $N$ non-interacting bosons with Hamiltonian $H_0 = \sum_i^N \frac{p_i^2}{2m}$.

From statistical mechanics, we know that the occupancy number of bosons is

$$n(\vec{k}) = \frac{1}{e^{\beta(\varepsilon_k - \mu)} - 1},$$

where $\beta = 1/k_B T$, $\varepsilon_k = \dfrac{\hbar^2 k^2}{2m}$, and $\mu$ is the chemical potential determined by the condition

$$n = \frac{N}{V} = \int \frac{d^d k}{(2\pi)^d} n(\vec{k}), \quad \text{where } d = \text{dimension of the system.}$$

At low temperature $T \to 0$, $e^{\beta(\varepsilon_k - \mu)} \to \infty$ for any $\varepsilon_k > \mu$, and $n(\vec{k}) \to 0$. The only possible solution where a finite density of bosons exists in this case has

$$\mu \to \varepsilon_{k=0} = 0 \text{ and } n(\vec{k}) = n\delta(\vec{k}), \tag{9.21}$$

corresponding to putting all bosons in the lowest-energy state with zero momentum. Note that this is impossible if the particles under consideration are fermions, because of the Pauli exclusion principle.

*A thermodynamic state where macroscopic numbers of bosons are put into a single one-particle eigenstate is called a state with Bose condensation, or a Bose-condensed state.*

It turns out that, at finite temperature, a thermodynamic state with $\mu = 0$ and

$$n(\vec{k}) = \begin{bmatrix} \dfrac{1}{e^{\beta\varepsilon_k} - 1}, & \vec{k} \neq 0, \\ n'(T), & \vec{k} = 0, \end{bmatrix} \tag{9.22}$$

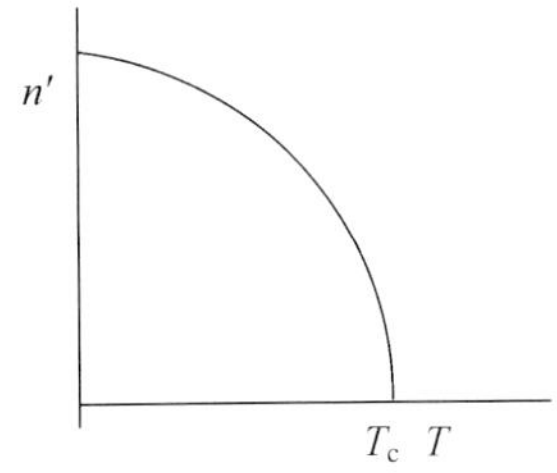


where $n'(T) \neq 0$, exists in 3D at temperatures $T < T_c$, where $T_c$ is the temperature where Bose condensation begins to occur (see above figure and e.g. Ref. [2] for details).

The Bose-condensation state is of interest because we believe that it is related to the phenomenon of *superfluidity*, which is a very special phenomenon observed in liquid $He^4$ (a liquid of boson atoms) at very low temperature.

### 9.2.2
### Superfluid $He^4$

Liquid $He^4$ undergoes a transition at $T_c = 2.18$ K. Above $T_c$, $He^4$ has properties like any normal liquid. Very spectacular changes happen below $T < T_c$. Of the many interesting behaviors, the most important one is perhaps the loss of viscosity at $T < T_c$. It was found that, below a critical velocity $v_c$, $He^4$ flows through a thin capillary tube with zero resistance. This state of zero viscosity is called a superfluid state. There are many other interesting properties associated with superfluidity, which we shall not discuss here. More details can be found in Refs. [1, 2].

### 9.2.3
### Landau's Analysis for Superfluidity

Landau worked out the first phenomenological theory for superfluidity. His discussion is very important because it points out the *necessary* conditions for the formation of a superfluid state. We shall follow his discussion in the following.

Before we go into the explanation of superfluidity, we have to understand why viscosity exists in normal liquid flow (or why when we stir up a normal liquid, the liquid will stop flowing after some time) in the first place.

First, we have to keep in mind that a system described by a Hamiltonian with translational invariance has no viscosity. This follows from the momentum conservation law that the total momentum of a system is conserved if there is translational invariance.

$$\text{Total current} = \frac{\text{total momentum}}{\text{mass of all particles}}$$

$\Rightarrow$ *total current is conserved in a system with translational invariance, that is current flow cannot stop by itself in systems with translational invariance* $\Rightarrow$ *no viscosity!*

In a real fluid flow, the current is stopped by frictional force between the fluid and the container. The existence of a container breaks translational symmetry and makes viscosity possible.

Microscopically, the current is stopped in a normal fluid because when the atoms in the fluid collide with the container's wall, they give up some of their kinetic energies to the wall of container and slow down. *In this process the energy carried by the fluid is lowered. Superfluidity appears at very low temperature when the fluid cannot give up its energy by colliding with the wall.*

To see how it can happen, let us consider temperature $T \to 0$ such that the liquid is in a quantum mechanical ground state with energy $E_G$.

Now, let us consider giving the liquid a uniform velocity $\vec{v}$. In the absence of interaction between the container's wall and the liquid, the energy of the liquid is

$$E_{\mathrm{G}}^{v} = E_{\mathrm{G}} + \frac{1}{2} M v^2, \tag{9.23a}$$

where $M$ is the total mass of the liquid.

Next, we analyze the excitation energies of this quantum liquid. Landau assumed that the low-energy excitations of the system at rest can be described by something like phonons in a solid with effective Hamiltonian

$$H_{\text{low energy}} = E_{\mathrm{G}} + \sum_{\vec{k}} \omega(\vec{k}) n_{\vec{k}}, \tag{9.23b}$$

where $\omega(\vec{k})$ is an unknown energy dispersion for the excitations and the $\hbar\vec{k}$ are the momenta of the excitations. $n_{\vec{k}} = 0, 1, 2, \ldots$ are integers. (Note that by allowing $n_{\vec{k}}$ to be arbitrary integers, we have assumed that these excitations are bosons.)

In the rest frame of a moving liquid, the energies of the excitations are $\omega(\vec{k})$. But, how about in the laboratory frame where the liquid is moving as a whole?

In the laboratory frame, the total momentum of the liquid when one excitation is present is $\vec{P}_{\mathrm{tot}} = M\vec{v} + \hbar\vec{k}$, and therefore the energy is now

$$E_1^{v} = \left[E_{\mathrm{G}} + \hbar\omega(\vec{k})\right] + \frac{1}{2M} {P_{\mathrm{tot}}}^2 \approx E_{\mathrm{G}}^{v} + \omega(\vec{k}) + \vec{v}.(\hbar\vec{k}), \tag{9.24}$$

where we have neglected the term $(\hbar k)^2/2M$. (Note that $M$ is the *total mass* of the liquid, whereas $\hbar k$ is the momentum of one excitation only.) Therefore, the excitation energy in the laboratory frame is $\omega(\vec{k}) + \vec{v}.(\hbar\vec{k})$. Note that the excitation energy is reduced and may become negative if $\vec{v}.(\hbar\vec{k}) < 0$.

We can now discuss the effect of the container and friction (viscosity). The presence of a wall breaks translational invariance and the momentum of the fluid is allowed to change. In particular, the fluid can lose energy by spontaneous creation of excitations if

$$\omega(\vec{k}) + \vec{v}.(\hbar\vec{k}) < 0. \tag{9.25}$$

Note that in this case we must have $\vec{v}.(\hbar\vec{k}) < 0$, indicating that the total momentum (current) of the system must be reduced by friction.

In particular, Landau pointed out that the liquid would be a superfluid if, for small enough $\vec{v}$, it is impossible to satisfy the inequality $\omega(\vec{k}) + \vec{v}.(\hbar\vec{k}) < 0$ since, in this case, the system cannot lower its energy by creating excitations.

#### 9.2.3.1 A Free Boson Condensate is Not a Superfluid

For free bosons, an excitation can be created by removing a boson from the condensate (momentum $=0$ state) to a finite-momentum state with $\hbar\vec{k}\neq 0$. The excitation energy is

$$\omega(\vec{k}) = (\hbar k)^2/2m.$$

Note that, for any velocity $\vec{v}\neq 0$, we can always have small enough $\hbar\vec{k}\neq 0$ such that $\hbar k < 2mv$, and the inequality $\omega(\vec{k})+\vec{v}.(\hbar\vec{k})<0$ is satisfied, that is a moving free boson liquid can always lower its energy by creating excitations. The velocity of the liquid is gradually reduced in the process. Therefore, a free boson condensate is *not* a superfluid.

#### 9.2.3.2 A Boson Fluid with Phonon-Like Excitation Spectrum is a Superfluid

Landau pointed out that a fluid system with a phonon-like dispersion at small $\vec{k}$, that is

$$\omega(\vec{k}) = c|\vec{k}|,$$

is a superfluid since in this case the inequality $c|\vec{k}|+\vec{v}.(\hbar\vec{k})<0$ can never be satisfied if $|\vec{v}|<v_c=c/\hbar$. In this case, the system cannot lower its energy by creating excitations. $v_c$ defines the critical velocity where superfluidity is destroyed. Unfortunately, the experimentally determined critical velocity for $He^4$ superfluid is much less than the one determined theoretically using this criterion. The discrepancy is still not fully understood. You can look at Ref. [1] to understand more about this problem.

Note that we have shown that bosons with short-range repulsive interaction have an excitation spectrum $\omega(\vec{k})=c|\vec{k}|$ at small enough wave vector $\vec{k}$. Therefore, it is believed that quantum liquids of repuling bosonic atoms are always superfluids at low enough temperature.

Note that $\omega(\vec{k})=\sqrt{\frac{k^2}{2m}\left(\frac{k^2}{2m}+2N_0 g\right)}$, and the phonon-like spectrum vanishes if $g=0$. The presence of repulsive interaction is said to give *rigidity* to the superfluid. Therefore, superfluidity is a result of Bose condensation (a large number of atoms remain in the same state) + repulsive interaction between bosons (phonon-like excitation spectrum). It would not exist without the two factors working together.

In reality, the energy dispersion of $He^4$ has a different form than our $\omega(\vec{k})$ at large $k$ because of the attractive part of the Lennard–Jones potential (see figure below). However, the small-$k$ behavior of the dispersion is well described by the point-interaction model.

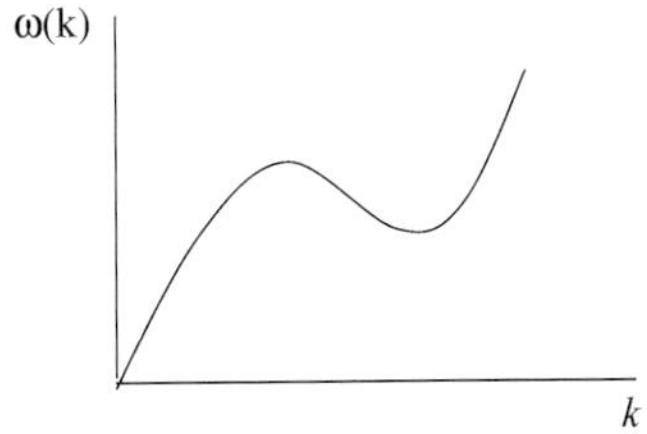

#### 9.2.3.3 Off-Diagonal Long-Range Order (ODLRO) and Collective Motion of Superfluid

A quantum state with a large number of bosons all occupying the same one-particle wavefunction $\phi_0(\vec{x})$ has a very special feature called off-diagonal long-range order (ODLRO). To understand this, we introduce boson annihilation and creation operators $a_0$ and $a_0^+$ which decrease and increase the number of bosons in the state $\phi_0(\vec{x})$ by one, respectively. A many-body wavefunction $|\Psi_N\rangle$ with $N$ bosons occupying the state $\phi_0(\vec{x})$ can be written as (see quantum mechanics textbooks on harmonic oscillators)

$$|\Psi_N> = \left[\frac{(a_0^+)^N}{\sqrt{N!}}\right]|0>, \tag{9.26a}$$

where $|0\rangle$ is the vacuum state. We also have

$$a_0|\Psi_N> = \sqrt{N}|\Psi_{N-1}>, \quad a_0^+|\Psi_N> = \sqrt{N+1}|\Psi_{N+1}>. \tag{9.26b}$$

Next, we consider boson creation and annihilation operators $\bar{\psi}(\vec{x})$ and $\psi(\vec{x})$ that add and remove a boson at position $\vec{x}$, respectively. By completeness, $\psi(\bar{\psi})(\vec{x})$ can be expanded in a complete and orthonormal set of wavefunctions $\phi_n(\vec{x})$,

$$\psi(\bar{\psi})(\vec{x}) = \sum_n \phi_n(\vec{x})a_n(a_n^+), \tag{9.27a}$$

where $a_n(a_n^+)$ decrease (increase) the number of bosons in state $\phi_n(\vec{x})$ by one.

ODLRO can be seen from the expectation value

$$\begin{aligned} G(\vec{x},\vec{x}') &= <\Psi_N|\psi(\vec{x})\bar{\psi}(\vec{x}')|\Psi_N> \\ &= \sum_{n,m}\phi_n(\vec{x})\phi_m(\vec{x}')<\Psi_N|a_n a_m^+|\Psi_N> = \phi_0(\vec{x})\phi_0(\vec{x}')(N+1) \end{aligned} \tag{9.27b}$$

for a Bose-condensed wavefunction with $N$ bosons all in state $\phi_0(\vec{x})$.

We observe that $G(\vec{x},\vec{x}')$ has the special property that it does not vanish (unless $\phi_0(\vec{x})(\phi_0(\vec{x}'))$ vanishes) when $\vec{x}-\vec{x}' \to \infty$. This is a property that does not occur for most macroscopic systems. Systems with this (or similar) properties are said to possess ODLRO.

For example, for free bosons, $\phi_0(\vec{x}) = \frac{1}{\sqrt{V}}(\vec{k}=0)$ and $G(\vec{x},\vec{x}') = \frac{N+1}{V} \cong n$ (density of bosons).

In quantum field theory, this property is usually represented as

$$\widehat{\psi}(\vec{x})|\Psi_N\rangle \sim \sqrt{N}\phi_0(\vec{x})|\Psi_N\rangle \tag{9.28}$$

(or $\widehat{\psi}(\vec{x}) = \sqrt{N}\phi_0(\vec{x}) + \delta\,\widehat{\psi}(\vec{x})$ if not all bosons occupy the same state).

Note that formally Equation (9.28) means that $|\Psi_N\rangle$ is a coherent state of the annihilation operator $\widehat{\psi}(\vec{x})$ and thus the number of particles is not mixed in the state. The situation is similar to thermodynamics when the grand canonical ensemble is used to describe closed systems with a fixed number of particles. The 'mistake' we made vanishes in the thermodynamic limit when the number of particles $N \to \infty$.

Equation (9.28) provides an understanding of the semiclassical approximation from a wavefunction point of view. The semiclassical approximation essentially assumes that the many-body boson wavefunction can be written as $|\Psi\rangle = e^J|\Psi_c\rangle$, where $e^J$ is the Jastrow factor we discussed before (Equation (9.13a-d)) and $|\Psi_c\rangle$ satisfying $\widehat{\psi}(\vec{x})|\Psi_c\rangle \sim \sqrt{n}|\Psi_c\rangle$ is a coherent state. Note that the broken global U(1) symmetry follows naturally in this language, since particle number is no longer a conserved quantity in the wavefunction (see Chapter 2 for the relation between particle number conservation and global U(1) phase).

Equation (9.28) suggests that $\sqrt{N}\phi_0(\vec{x})$ can be interpreted as *a wavefunction of the whole condensate*. It describes the collective motion of all atoms and was first used by Ginsburg and Landau to study the effect of external electromagnetic fields on superconductors. Basically, the physical picture behind it is that the single-particle wavefunction $\phi_0(\vec{x},t)$ changes in the presence of (weak) external perturbations. However, all the bosons are still condensed in the (changing) state $\phi_0(\vec{x},t)$, and the effect of external perturbations on the $N$-body system can be described by a time-dependent condensate wavefunction $\sqrt{N}\phi_0(\vec{x},t)$. For the contact interaction model (9.3), the equation governing the time evolution of the condensate is called the Gross–Pitaevskii equation

$$i\hbar\frac{\partial}{\partial t}\phi_0 = -\frac{\hbar^2}{2m}\nabla^2\phi_0 + v(\vec{x},t)\phi_0 + g|\phi_0|^2\phi_0, \tag{9.29}$$

where $v(\vec{x},t)$ is the external time-dependent potential. This equation is applied very frequently to describe the behavior of cold atom condensates nowadays.

#### 9.2.3.4 **Two-Fluid Picture**

At finite temperatures below $T_c$, some of the atoms in the liquid are in excited states (i.e. not a part of the condensate) and the liquid can be considered as composed of two components: the first component is the condensate, which describes the collective motion of the bosons and is called the superfluid component. The second component is formed by the excited atoms which can exchange energies with the surroundings easily by collision and is called the normal component. The dynamics of the whole liquid can be described by a two-fluid model composed of these two components of liquids.

As explained above, the dynamics of the superfluid component is described by the Gross–Pitaevskii equation. The question is how about the normal component? What is the appropriate 'thermodynamic condition' that determines the distribution of the normal component atoms? Would the normal component have an effect also on the superfluid dynamics?

A complete description of the dynamics and thermodynamics of a non-equilibrium superfluid system is so far not available and physicists usually attack this problem using the so-called 'local-equilibrium' approximation. The idea of this approximation is that locally we can treat the superfluid component as a superfluid with uniform density moving with a constant velocity $\vec{v}_s$. In this case, we can make a Galilean transformation to the rest frame of the superfluid and determine the distribution of normal component atoms, assuming that usual thermodynamics can be applied in this rest frame.

If the center of mass of the normal fluid component is at rest with the superfluid, we expect that the number density of excited atoms with momentum $\vec{p}$ is

$$n_B(\vec{p}) = \frac{1}{e^{\frac{\omega(\vec{p})}{k_B T}} - 1}, \tag{9.30a}$$

where $\omega(\vec{p})$ is the phonon excitation spectrum we determined before. It is more interesting and important to consider the case when the center of mass (CM) of the normal fluid component is moving with a relative velocity $\vec{v} = \vec{v}_n - \vec{v}_s$ to the superfluid ($\vec{v}_n$ is the CM velocity of the normal component in the laboratory frame). In this case, we expect as a result of Galilean transformation (see also Equation (9.24)) that

$$\omega(\vec{p}) \to \omega(\vec{p}) - \vec{p}.\vec{v} \tag{9.30b}$$

in the rest frame of the superfluid. By inserting this into Equation (9.30a), we obtain for the (momentum) current carried by the normal fluid

$$\vec{j}_n = \sum_{\vec{p}} \vec{p} \frac{1}{e^{\frac{\omega(\vec{p}) - \vec{p}.\vec{v}}{k_B T}} - 1} = \vec{v} \sum_{\vec{p}} \left( \frac{\vec{p}.\vec{v}}{v^2} \right) \frac{1}{e^{\frac{\omega(\vec{p}) - \vec{p}.\vec{v}}{k_B T}} - 1} = \rho_n(v) \vec{v} \tag{9.31a}$$

in the rest frame of the superfluid. Therefore, the current in the laboratory frame is

$$\vec{j} = \rho \vec{v}_s + \rho_n (\vec{v}_n - \vec{v}_s) = \rho_s \vec{v}_s + \rho_n \vec{v}_n, \tag{9.31b}$$

where $\rho_s = \rho - \rho_n$ is the superfluid density. Thus, at non-zero temperature current flow in the system is in general a sum of supercurrent and normal current.

Following this line, one can also obtain a hydrodynamic theory for superfluid flow starting from general conservation laws and thermodynamic considerations, similar to what we have done in Chapter 7 in deriving the Navier–Stokes equation. We shall not go into these details here. Interested readers can look at Ref. [4] for more details.

**Homework Q5**

Show that, in the limit of small $\vec{v}$, the normal fluid density is given by

$$\rho_n \sim \sum_{\vec{p}} \left( \frac{p^2}{3} \right) \left( -\frac{\partial n_B(\omega)}{\partial \omega} \right)_{\omega = \omega(\vec{p})}.$$

Show that $\rho_n \sim T^4$ at low temperature in 3D.

## 9.3 Charged Superfluids: Higgs Mechanism and Superconductivity

### 9.3.1 Goldstone Theorem and Higgs Mechanism

You may wonder why we are getting bosonic excitations with energy dispersion $\omega_q = cq$ all the time. The sound wave in solids has this dispersion, and so does the density wave in a (quantum) liquid of bosons. Actually, there is a reason behind it – the Goldstone theorem, which states the following.

> *If the ground state breaks a continuous symmetry possessed by the Hamiltonian, there exists always low-energy excitations with dispersion $\omega_q = cq$.*

The continuous symmetry being broken for solids is translational symmetry. The energy of the solid does not change if the location of the *whole* crystal is shifted by an arbitrary amount, that is there is huge ground-state degeneracy. However, once a solid is formed, the positions of all atoms are fixed (or a particular ground state is being chosen). Instead of translational symmetry, the continuous symmetry being broken in a quantum liquid of bosons is gauge symmetry, the symmetry with respect to a global change in phase of the boson system $\psi(\vec{r}) \rightarrow e^{i\theta}\psi(\vec{r})$. The energy of the system remains unchanged when $\theta$ is varied. However, a particular value of $\theta$ is chosen in the ground state.

We shall not prove the Goldstone theorem here except to give you a qualitative picture of how the theorem works. Let us take the quantum liquid as example. Imagine a quantum state where the wavefunction is essentially the ground-state wavefunction except that the overall phase $\theta$ of the wavefunction is slowly varying as a function of position in the liquid, that is

$$|\psi\rangle \sim e^{i\sum_i \theta(\vec{r}_i)}|\psi_G\rangle$$

with, for example, $\theta(\vec{r}) \sim \theta_0 \sin(\vec{q}.\vec{r})$, where $\vec{q}$ is small. As $\vec{q} \rightarrow 0$, the phase is fixed and the system is in the ground state. We also expect that $|\psi\rangle$ is orthogonal to $|\psi_G\rangle$ when $\vec{q} \neq 0$. Therefore, the excitation energy of this state should have the property $\omega(\vec{q}) \rightarrow 0$ when $\vec{q} \rightarrow 0$. It is much harder to show that $\omega_q \sim cq$ at small $\vec{q}$, and we shall not prove this here.

Historically, the Goldstone theorem posed a problem in high-energy physics when physicists were building up the theory unifying weak and electromagnetic interactions. Roughly speaking, both electromagnetic and weak interactions are described by gauge theories and are expected to support massless photon-like excitations. However, the particle mediating the weak interaction is massive and short range. To explain this disparity between electromagnetic and weak interactions, a boson field (with Bose condensation) that couples with the weak interaction has to be introduced. However, such a boson field would introduce a massless (or gapless) excitation ($\omega_q \sim cq$) according to the Goldstone theorem. Unfortunately, this massless bosonic excitation was never observed experimentally.

A solution to this problem was pointed out by Anderson and Higgs in different contexts and is now called the Anderson–Higgs mechanism. The mechanism says that a massless Goldstone boson will become massive ($\omega_q^2 \rightarrow m^2 + c^2q^2$) if the boson system couples with a gauge field. The mechanism provides a natural explanation of why there is no Goldstone mode associated with the bosons coupling to the weak interaction. We shall illustrate the Anderson–Higgs mechanism in the following using a simple model of charged bosons coupling to a U(1) electromagnetic field in three dimensions,

$$S=\int d\vec{r}\int_0^\beta d\tau\left\{\bar{\psi}(\vec{r})\left[\frac{\partial}{\partial\tau}-\frac{\hbar^2}{2m}(\nabla-ie^*\vec{A})^2-\mu\right]\psi(\vec{r})+\frac{1}{2}g[\bar{\psi}(\vec{r})\psi(\vec{r})]^2\right\}$$

$$+\frac{1}{2}\int d\vec{r}\int d\vec{r}'\int d\tau\bar{\psi}(\vec{r},\tau)\bar{\psi}(\vec{r}',\tau)\frac{e^{*2}}{|\vec{r}-\vec{r}'|}\psi(\vec{r}',\tau)\psi(\vec{r},\tau)+S(\vec{A}), \tag{9.32}$$

where we have written the action in the Coulomb gauge $\nabla.\vec{A}=0$. $S(\vec{A})$ is the action for the transverse electromagnetic field. The scalar field $A_0$ is eliminated in favor of the Coulomb interaction $e^{*2}/r$ between charges. Note that $e^*$ may not be equal to ordinary electric charge $e$ in Equation (9.32).

We proceed as before and let $\psi=\psi_0+\delta\psi$ ($\bar{\psi}=\bar{\psi}_0+\delta\bar{\psi}$), where $\psi_0=\sqrt{N_0}e^{i\theta}$, $\bar{\psi}_0=\sqrt{N_0}e^{-i\theta}$, and $N=N_0+n$.

Expanding to second order in $n$ and $\theta$ as before, we obtain

$$S=S_{\text{cl}}+\delta S+S(\vec{A}),\quad \text{where}$$

$$\delta S=\int d\vec{r}\int_0^\beta d\tau\left\{i\hbar n\cdot\partial_\tau\theta+\frac{\hbar^2 N_0}{2m}(\nabla\theta-e^*\vec{A})^2+\frac{1}{2}g\cdot n^2\right\}$$
$$+\int d\tau\int d\vec{r}\int d\vec{r}'\frac{e^2 n(\vec{r})n(\vec{r}')}{|\vec{r}-\vec{r}'|}. \tag{9.33}$$

Fourier transforming and integrating out $n$, we obtain a phase action

$$\delta S=\sum_{\vec{q}}\frac{1}{2g}\left[\omega(\vec{q})^2-\frac{g(i\omega_n)^2}{g+V(q)}\right]\cdot\left[\theta(q)-\frac{e^*A(q)}{iq}\right]\cdot\left[\theta(-q)-\frac{e^*A(-q)}{iq}\right], \tag{9.34}$$

where $\omega(\vec{q})=cq$, $q=(\vec{q},i\omega)$, and $V(q)=\frac{4\pi e^{*2}}{q^2}$ is the Fourier transform of the Coulomb potential.

The excitation spectrum is given by the pole of the action (see also Chapter 10), determined by

$$\frac{g\omega^2}{g+V(q)}=\omega(q)^2,\quad \text{where}\quad V(q)=\frac{4\pi e^{*2}}{q^2}. \tag{9.35a}$$

Note that $e^*=0\Rightarrow\omega=\omega(q)=cq$, which is the original Goldstone mode.

However, for $e^*\neq 0$, we obtain $\frac{gN_0q^2}{m}=\frac{g\omega^2q^2}{4\pi e^{*2}}=0$ in the limit $q\to 0$

$$\Rightarrow\omega^2=\frac{4\pi e^{*2}N_0}{m}=\omega_{\text{p}}^2\ \text{(called the plasma frequency)} \tag{9.35b}$$

and the original Goldstone mode acquires a gap! This is an example of the Anderson–Higgs mechanism.

**Homework Q6a**

Derive the action $\delta S$(9.34) and then the corresponding Hamiltonian. Show that the excitation energy is $\omega(q) \to \omega_p$ in the limit $q \to 0$.

## 9.3.2 Higgs Mechanism and Superconductivity

A quantum liquid of charged bosons (coupling to a U(1) gauge field) with a ground state that breaks gauge symmetry is automatically a superconductor.

To see this, we consider the action (9.34) at small $\vec{q}$ and $\omega$. Neglecting terms of order $q^2\omega^2$, we obtain

$$\delta S = \int dt d\vec{r} \frac{N_0 g^2}{2m}(\nabla\theta - e^*\vec{A})^2 = \int dt\, d\vec{r} \rho_s (\nabla\theta - e^*\vec{A})^2, \tag{9.36}$$

where $\rho_s = \frac{N_0 g^2}{2m} > 0$ is called the superfluid density.

The total action for an electromagnetic field (in Coulomb gauge) in the small-$\vec{q}$ limit is thus

$$S = \delta S + S(\vec{A}) = \int dt d\vec{r} \left[ (e^2\rho_s)\vec{A}^2 + \frac{1}{8\pi}\left(\frac{\partial \vec{A}(\vec{x})}{\partial t}\right)^2 - \frac{1}{8\pi}(\nabla \times \vec{A})^2 \right]. \tag{9.37}$$

Note that $\int d^d r \vec{A}.\nabla\theta = 0$. We have also set $c = 1$.

We shall show that this action reproduces the Meissner effect, which is a trade mark for superconductivity.

### 9.3.2.1 Meissner Effect (Higgs Mechanism on Gauge Field)

We shall study the effect classically. Minimizing the action, we obtain the equation of motion for a gauge field,

$$\frac{\partial^2 \vec{A}}{\partial t^2} - \nabla \times \nabla \times \vec{A} + \frac{1}{\lambda^2}\vec{A} = 0, \tag{9.38a}$$

where $\lambda = \frac{1}{8\pi e^{*2}\rho_s}$ is the penetration depth.

Taking curl on both sides of the equation, we obtain

$$\frac{\partial^2 \vec{B}}{\partial t^2} - \nabla^2 \vec{B} + \frac{1}{\lambda^2}\vec{B} = 0. \tag{9.38b}$$

For a time-independent magnetic field we obtain $\vec{B}(x) \sim \vec{B}(0)e^{-x/\lambda}$, meaning that a constant magnetic field cannot exist inside a Bose-condensed charged boson liquid and must decay inside the liquid with penetration length $\lambda$. This is precisely the Meissner effect.

Solving the corresponding time-dependent equation with a solution $\vec{B}(\vec{x},t) \sim \vec{B}_0 e^{i(\vec{k}.\vec{x}-\omega t)}$, we find that the energy dispersion of a photon changes from $\hbar\omega(\vec{k}) = c|\vec{k}|$ to

$$\hbar\omega(\vec{k}) = \sqrt{m^2 + (ck)^2} \tag{9.38c}$$

in a superconductor, where $m \propto \lambda^{-1}$. This phenomenon where the photon acquires a mass gap is the Anderson–Higgs mechanism acting on an electromagnetic field. We shall discuss superconductivity in more detail in Chapter 11.

**Homework Q6b**

Derive Equation (9.38c). Explain what happens when the action (9.37) is quantized. (See Chapter 6 on quantization of the electromagnetic field for reference.)

## 9.4 Supersolids

Imagine liquid $He^4$ (or a liquid of other boson atoms) put under pressure. When the pressure is large enough, the liquid solidifies. The resulting quantum solid state may have very peculiar properties. Besides a normal solid state, at low enough temperature and not too high a pressure, it may display so-called supersolid behavior, meaning that it is both a solid and a superfluid. This unusual state is possible only because of quantum mechanics!

To show the plausible existence of a supersolid state we have to show two things: (1) we have to show that, even in the presence of a periodic lattice structure, a quantum state with a uniform current flow could exist; and (2) this state of uniform current flow is not destroyed by excitation if the flow velocity is low enough (Landau criterion). We shall concentrate on (1) in the following.

In the absence of lattice structure, a state corresponding to uniform fluid flow exists because of translational invariance. However, in the presence of the periodic lattice structure the translational invariance is broken and the existence of a uniform current flow state is no longer automatic. In fact, the plausible existence of a uniform current flow state is a consequence of quantum mechanics and is reflected in a (generalized) Bloch theorem.

Let us consider a Hamiltonian

$$H = -\frac{\hbar^2}{2m}\sum_i \nabla_i^2 + \frac{1}{2}\sum_{i\neq j} U(\vec{r}_i - \vec{r}_j) + \varepsilon \sum_i V(\vec{r}_i), \tag{9.39}$$

where $\vec{r}_i$ is the position of the $i$th particle. $U(\vec{r})$ is the interaction potential between particles and $V(\vec{r}) = V(\vec{r} + \vec{R})$ is a periodic potential which the particles see. $V(\vec{r})$ is chosen in a way such that the $\vec{R}$ are Bravais lattice vectors of the underlying boson crystal lattice in the physical limit $\varepsilon \to 0$.

Because the Hamiltonian is invariant under translation by the $\vec{R}$, that is $H[\vec{r}+\vec{R}] = H[\vec{r}]$, by Bloch's theorem the boson wavefunction $\psi(\vec{r}_1, \vec{r}_2, \ldots, \vec{r}_N) = \psi([\vec{r}])$ must have the form

$$\begin{aligned}\psi(\vec{r}_1, \vec{r}_2, \ldots, \vec{r}_N) &= e^{i\vec{k}.(\vec{r}_1+\vec{r}_2+\ldots+\vec{r}_N)}\phi(\vec{r}_1, \vec{r}_2, \ldots, \vec{r}_N) \\ &= e^{iN\vec{k}.\vec{r}_{CM}}\phi(\vec{r}_1, \vec{r}_2, \ldots, \vec{r}_N),\end{aligned} \tag{9.40}$$

where $\vec{r}_{CM} = \frac{1}{N}\sum_i \vec{r}_i$ is the center of mass coordinate of the system and $\phi(\vec{r}_1+\vec{R}, \vec{r}_2+\vec{R}, \ldots, \vec{r}_N+\vec{R}) = \phi(\vec{r}_1, \vec{r}_2, \ldots, \vec{r}_N)$ is a periodic function with period of the $\vec{R}$. Therefore, the boson eigenstate wavefunctions can be classified by 'bands' associated with the center of mass motion,

$$\psi(\vec{r}_1, \vec{r}_2, \ldots, \vec{r}_N) = \psi[\vec{r}] = e^{iN\vec{k}.\vec{r}_{CM}}\phi_{n\vec{k}}[\vec{r}].$$

Let us consider the lowest band of the system such that $\phi_{00}[\vec{r}]$ is the ground state of the system (note that the ground-state wavefunction of a boson system must be real [1]) and evaluate the current carried by the state $\psi[\vec{r}] = e^{iN\vec{k}.\vec{r}_{CM}}\phi_{0\vec{k}}[\vec{r}]$, which, by definition, is given by

$$\vec{J}_{0\vec{k}} = \sum_i <\dot{\vec{r}}_i> = -\frac{i\hbar}{m}\sum_i <\psi_{0\vec{k}}|\nabla_i|\psi_{0\vec{k}}> = -\frac{i\hbar}{m}\sum_i <\phi_{0\vec{k}}|(\nabla_i + i\vec{k})|\phi_{0\vec{k}}>. \tag{9.41a}$$

The expectation value can be evaluated most easily by noting that the Schrödinger equation determining $\phi_{0\vec{k}}[\vec{r}]$ is $H'_{\vec{k}}\phi_{0\vec{k}} = E_{0\vec{k}}\phi_{0\vec{k}}$, where

$$H'_{\vec{k}} = -\frac{\hbar^2}{2m}\sum_i(\nabla_i + i\vec{k})^2 + \frac{1}{2}\sum_{i\neq j} U(\vec{r}_i - \vec{r}_j) + \varepsilon\sum_i V(\vec{r}_i). \tag{9.41b}$$

Therefore,

$$H'_{\vec{k}+\vec{q}} = H'_{\vec{k}} - \frac{\hbar^2}{m}i\vec{q}.\sum_i(\nabla_i + i\vec{k}) + O(\vec{q}^2). \tag{9.41c}$$

Taking the expectation value, we obtain

$$E_{0\vec{k}+\vec{q}} = E_{0\vec{k}} - \hbar\vec{q}.\vec{J}_{0\vec{k}} + O(\vec{q}^2). \tag{9.42a}$$

Therefore,

$$\vec{J}_{0\vec{k}} = \frac{1}{\hbar}\nabla_{\vec{k}}E_{0\vec{k}} \tag{9.42b}$$

is non-zero as long as $E_{0\vec{k}}$ has a finite dispersion. Since we expect that the ground state is a state with zero current, $E_{0\vec{k}} \sim E_{00} + N\frac{\hbar^2\vec{k}^2}{2m^*} + O(k^3)$ at small $\vec{k}$ and

$$\vec{J}_{0\vec{k}} \sim \left(\frac{1}{m^*}\right)N\hbar\vec{k} = \left(\frac{m}{m^*}\right)\vec{J}^{(0)}_{0\vec{k}}, \tag{9.42c}$$

where $\vec{J}^{(0)}_{0\vec{k}}$ is the supercurrent carried by the corresponding non-crystalline superfluid state. Generally speaking, $m^* > m$ in the presence of a periodic potential and the superfluid density defined by $\vec{j}_{0\vec{k}} = \frac{\vec{J}_{0k}}{V} = \rho_s \frac{h\vec{k}}{m}$ is reduced in the supersolid state by $\rho_s = \left(\frac{m}{m}\right)\rho_s^{(0)}$. The (crystal) momentum carried by the supersolid state is $\vec{P}_{0\vec{k}} \sim N\hbar\vec{k}$. This analysis also suggests that a normal solid state is one with effective mass $m^* \longrightarrow \infty$.

Landau's argument applies if we replace the momentum of the translational motion by the crystal momentum and if excitations in the supersolid state have dispersions $\omega \sim ck$. In this case a superfluid flow exists if the velocity of the superfluid flow is below a certain critical value. Supersolid states in $He^4$ and $H^2$ under high pressure were observed a few years ago by Kim and Chan [ref. (5)], but so far no other experimental groups have confirmed their observations. There are still lots of controversies about the supersolid state and it remains a topic of frontier research.

## 9.5 A Brief Comment Before Ending

We have introduced in this chapter some basic concepts and techniques commonly used in studying simple boson quantum liquids. It should be emphasized that boson quantum liquids display a very rich variety of phenomena that go far beyond the scope of this book, for example bosons with spins and rotating boson liquids. The important topic of vortices is also not covered in this chapter and will be discussed in Chapters 11 and 12. A similar situation also applies in later chapters. The existence of rich and diversified phenomena that cannot be covered by a few simple rules is an important feature of condensed matter physics, which readers should keep in mind when entering this field.

## References


1 Feynman, R.P. *Statistical Mechanics*, The Benjamin/Cummings Publishing Company, Inc. (Reading, Massachusetts) (1982).
2 Huang K., *Statistical Mechanics*, John Wiley, (New York) (1987).
3 Nagaosa, N., *Quantum Field Theory in Condensed Matter Physics*, Springer – Verlag Berlin Heidelberg New York (1999).
4 Khalatnikov, I.M. *An Introduction to the Theory of Superfluidity*, Addison-Wesley (1965).
5 E. Kim and M.H.W. Chan, *Nature*, **427**, 225 (2004).

# 10
# Simple Fermion Liquids: Introduction to Fermi Liquid Theory

## 10.1
## Single-Particle and Collective Excitations in Fermi Liquids

### 10.1.1
### The Spectrum of a Free Fermi Gas

The spectrum of a free Fermi gas differs from the spectrum of free bosons considerably. All bosons condense in the zero-momentum single-particle state in the ground state, while this is not allowed for fermions because of the Pauli exclusion principle. The single-particle–hole-pair excitation spectrum of Bose-condensed bosons with momentum $\vec{q}$ is

$$E(\vec{q}) = \varepsilon_{\vec{q}} - \varepsilon_0, \tag{10.1}$$

where $\varepsilon_q = \hbar^2\vec{q}^2/2m$ is the single-particle spectrum for the particles. We note that, *for each value of the momentum, there is only one possible excitation energy.* This is a unique character for the boson excitation spectrum which is more or less kept even when interaction between bosons exists – at least in the saddle-point approximation we discussed in Chapter 9.

Because of the Pauli exclusion principle, a system of $N$ free fermions occupies the lowest $N$ single-particle states of the single-particle spectrum. The single-particle–hole-pair excitation spectrum of fermions with momentum $\vec{q}$ is therefore

$$E_{\vec{k}}(\vec{q}) = \varepsilon_{\vec{k}+\vec{q}} - \varepsilon_{\vec{k}}, \tag{10.2}$$

where any value of $\vec{k}$ with $|\vec{k}| < k_F$, $|\vec{k}+\vec{q}| > k_F$ is allowed, where $k_F$ is the Fermi momentum determined by $n = \frac{1}{(2\pi)^3}\left(\frac{4\pi}{3}k_F^3\right)$ in 3D ($n$ = density of electrons) (see figure next page).

*Introduction to Classical and Quantum Field Theory.* Tai-Kai Ng

ISBN: 978-3-527-40726-2

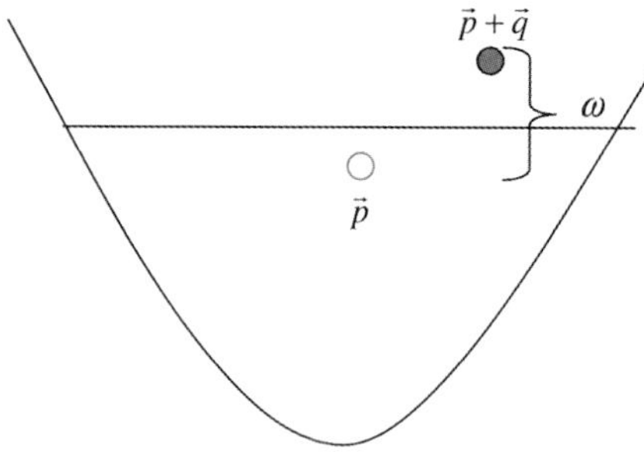


When compared with the boson excitation spectrum, we see that the main difference is that for a single momentum $\vec{q}$ there are many possible excitations with different energies. In fact, it is not difficult to see that all excitations with energy

$$qk_F/m+\varepsilon_{\vec{q}}>\varepsilon_{\vec{k}+\vec{q}}-\varepsilon_{\vec{k}}>0 \text{ are allowed for } q<2k_F. \tag{10.3}$$

**Homework Q1**

Derive Equation (10.3). How about $q>2k_F$?

The excitation spectrum for single-particle–hole excitations is often shown in a $\omega-q$ diagram as follows. First, we consider bosons.

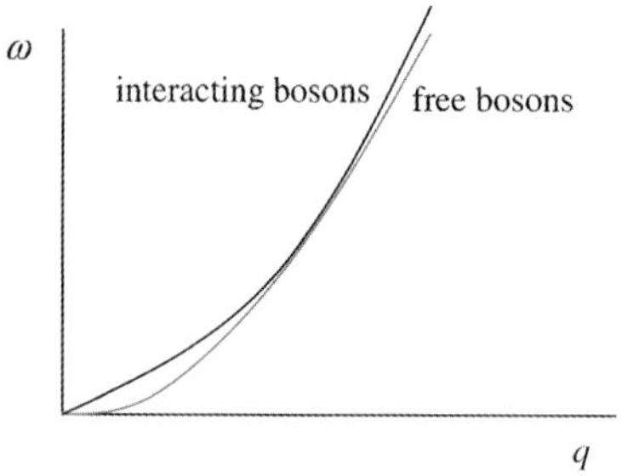


Note that $\omega\sim cq$ at small $q$ for interacting bosons (Chapter 9).
Then, for fermions, we have the following diagram:

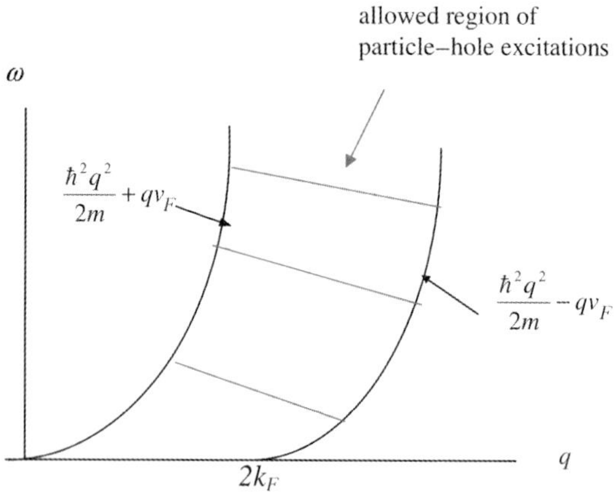

Landau proposed that, unlike bosons, where the excitation spectrum changes from $E(q)=\varepsilon_{\vec{q}}$ to $E(\vec{q})=\sqrt{\frac{q^2}{2m}\left(\frac{q^2}{2m}+2N_0g\right)}$ in the presence of interaction, the (particle–hole) excitation spectrum of fermions is very robust against interaction. His proposal is called 'Fermi liquid theory' nowadays and remains the foundation of the theory of fermion gases in two and higher dimensions. (In 1D, a Landau Fermi liquid is replaced by a different kind of liquid called a Luttinger liquid.) (ref. [1], [2]). We shall discuss Landau's Fermi liquid theory in more detail in the second part of this chapter.

### 10.1.2 Collective Modes in Fermi Liquid

Besides particle–hole excitations where individual particles are excited from below the Fermi sea to above, collective excitations where a macroscopic number of particles move coherently together is also allowed in fermion liquids when interaction is present. Here, we introduce an important technique in quantum field theory (QFT) that is often applied to study collective motions in fermionic systems. We start with a simple model of interacting fermions with repulsive interaction $V(\vec{r})$. We shall consider two kinds of interactions in the following: (i) short-range interaction $V(\vec{r})\sim g\delta(\vec{r})$ ($g>0$) and (ii) $V(\vec{r})=\frac{e^2}{|\vec{r}|}$ (Coulomb interaction). The first interaction is a good model for fermionic atoms ($He^3$, for example), whereas the second interaction is a good model for electrons in metals.

The Hamiltonian of the system is

$$\hat{H}=\sum_{i=1}^{N}\frac{\hat{p}_i^2}{2m}+\sum_{i<j}V(\vec{r}_i-\vec{r}_j). \tag{10.4a}$$

For electrons in metals (Coulomb interaction), an additional term should be added to ensure charge neutrality in the system:

$$\hat{H}=\sum_{i=1}^{N}\frac{\hat{p}_i^2}{2m}+\sum_{i<j}\frac{e^2}{|\vec{r}_i-\vec{r}_j|}+\text{background}. \tag{10.4b}$$

The background term represents the potential energy from a uniform positive charge background with density the same as the average density of electrons.

Fourier transforming in second quantization representation, we obtain

$$\hat{H}=\sum_{\vec{k},\sigma}\left(\frac{\hbar^2k^2}{2m}-\mu\right)\hat{C}^{+}_{\vec{k},\sigma}\hat{C}_{\vec{k},\sigma}+\frac{1}{2}\sum_{\vec{q}}V(\vec{q})\hat{\rho}_{\vec{q}}\hat{\rho}_{-\vec{q}}\ \text{for neutral atoms} \tag{10.5a}$$

and

$$\hat{H}=\sum_{\vec{k},\sigma}\left(\frac{\hbar^2k^2}{2m}-\mu\right)\hat{C}^{+}_{\vec{k},\sigma}\hat{C}_{\vec{k},\sigma}+\frac{1}{2}\sum_{\vec{q}\neq 0}\frac{4\pi e^2}{q^2}\hat{\rho}_{\vec{q}}\hat{\rho}_{-\vec{q}}\ \text{for electrons.} \tag{10.5b}$$

Note that the $\vec{q}=0$ component of the interaction is cancelled by the positive charge background.

$$\hat{\rho}_{\vec{q}} = \int d\vec{r}\hat{\rho}(\vec{r})e^{-i\vec{q}\cdot\vec{r}} = \sum_{\vec{k},\sigma} \hat{C}^{+}_{\vec{k},\sigma}\hat{C}_{\vec{k}+\vec{q},\sigma} \tag{10.5c}$$

is the density operator.

In the following, we shall not emphasize the differences between neutral and charged liquids except keeping in mind that the $\vec{q} = 0$ term of the interaction will be absent for the charged liquid.

The corresponding action at imaginary time is

$$S(\beta) = \sum_{\vec{k}} \int d\tau \left[ \bar{\psi}_\sigma(\vec{k},\tau)\left[\frac{\partial}{\partial\tau} + \frac{\hbar^2 k^2}{2m} - \mu\right]\psi_\sigma(\vec{k},\tau) - \frac{1}{2}V(\vec{k})\rho_k(\tau)\rho_{-k}(\tau)\right], \tag{10.6a}$$

where

$$\rho_{\vec{q}}(\tau) = \sum_{\vec{k},\sigma} \bar{\psi}_\sigma(\vec{k},\tau)\psi_\sigma(\vec{k}+\vec{q},\tau) \tag{10.6b}$$

and $Z(\beta) = \int D\psi \int D\bar{\psi} e^{-S(\beta)}$. Note that $\psi$, $\bar{\psi}$ are Grassmann variables.

Following our experience in working with bosons, we may like to make the semiclassical approximation and minimize the action with respect to $\psi$, $\bar{\psi}$. However, this is not meaningful because $\psi$, $\bar{\psi}$ are Grassmann numbers. For example, an equation of the form

$$-ax + bx^3 = 0$$

where $a$, $b$ are ordinary numbers but $x$ is a Grassmann number has no solution in general. (Note that $x = 0$ is not a solution since '0' is not a Grassmann number.) Physically, this is because we cannot put more than one fermion into a single state to obtain a coherent state wavefunction as in the case of bosons.

However, physically there is also no reason why macroscopic motions in fermionic systems where many fermions move together in similar ways cannot exist (an example is electric current flow). The question is: how can we describe these macroscopic (or collective) behaviors of fermionic systems?

We shall consider as example density fluctuations in fermionic systems in the following. To begin with, we note that the density operator $\rho$ is a product of two fermion operators, and thus has bosonic character. Thus, to understand density fluctuations it would be useful if we can write the Hamiltonian or action of the system in terms of $\rho$ instead of the Grassmann variables $\psi$, $\bar{\psi}$ that represent individual atoms/electrons. This is what the Hubbard–Stratonovich transformation does.

#### 10.1.2.1 Hubbard–Stratonovich Transformation

The Hubbard–Stratonovich (HS) transformation is a simple mathematical trick involving Gaussian integrals we discussed in Chapter 1. For the Coulomb interaction we note that, aside from a constant of integration,

$$\exp\left[-2\pi e^2\int d\tau\sum_{q\neq 0}\frac{\rho_q(\tau)\rho_{-q}(\tau)}{q^2}\right]$$

$$=\int D\varphi(\vec{q},\tau)\exp\left[-\frac{1}{8\pi}\int d\tau\left(\sum_{\vec{q}\neq 0}q^2\varphi(\vec{q},\tau)\varphi(-\vec{q},\tau)+i4\pi e(\varphi(\vec{q},\tau)\rho_q(\tau)\right.\right.$$
$$\left.\left.+\rho_q(\tau)\varphi(-\vec{q},\tau)\right)\right], \tag{10.7}$$

where $\rho_q(\tau)$ and $\varphi(\vec{q},\tau)$ are Fourier transforms of real number fields $\rho(\vec{r},\tau)$ and $\varphi(\vec{r},\tau)$. $\varphi(\vec{r},\tau)$ is a fictitious field (the Hubbard–Stratonovich field) similar to the $\vec{J}$ field we introduced in Chapter 8 when studying vortices. Therefore, after Fourier transformation, we may write

$$S(\psi,\bar{\psi})\Rightarrow S(\psi,\bar{\psi},\varphi)=\sum_{\vec{k}}\int d\tau\int d\vec{r}\left\{\frac{1}{8\pi}(\nabla\varphi)^2+\bar{\psi}_\sigma\left[\frac{\partial}{\partial\tau}+\frac{\nabla^2}{2m}-\mu+ie\varphi(\vec{r},\tau)\right]\psi_\sigma\right\}$$

$$\text{and } Z(\beta)=\int D\psi\int D\bar{\psi}\int D\varphi e^{-S(\psi,\bar{\psi},\varphi)}. \tag{10.8}$$

Note that by minimizing the action with respect to $\varphi$ we obtain

$$\nabla^2\varphi=-4\pi i\rho\,(\text{Gauss' law})$$

and $i\varphi$ is just the scalar potential in electrostatics. A similar result can be obtained for general positive $V(\vec{q})>0$.

**Homework Q2a**

Derive the HS transformation (10.7) and effective action (10.8) for general $V(\vec{q})>0$.

To proceed further, we Fourier transform the action to obtain

$$S=\sum_{k,k',\sigma}\left[\frac{(\vec{k}-\vec{k}')^2}{8\pi}\varphi(k-k')\varphi(k'-k)+\bar{\psi}_\sigma(k)M_{k,k'}[\varphi]\psi_\sigma(k')\right], \tag{10.9a}$$

where

$$k=(\vec{k},i\omega_n),\ k'=(\vec{k}',i\omega'_n),\ \text{and}$$
$$M_{k,k'}[\varphi]=\delta_{k,k'}(-i\omega_n+\varepsilon_k-\mu)+ie\varphi(k-k'). \tag{10.9b}$$

Note that the action is now quadratic in the fermion variables. Next, we use a useful formula for Grassmann variable integration:

$$\int D\psi\int D\bar{\psi}\exp\left[-\sum_{i,j}\bar{\psi}_i M_{ij}\psi_j\right]=\det M=e^{\ln\det M}. \tag{10.10}$$

**Proof**

*Introduce a unitary transformation to diagonalize the matrix M. The path integral can be performed over each eigenstate Grassmann variable separately to obtain the result (see also Chapter 3).*

Therefore, the integration over Grassmann variables can be performed to obtain

$$Z(\beta) = \int D\varphi e^{-\left\{\sum_q \frac{q^2}{8\pi}\varphi\varphi\right\} + 2\ln\det M_{k,k'}}. \tag{10.11}$$

The factor of two is coming from spin. Note that $Z$ is now a functional of only $\varphi(\vec{r}, \tau)$ which is an ordinary number field, and we can use techniques we discussed in Chapter 9 to treat $Z$ if we know how to evaluate $\ln\det M[\varphi]$.

As in the case of bosons, we expect that the action has a minimum for a particular configuration $\varphi(\vec{r}, \tau) = \varphi_0(\vec{r})$. For a translationally invariant system, we expect that $\varphi_0(\vec{r}) = i\varphi_0$ is independent of $\vec{r}$. In this case,
$M_{k,k'}[\varphi] \to \delta_{k,k'} M_0([\varphi_0], k)$ is a diagonal matrix and

$$\ln\det M_0 = \ln\left(\prod_{\vec{k}, i\omega_n} (i\omega_n - \varepsilon_k + \mu + e\varphi_0)\right) = \sum_{\vec{k}, i\omega_n} \ln(i\omega_n - \varepsilon_k + \mu') = \mathrm{Tr}(\ln M_0), \tag{10.12}$$

where we find that $e\varphi_0$ can be absorbed by a redefinition of chemical potential.

At zero temperature, we find that, after minimizing the free energy with respect to $\varphi_0$ (recall that we have learnt how to evaluate the sum over $i\omega_n$ in Chapter 4),

$\varphi_0 = V(q = 0)\, n$ for (neutral) atoms (Hartree energy)

and $\varphi_0 = 0$ for electrons in metals.

**Homework Q2b**

Derive the above results.

We next expand the $\varphi(\vec{r}, \tau)$ field around the classical minimum (semiclassical approximation), that is we write

$$\varphi(\vec{r}, \tau) = i(\varphi_0 + \varphi'(\vec{r}, \tau)), \tag{10.13}$$

and expand $\ln\det M = \mathrm{Tr}(\ln M)$ to second order in $\varphi'(\vec{r}, \tau)$.

*Perturbation expansion for (lnM):*

We note that, for general matrix $M_{k,k'} = M_{0k}\delta_{k,k'} + \lambda \cdot M_{1k,k'}$,

$$\begin{aligned} \ln M &= \ln(M_0 + \lambda M_1) = \ln M_0(1 + \lambda M_0^{-1} M_1) \\ &= \ln M_0 + \ln(1 + \lambda M_0^{-1} M_1) \\ &= \ln M_0 + \lambda M_0^{-1} M_1 + \frac{1}{2}\lambda^2 M_0^{-1} M_1 M_0^{-1} M_1 + O(\lambda^3). \end{aligned} \tag{10.14a}$$

Therefore,

$$\mathrm{Tr}(\ln M) \sim \mathrm{Tr}(\ln M_0) - \lambda \mathrm{Tr}(G_0 M_1) + \frac{1}{2}\lambda^2 \mathrm{Tr}(G_0 M_1 G_0 M_1) + O(\lambda^3), \tag{10.14b}$$

where we have defined the matrix $G_{0k} = -M_{0k}^{-1}$ (see ref. [2]).

Now, let us apply the expansion to an electron gas, assuming that the charge $e$ is 'small'.

We identify

$$M_{0kk'} = -\delta_{k,k'}(i\omega_n - \varepsilon_k + \mu) = \delta_{k,k'} G_0(\vec{k}, i\omega_n)^{-1}$$
$$M_{1kk'} = ie\varphi(\vec{k}-\vec{k}', i\omega_n - i\omega'_n), \tag{10.15}$$

$$\mathrm{Tr}(\ln M) \to \mathrm{Tr}(\ln M_0) - \mathrm{Tr}(G_0 M_1) + \frac{1}{2}\mathrm{Tr}(G_0 M_1 G_0 M_1) + O(e^3), \tag{10.16a}$$

$$\mathrm{Tr}(G_0 M_1) = \sum_{k,k'} G_{0kk'} M_{1k'k} = \sum_{\vec{k},i\omega_n} G_0(\vec{k}, i\omega_n) \cdot ie\varphi(0) = 0, \tag{10.16b}$$

$$\begin{aligned}
\mathrm{Tr}(G_0 M_1 G_0 M_1) &= \frac{1}{\beta V} \sum_{k_1,k_2,k_3,k_4} G_{0k_1k_2} M_{1k_2k_3} G_{0k_3k_4} M_{1k_4k_1} \\
&= \frac{1}{\beta V} \sum_{k_1,k_3} G_0(k_1) M_{1k_1k_3} G_0(k_3) M_{1k_3k_1} \\
&= -\frac{e^2}{\beta V} \sum_{\substack{\vec{k}, i\omega_n \\ \vec{k}', i\omega''_n}} G_0(\vec{k}, i\omega_n)\varphi(\vec{k}-\vec{k}', i\omega_n - i\omega'_n) G_0(\vec{k}', i\omega'_n) \\
&\quad \times \varphi(\vec{k}'-\vec{k}, i\omega'_n - i\omega_n) \\
&= \sum_{\vec{q},i\omega_n} \frac{e^2}{2} \chi_0(\vec{q}, i\omega_n)\varphi(\vec{q}, i\omega_n)\varphi(-\vec{q}, -i\omega_n),
\end{aligned} \tag{10.16c}$$

where $q = k - k'$ and

$$\chi_0(\vec{q}, i\omega_n) = \frac{-2}{\beta V} \sum_{\vec{k},i\Omega_n} G_0(\vec{k}+\vec{q}, i\omega_n + i\Omega_n) G_0(\vec{k}, i\Omega_n) \tag{10.17}$$

is called the Lindhard function. The Lindhard function is usually represented by a 'bubble' diagram in perturbation theory (see below)

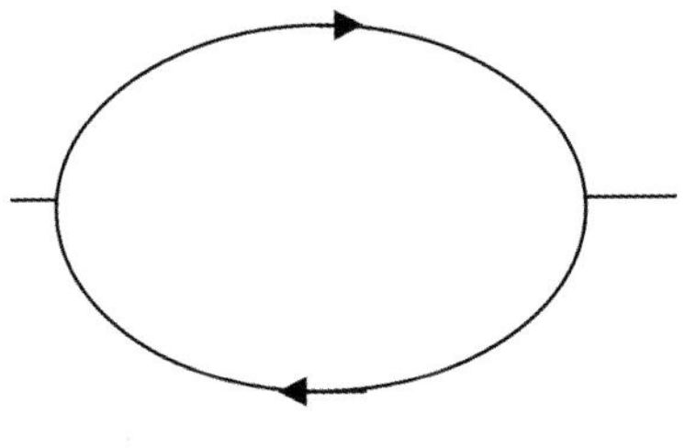

Therefore, to second order in $e\varphi$,

$$S_{\mathrm{eff}}(\varphi) = \frac{1}{2} \sum_{\vec{q},i\omega_n} \left[\frac{\vec{q}^2}{4\pi} - e^2 \chi_0(\vec{q}, i\omega_n)\right] \cdot \varphi(\vec{q}, i\omega_n)\varphi(-\vec{q}, -i\omega_n) + O(e^3). \tag{10.18}$$

This is called the RPA (random-phase approximation) effective action for density fluctuations. With this expression, one can evaluate (approximately) the free energy for the fermion system (see e.g. Ref. [3]).

**Homework Q3**
Introducing a source term as in Chapter 9, show that the density–density response function for an electron gas in RPA is $\chi(\vec{q},\omega)=\frac{\chi_0(\vec{q},\omega)}{1-V(q)\chi_0(\vec{q},\omega)}$, where $v(q)=\frac{4\pi e^2}{q^2}$.

#### 10.1.2.2 Excitation Spectrum of Electron Gas in RPA

We can obtain information on the excitation spectrum of an electron gas by looking at the imaginary part of the response function (recall the spectral representation of correlation functions in Chapter 5)

$$\chi(\vec{q},\omega)=\frac{\chi_0(\vec{q},\omega)}{1-\frac{4\pi e^2}{q^2}\chi_0(\vec{q},\omega).} \tag{10.19a}$$

The imaginary part is given by

$$\mathrm{Im}\chi(\vec{q},\omega)=\frac{\mathrm{Im}\chi_0(\vec{q},\omega)}{\left(1-\frac{4\pi e^2}{q^2}\mathrm{Re}\chi_0(\vec{q},\omega)\right)^2+\left(\frac{4\pi e^2}{q^2}\mathrm{Im}\chi_0(\vec{q},\omega)\right)^2} \tag{10.19b}$$

and is non-zero when either

$$\text{(i)}\ \mathrm{Im}\,\chi_0(\vec{q},\omega)\neq 0 \text{ or}$$

$$\text{(ii)}\ \mathrm{Im}\,\chi_0(\vec{q},\omega_q)=0 \quad\text{and}\quad 1-\frac{4\pi e^2}{q^2}\mathrm{Re}\chi_0(\vec{q},\omega_q)=0$$

$$(\text{or}\,1-V(\vec{q})\chi_0(\vec{q},\omega_q)=0 \text{ for general interaction}). \tag{10.20}$$

It can be shown that the first possibility describes particle–hole excitations which exist even in the absence of electron–electron interaction, whereas the second term corresponds to collective modes which exist only in the presence of interaction.

**Homework Q4**
Show that $\mathrm{Im}\chi_0(\vec{q},\omega)\neq 0$ only when $qk_F/m+\varepsilon_{\vec{q}}>\varepsilon_{\vec{k}+\vec{q}}-\varepsilon_{\vec{k}}>0$ for $q<2k_F$. What is the corresponding condition when $q>2k_F$? What are the effects of interaction on the particle–hole excitations? (Hint: the spectral representation of the density–density response function may help.)

To see why the second possibility represents a collective mode, we expand the action $S_{\mathrm{eff}}(\varphi)$ (at real time) around $\omega\sim\omega_q$,

$$S_{\text{eff}}(\varphi) = \frac{1}{2}\sum_{\vec{q},i\omega_n}\left[\frac{\vec{q}^2}{4\pi} - e^2\chi_0(\vec{q},\omega_q) - e^2\left.\frac{\partial\chi_0}{\partial\omega}\right|_{\omega=\omega_q}(i\omega-\omega_q)\right]\cdot\varphi(\vec{q},i\omega)\varphi(-\vec{q},-i\omega)$$

$$\sim\frac{-1}{2}\sum_{\vec{q},i\omega_n}\left[e^2\left.\frac{\partial\chi_0}{\partial\omega}\right|_{\omega=\omega_q}(i\omega-\omega_q)\right]\cdot\varphi(\vec{q},\omega)\varphi(-\vec{q},-\omega), \tag{10.21}$$

which has the form of an action for free bosons with energy $\omega_q$ (see Chapter 4). This suggests that $\omega_q$ represents the excitation energy of the system associated with collective bosonic excitations. To understand the nature of the excitations, we consider in the following an alternative derivation for RPA.

**Homework Q5a**
By performing the sum over $i\Omega$, show that the Lindhard function

$$\chi_0(\vec{q},i\omega_n) = \frac{-2}{\beta V}\sum_{\vec{k},i\Omega_n} G_0(\vec{k}+\vec{q},i\omega_n+i\Omega_n)G_0(\vec{k},i\Omega_n)$$

is given by $\chi_0(\vec{q},i\omega_n) = \frac{2}{V}\sum_{\vec{k}}\frac{n_F(\varepsilon_{\vec{k}-\vec{q}/2}) - n_F(\varepsilon_{\vec{k}+\vec{q}/2})}{i\omega_n - \varepsilon_{\vec{k}+\vec{q}/2} + \varepsilon_{\vec{k}-\vec{q}/2}}$.

**Homework Q5b**
Show that $\chi_0(\vec{q},\omega \gg \frac{k_F q}{m}) \sim -\frac{nq^2}{m\omega}$, where $n=$ electron density and $\chi_0(\vec{q}\to 0, \omega=0) = -2N(0)$, where $N(0) = \frac{2mk_F}{(2\pi)^2}$ is the density of states on the Fermi surface (3D).

### 10.1.3
### Alternative Derivation for RPA

Let $\chi_0(\vec{q},\omega)$ be the density–density response function of a non-interacting electron gas to an external potential $\Phi(\vec{q},\omega)$, that is the induced density fluctuation is

$$\rho(\vec{q},\omega) = \chi_0(\vec{q},\omega)\Phi(\vec{q},\omega). \tag{10.22}$$

RPA assumes that, in the presence of interaction, the system still responds like a non-interacting electron gas, except that the electrons 'see' a total potential $\Phi_{\text{tot}}(\vec{q},\omega)$ which is a sum of the external potential $\Phi(\vec{q},\omega)$ and an induced potential $\Phi_{\text{ind}}(\vec{q},\omega)$ coming from the charge fluctuations, that is

$$\rho(\vec{q},\omega) = \chi_0(\vec{q},\omega)\Phi_{\text{tot}}(\vec{q},\omega). \tag{10.23a}$$

The induced potential is just $\Phi_{\text{ind}}(\vec{r},t) = \int d^d r' V(\vec{r}-\vec{r}')\rho(\vec{r}',t)$ from classical electrostatics. Therefore, after Fourier transforming,

$$\Phi_{\text{tot}}(\vec{q},\omega) = \Phi(\vec{q},\omega) + V(q)\rho(\vec{q},\omega) \tag{10.23b}$$

and we obtain, by combining Equations (10.23a) and (10.23b),

$$\rho(\vec{q},\omega) = \frac{\chi_0(\vec{q},\omega)}{1-V(q)\chi_0(\vec{q},\omega)}\Phi(\vec{q},\omega), \tag{10.23c}$$

which is the RPA result for the density–density response function.

Equations (10.23a) and (10.23b) can also be used to examine under what condition the system can self sustain a collective density fluctuation, that is under what condition can we have solution with $\rho(\vec{q},\omega) \neq 0$ but $\Phi(\vec{q},\omega) = 0$. This can occur if

$$\rho(\vec{q},\omega) = \chi_0(\vec{q},\omega)\Phi_{\text{tot}}(\vec{q},\omega) = \chi_0(\vec{q},\omega)[V(q)\rho(\vec{q},\omega)], \tag{10.24}$$

and has a solution if $1 = V(\vec{q})\chi_0(\vec{q},\omega_q)$, which is precisely the condition we derived for the appearance of a collective mode in $S_{\text{eff}}(\varphi)$.

In the case of Coulomb interaction, the self-sustained mode is called a plasma mode and it can be shown that, at small $\vec{q} \to 0$, the plasma oscillation frequency is $\omega_{q\to 0} \sim \omega_p = \frac{4\pi e^2 n}{m}$.

**Homework Q5c**

Show using the result of Homework Q5b that for Coulomb interaction $\omega_{q\to 0} \sim \omega_p = \frac{4\pi e^2 n}{m}$. Show that if the Coulomb interaction is replaced by a point-contact interaction with $V(q \to 0)$ being a finite, positive number, then $\omega_q \sim cq$ at small $q$.

The following figures show the excitation spectra for a fermion gas with (1) long-range Coulomb interaction and (2) short-range repulsive interaction. We assume that the space dimension = 3 in the calculation.

1. Coulomb interaction

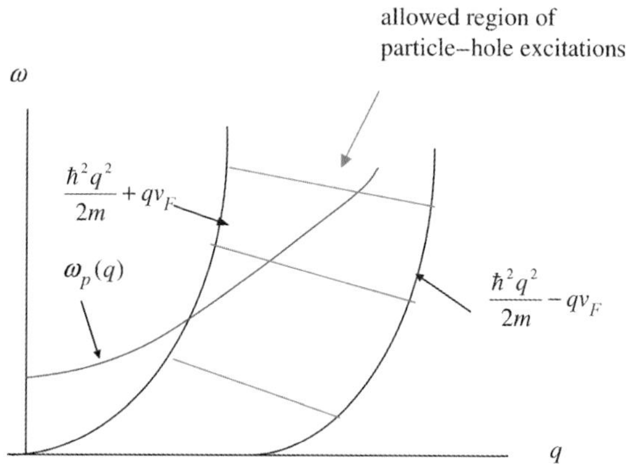


For short-range interaction, $V(q \to 0) \sim g\ (>0)$, the excitation spectrum is similar for particle–hole pairs, but the collective mode is different. It corresponds to the sound waves (phonons) in bosonic liquids with $\omega_s(q) \sim cq$, same as what we found in chapter 9.

2. Short-range interaction

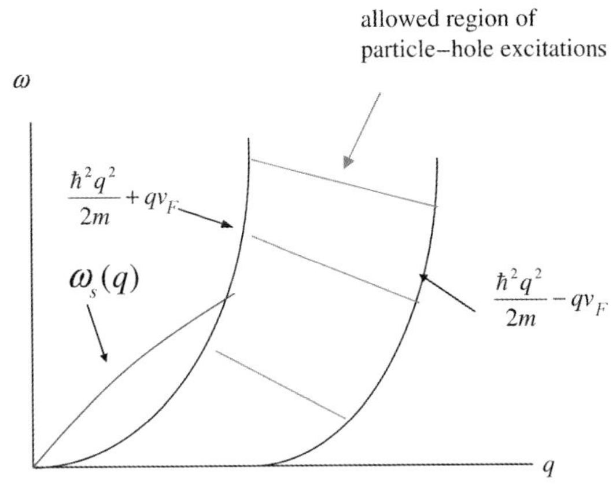


## 10.1.4 Screening

The collective motion of electrons also leads to the effect of screening in charged systems. An external potential $\Phi(\vec{q}, \omega)$ is 'screened' because the 'real' potential that charges 'see' is the total potential

$$\Phi_{\text{tot}}(\vec{q}, \omega) = \Phi(\vec{q}, \omega) + V(q)\rho(\vec{q}, \omega) = \frac{\Phi(\vec{q}, \omega)}{1 - V(q)\chi_0(\vec{q}, \omega)} = \frac{\Phi(\vec{q}, \omega)}{\varepsilon(\vec{q}, \omega)}. \tag{10.25}$$

The last equality defines the dielectric function $\varepsilon(\vec{q}, \omega) = 1 - V(q)\chi_0(\vec{q}, \omega)$ in RPA. With the result of Homework Q5b, it is easy to see that

$$\varepsilon(\vec{q}, 0) \sim 1 + V(q)N(0) \sim 1 + \frac{q_{\text{TF}}^2}{q^2} \tag{10.26}$$

at small wave vectors $q$, where $q_{\text{TF}}^2 = 4\pi e^2 N(0)$; $q_{\text{TF}}^{-1}$ is called the Thomas–Fermi screening length. Note that $\varepsilon(\vec{q}, 0) \to \infty$ at $q \to 0$ and $\Phi_{\text{tot}}(\vec{q} \to 0, 0) \to \frac{\Phi(\vec{q} \to 0,0)}{\varepsilon(q,0)} \to 0$, that is we cannot set up a finite electrostatic potential difference in a system with freely moving charges – which is what classical electrodynamics told us about metals.

Next, we imagine that $\Phi(\vec{q}, \omega) = 4\pi e^2/q^2$ is coming from an electron in the metal. In this case, the total potential seen by other electrons is in the static limit

$$V_{\text{eff}}(q) = \frac{4\pi e^2}{q^2 + q_{\text{TF}}^2}. \tag{10.27a}$$

Fourier transforming (3D), we find that

$$V_{\text{eff}}(r) = \frac{e^2}{r} e^{-q_{\text{TF}} r}. \tag{10.27b}$$

The long-range Coulomb potential is 'screened' and becomes a short-range potential with decaying length $\lambda \sim q_{\mathrm{TF}}^{-1}$ in a metal. This is an important reason why the random-phase approximation where electrons are assumed to be behaving more or less like free particles can be justified. By simply allowing collective motion of the whole electron liquid we find that the effective interaction between individual electrons becomes much weakened!

## 10.2
## Introduction to Fermi Liquids and Fermi Liquid Theory

Fermi liquid theory is a general theoretical framework (an effective field theory) describing the behavior of interacting fermion systems at low temperature. The random-phase approximation (RPA) and the corresponding excitation spectra we discussed in the last section are consistent with Fermi liquid behavior. We recall that, in RPA, the electronic excitation spectrum consists of two parts: collective and single-particle–hole-pair excitations. The particle–hole-pair excitation spectrum is basically the same as the corresponding spectrum for non-interacting particles.

Fermi liquid theory is a theoretical framework which justifies this qualitative behavior, at least when the excitation energy is low. The basic assumption of Fermi liquid theory is adiabaticity, meaning that there is a one-to-one correspondence between the ground state and the low-energy excitation spectrum of fermions in the absence and in the presence of (repulsive) interactions.

As a consequence of this assumption, the low-energy particle–hole excitation spectrum can basically be 'labeled' by the same quantum number as for non-interacting fermions and the natures of the excitations are qualitatively similar. To capture this physics, Landau introduced the concept of 'quasi-particles'.

Landau imagined that the ground state of an interacting fermion liquid can be labeled as a filled Fermi gas of non-interacting fermions. In the presence of interaction these fermions are not the same as the original fermions in the problem. Landau imagined them as the original (free) fermions 'dressed' by the interaction and called them quasi-particles. The energy of the system can be written in terms of the occupation number of quasi-particles $n_{\vec{k}}$. In particular, the ground-state energy is

$$E_G = E[\{n_{\vec{k}} = \theta(k_F - |\vec{k}|)\}], \tag{10.28a}$$

Corresponding to a filled Fermi sea.

Low-energy excitations can be labeled by small changes in the occupation number of quasi-particles, $n_{\vec{k}} = \theta(k_F - |\vec{k}|) + \delta n_{\vec{k}}$. Assuming that $\delta n_{\vec{k}}$ is small, Landau expanded the energy functional to second order,

$$E[n_{\vec{k}}] \sim E_G + \sum_{\vec{k}} \xi^*_{\vec{k}} \delta n_{\vec{k}} + \frac{1}{2V} \sum_{\vec{k},\vec{k}\prime} f_{\vec{k}\vec{k}\prime} \delta n_{\vec{k}} \delta n_{\vec{k}\prime}, \tag{10.28b}$$

where $\xi^*_{\vec{k}} \sim \frac{\hbar k_F}{m^*}(|\vec{k}| - k_F)$. $m^*$ is called the effective mass and $f_{\vec{k}\vec{k}\prime}$ are called the Landau parameters. The form of $\xi^*_{\vec{k}}$ ensures that the ground state (filled Fermi sea) is stable.

For free fermions $m^* = m$ and $f_{\vec{k}\vec{k}'} = 0$. We have neglected spin to keep the discussion simple.

Landau next derived the (low-energy) dynamics of the quasi-particles. This is quite impossible from a microscopic point of view since the nature of the quasi-particles is not specified. Landau realized that he could bypass this problem if he interpreted the above energy functional as a 'local' Hamiltonian functional for 'classical' particles, that is

$$E[n_{\vec{k}}] \to E_G + \sum_{\vec{k}} \xi^*_{\vec{k}} \int \delta n_{\vec{k}}(\vec{r}) d^d r + \frac{1}{2} \sum_{\vec{k},\vec{k}'} f_{\vec{k}\vec{k}'} \int \delta n_{\vec{k}}(\vec{r}) \delta n_{\vec{k}'}(\vec{r}) d^d r, \qquad (10.29)$$

where the dynamics of the distribution function $\delta n_{\vec{k}}(\vec{r})$ is determined by the classical equation of motion that followed from the Hamiltonian (essentially a Boltzmann equation). This assumption is justifiable in the $\vec{q}, \omega \to 0$ limit where the system should become 'classical like' (correspondence principle). Recall that the uncertainty principle forbids simultaneous identification of position and momentum of particles. Therefore, for this description to be valid, the distribution function $\delta n_{\vec{k}}(\vec{r})$ should not be changing too rapidly in space so that the uncertainty principle is not violated (i.e. $\Delta r \Delta k \geq 1$).

As see shall see later, we shall be interested in states $\vec{k}, \vec{k}'$ in the vicinity of the Fermi surface. In this regime we may replace $f_{\vec{k}\vec{k}'}$ by $f_{\vec{k}_F\vec{k}'_F}$ where the $\vec{k}_F$ are wave vectors on the Fermi surface. For a rotationally symmetric system, we also expect that $f_{\vec{k}_F\vec{k}'_F} \to f(\theta)$, where $\theta$ is the angle between the directions $\vec{k}_F$ and $\vec{k}'_F$. This form of Landau interaction is assumed in most studies of Fermi liquid theory and will be assumed in the following.

Following Chapter 7, if a particle with momentum $\vec{k}_i$ is located at $\vec{r}_i$ at time $t$ then, at time $t + dt$, it is located at $\vec{r}_i + \delta\vec{r}_i = \vec{r}_i + \vec{v}(\vec{r}_i)dt$, with momentum $\vec{k}_i + \delta\vec{k}_i = \vec{k}_i + \vec{F}_i(t)dt$

$$\Rightarrow n(\vec{r} + \delta\vec{r}, \vec{k} + \delta\vec{k}, t + \delta t) = n(\vec{r}, \vec{k}, t) \Rightarrow \frac{\partial n}{\partial t} + \vec{v}(\vec{k}) \cdot \nabla_{\vec{r}} n + \vec{F} \cdot \nabla_{\vec{k}} n = \frac{dn}{dt} = 0. \qquad (10.30a)$$

To complete the equation, we have to determine the velocities $\vec{v}(\vec{r}_i)$ and forces $\vec{F}_i(t)$ from Landau Fermi liquid theory. They are given in classical mechanics by

$$\vec{v}(\vec{k}_i) = \frac{\partial \varepsilon_{\vec{k}}}{\partial \vec{k}_i}, \quad \vec{F}_i(t) = -\frac{\partial \varepsilon_{\vec{k}}}{\partial \vec{r}_i}, \qquad (10.30b)$$

where $\varepsilon_{\vec{k}}(\vec{r}) = \xi^*_{\vec{k}} + \sum_{\vec{k}'} f_{\vec{k}\vec{k}'} \delta n_{\vec{k}'}(\vec{r})$ from Equation (10.29).

Therefore, to leading order where $f_{\vec{k}\vec{k}'} \to f_{\vec{k}_F\vec{k}'_F}$,

$$\vec{v}(\vec{r}_i) = \frac{\partial \xi^*_{\vec{k}}}{\partial \vec{k}_i} = \frac{\hbar \vec{k}_F}{m^*} = \vec{v}_{\vec{k}_F} \quad \text{and} \quad \vec{F}_i(t) = -\sum_{\vec{k}'} f_{\vec{k}_F\vec{k}'_F} \nabla n_{\vec{k}'}(\vec{r}). \qquad (10.30c)$$

Writing $n_{\vec{k}}(\vec{r},t) = n_F(\vec{k}) + \delta n_{\vec{k}}(\vec{r},t)$, where $n_F$ is the equilibrium Fermi distribution, the Boltzmann equation becomes

$$\frac{\partial \delta n_{\vec{k}}(\vec{r},t)}{\partial t} + \vec{v}_{\vec{k}} \cdot \nabla_{\vec{r}} \delta n_{\vec{k}}(\vec{r},t) + \delta(\xi^*_{\vec{k}}) \sum_{\vec{k}'} f_{\vec{k}\vec{k}'} \vec{v}_{\vec{k}} . \nabla \delta n_{\vec{k}'}(\vec{r},t) = 0 \qquad (10.30d)$$

to linear order in $\delta n$, where we considered $T \to 0$ so that $n_F(\vec{k}) \sim \theta(-\xi^*_{\vec{k}})$. Note that, in this limit, only states in the vicinity of the Fermi surface are involved in the transport equation.

In this way Landau derived results that cannot be obtained directly from the 'quantum' energy functional (10.28a,b). Note that interactions between quasi-particles represented by the Landau interaction $f_{\vec{k}\vec{k}'}$ exist in the transport equation. We shall see later that the interactions are not 'small' and cannot be neglected in general. Therefore, Landau theory predicts that the low-energy effective theory of Fermi liquids is not a theory of free fermions. Interaction between quasi-particles persists down to zero temperature. The validity and meaning of Landau's approach were not fully understood until the 1990s, when the mathematical techniques of renormalization groups and bosonization had matured (see Ref. [4] and the last section of this chapter for a brief introduction to bosonization). In the following, we shall discuss some basic features of Fermi liquid theory.

### 10.2.1
### Quasi-particles and Single-Particle Green's Function

Fermi liquid theory is based on adiabaticity, and is valid only when adiabaticity is a good assumption. What Landau observed was that adiabaticity is, strictly speaking, valid only for quasi-particles on the Fermi surface, and would be violated gradually when we look at quasi-particle states away from the Fermi surface.

A measure of adiabaticity is the one-particle Green's function for the 'bare' fermions in the system. If adiabaticity is obeyed, there will be a non-zero probability that adding (or removing) a 'bare' particle with momentum $\vec{k}$ in the system will generate (remove) a quasi-particle with the same momentum in the system. Moreover, if this quasi-particle state is a good eigenstate of the system, it should have an infinite lifetime. This should also be reflected in the one-particle Green's function. Thus, although we may not have a precise wavefunction description of quasi-particles, the stability of a quasi-particle can be inferred by studying the one-particle Green's function of the 'bare' fermions.

We have seen in Chapter 5 (and see Appendix A) that the one-particle Green's function has a general form

$$G(\vec{k};\omega) = \frac{1}{\omega + i\delta - \varepsilon_{\vec{k}} - \Sigma(\vec{k}, \omega + i\delta) + \mu}. \qquad (10.31a)$$

Around the pole given by $E_{\vec{k}} - \varepsilon_{\vec{k}} - \mathrm{Re}\Sigma(\vec{k}, E_{\vec{k}}) + \mu = 0$, we may expand the self energy to obtain

$$G(\vec{k};\omega\sim E_{\vec{k}})=\frac{Z_{\vec{k}}}{(\omega+i\delta-E_{\vec{k}})-iZ_{\vec{k}}\,\mathrm{Im}\Sigma(\vec{k},E_{\vec{k}})}, \qquad (10.31b)$$

where $Z_{\vec{k}}^{-1}=1-\frac{\partial \mathrm{Re}\Sigma(\vec{k},\omega)}{\partial\omega}\Big|_{\omega=E_{\vec{k}}}$ measures the overlap between the 'bare' and 'quasi'-particle states.

Defining energy zero to be on the Fermi surface (i.e. $E_{k=k_F}=0$) and performing perturbation theory to second order, it is found that (see Appendix A)

$$\mathrm{Im}\Sigma(\vec{k},\omega)\sim\frac{\omega^2}{E_F}. \qquad (10.32)$$

Therefore, we expect that quasi-particles have lifetime $\sim\left(\frac{E_{\vec{k}}^2}{E_F}\right)^{-1}$ and adiabaticity is, strictly speaking, valid only for quasi-particle states on the Fermi surface provided that $Z_{\vec{k}}$ is a non-zero number. The behavior of $Z_{\vec{k}}$ can be studied by using the Kramers–Kronig relation, where we obtain

$$\mathrm{Re}\Sigma(\vec{k},\omega)=-\frac{1}{\pi}\int d\omega' P\frac{\mathrm{Im}\Sigma(\vec{k},\omega')}{\omega-\omega'}. \qquad (10.33a)$$

The small-$\omega'$ contributions can be estimated using Equation (10.32), and we obtain

$$\mathrm{Re}\frac{\partial\Sigma(\vec{k},\omega)}{\partial\omega}\Bigg|_{\omega=E_{\vec{k}}}\sim\frac{1}{\pi E_F}\int_{-\Lambda}^{\Lambda}d\omega'\frac{\omega'^2}{(E_{\vec{k}}-\omega')^2}\sim A+BE_{\vec{k}}, \qquad (10.33b)$$

where $A$, $B$ are constants and $\Lambda$ is a high-energy cutoff below which Equation (10.32) is valid. Thus, $Z_{\vec{k}}$ is a finite number if there is no singular high-energy contribution to the integral.

Note that, although strictly speaking adiabaticity is obeyed only for states on the Fermi surface, the inverse lifetime of the quasi-particles $\sim\frac{E\vec{k}^2}{E_F}$ is still much smaller than $E_{\vec{k}}$ itself for small $E_{\vec{k}}$. Therefore, at a small regime of energy above and below the Fermi surface, adiabaticity is still a good approximation and Fermi liquid theory remains valid.

**Homework Q6**

Discuss the validity of the adiabatic approximation and Fermi liquid theory for $\mathrm{Im}\Sigma(\vec{k},\omega)\sim\frac{\omega^{n+1}}{E_F^n}$ for $-1<n<1$. The case $n=0$ is called a marginal Fermi liquid in the literature. Explain why this is the case.

### 10.2.2 Charge and Current Carried by Quasi-particles

Next, we study the charge and current carried by quasi-particles. We expect that the charge carried by a quasi-particle is exactly equal to the charge carried by the original particle because of the basic assumption that there is a one-to-one correspondence between 'bare' particles and quasi-particles and that total charge is conceived. It turns out that the same conclusion cannot be drawn so readily for the current. To see why this is so, we examine the transport equation (10.30d). Summing the equation over $\vec{k}$, we obtain the continuity equation

$$\frac{\partial \rho(\vec{r},t)}{\partial t} + \nabla . \vec{j}(\vec{r},t) = 0, \tag{10.34a}$$

where

$$\rho(\vec{r},t) = \sum_{\vec{k}} \delta n_{\vec{k}}(\vec{r},t) \tag{10.34b}$$

is the total charge fluctuation and

$$\vec{j}(\vec{r},t) = \sum_{\vec{k}} \vec{v}_{\vec{k}} \delta n_{\vec{k}}(\vec{r},t) + \sum_{\vec{k},\vec{k}'} \delta(\xi^*_{\vec{k}'}) f_{\vec{k}'\vec{k}} \vec{v}_{\vec{k}'} \delta n_{\vec{k}}(\vec{r},t) = \sum_{\vec{k}} \vec{j}_{\vec{k}} \delta n_{\vec{k}}(\vec{r},t), \tag{10.34c}$$

where

$$\vec{j}_{\vec{k}} = \vec{v}_{\vec{k}} + \sum_{\vec{k},\vec{k}'} \delta(\xi^*_{\vec{k}'}) f_{\vec{k}'\vec{k}} \vec{v}_{\vec{k}'} \tag{10.34d}$$

is by definition the current carried by a quasi-particle. This is quite different from the naïve expectation $\vec{j}_{\vec{k}} = \vec{v}_{\vec{k}}$ for a single quasi-particle. Note that this result is a consequence of having interaction terms in the transport equation, and is rather independent of quantum mechanics. The reason why $\vec{j}_{\vec{k}} \neq \vec{v}_{\vec{k}}$ is that quasi-particles are never really independent from one another when $f_{\vec{k}\vec{k}'} \neq 0$. It is impossible to excite just one quasi-particle without creating a 'cloud' of other quasi-particles surrounding it. You can imagine the situation of a man trying to push his way through a crowd of people. When he pushes through the crowd, people in his way will all be affected and have to move away a little bit. This phenomenon is called 'back flow' in the study of fluids. In mathematical terms, one has to distinguish a quasi-particle from an 'elementary excitation', which is what is physically excited in the system. An 'elementary excitation' = a 'quasi-particle' + 'crowd of other quasi-particles around it'. The 'crowd' also contributes to the total current carried by the elementary excitation and is responsible for the $\sum_{\vec{k},\vec{k}'} \delta(\xi^*_{\vec{k}'}) f_{\vec{k}'\vec{k}} \vec{v}_{\vec{k}'}$ factor in Equation (10.34d).

For a translationally invariant system, another interesting result occurs. First, we recall that quasi-particles and bare particles carry the same charge, and the quasi-particle density $\rho(\vec{r},t)$ in Equation (10.34b) is equal to the density of bare particles. Therefore, the current $\vec{j}(\vec{r},t)$ is equal to the current of bare particles. For fermion systems with Hamiltonian (10.4a,b), it is easy to show that the expression for the current is

$$\vec{j}(\vec{r},t) = \sum_{\vec{k}} \vec{v}^{(0)}_{\vec{k}} \delta n_{\vec{k}B}(\vec{r},t) = \sum_{\vec{k}} \vec{v}^{(0)}_{\vec{k}} \delta n_{\vec{k}}(\vec{r},t), \tag{10.35}$$

where $\vec{v}^{(0)}_{\vec{k}} = \frac{\hbar \vec{k}}{m}$, $m$ is the bare mass of particles (the mass that enters Equation (10.4a,b)), and $\delta n_{\vec{k}B}(\vec{r},t) = \delta n_{\vec{k}}(\vec{r},t)$ is the density of bare particles. Comparing this with Equation (10.34d), we obtain the result

$$\vec{v}^{(0)}_{\vec{k}} = \vec{v}_{\vec{k}} + \sum_{\vec{k},\vec{k}'} \delta(\xi^*_{\vec{k}'}) f_{\vec{k}'\vec{k}} \vec{v}_{\vec{k}'}. \tag{10.36a}$$

Note that in this case $f_{\vec{k}_F\vec{k}'_F} \to f(\theta)$ and by symmetry

$$\sum_{\vec{k},\vec{k}'} \delta(\xi^*_{\vec{k}'}) f_{\vec{k}'\vec{k}} \vec{v}_{\vec{k}'} = \frac{1}{(2\pi)^3} \int k' dk' \sin\vartheta d\vartheta d\phi \delta(\xi^*_{\vec{k}'}) f(\vartheta) \frac{\vec{k}_F \cos\vartheta}{m^*}.$$

The result (10.36a) is often written as a relation between the effective mass $m^*$ and the Landau parameter $F_{1s}$ [1]:

$$\frac{1}{3} F_{1s} = \frac{1}{(2\pi)^3} \int k' dk' \sin\vartheta d\vartheta d\phi \delta(\xi^*_{\vec{k}'}) f(\vartheta) \cos\vartheta, \tag{10.36b}$$

where

$$\frac{1}{m} = \frac{1}{m^*} \left( 1 + \frac{F_{1s}}{3} \right). \tag{10.36c}$$

### 10.2.3
### Two Examples of Applications

In the following, we shall study two applications of Fermi liquid theory. First, we consider thermodynamics and evaluate the specific heat. In this case, we have to evaluate the system's free energy starting from the energy function (10.28b).

It is easier to start with the microcanonical ensemble of the system in this case. We consider the system with fixed energy

$$E = E[n_{\vec{k}}] \sim E_G + \sum_{\vec{k}} \xi^*_{\vec{k}} \delta n_{\vec{k}} + \frac{1}{2} \sum_{\vec{k},\vec{k}'} f_{\vec{k}\vec{k}'} \delta n_{\vec{k}} \delta n_{\vec{k}'} \tag{10.37}$$

and fixed particle number $N$. The particles (fermions) are distributed with equal probability to each available state $\vec{k}$ provided that the above constraint is satisfied. To derive the free energy, we first note that, for $N_j$ particles (fermions) distributed in $G_j$ states, the total number of possibilities is $\Gamma_j = \frac{G_j!}{N_j!(G_j - N_j)!}$, and the corresponding entropy is

$$\ln\Gamma_j \sim G_j \ln G_j - N_j \ln N_j - (G_j - N_j)\ln(G_j - N_j) = -G_j(n_j \ln n_j + (1 - n_j)\ln(1 - n_j)), \tag{10.38a}$$

where $n_j = N_j/G_j$; we have made the Stirling approximation $\ln N! \sim N(\ln N - 1)$ in Equation 10.38a. Applying this to our Fermi liquid system, we identify '$j$' as a group of states with momentum $\vec{p}$ between $\vec{k}$ and $\vec{k} + \Delta\vec{k}$ with $\Delta\vec{k}$ small enough so that $n_{\vec{p}} \sim n_{\vec{k}}$ and $G_j \to \frac{V}{(2\pi)^d} (\Delta k)^d$. The entropy per unit volume is thus

$$S \sim -k \int \frac{d^d k}{(2\pi)^d} \left( n_{\vec{k}} \ln n_{\vec{k}} + (1 - n_{\vec{k}}) \ln(1 - n_{\vec{k}}) \right), \tag{10.38b}$$

where $n_{\vec{k}}$ can be determined by maximizing the entropy with the energy constraint (10.37) and $\sum n_{\vec{k}} = N$. The constraints can be handled by introducing Lagrange multipliers $\alpha, \beta$, where we maximize $\frac{S}{k} - \alpha N - \beta E$ with respect to $n_{\vec{k}}$. It is easy to obtain

$$n_k = \frac{1}{1+e^{\frac{\bar{\varepsilon}(\vec{k})-\mu}{kT}}} \quad \text{or} \quad \delta n_{\vec{k}} = \frac{1}{1+e^{\frac{\bar{\varepsilon}(\vec{k})-\mu}{kT}}} - \theta(-\xi^*_{\vec{k}}), \tag{10.39a}$$

where

$$\bar{\varepsilon}(\vec{k}) = \frac{\delta E[n]}{\delta n_{\vec{k}}} = \xi^*_{\vec{k}} + \sum_{\vec{k}'} f_{\vec{k}\vec{k}'} \delta n_{\vec{k}'}. \tag{10.39b}$$

We have made the usual identification $\beta = (kT)^{-1}$ and $\alpha = -\beta\mu$ in Equation (10.39a). Note that in principle Equation (10.39a,b) has to be solved self consistently to determine $\delta n_{\vec{k}}$.

At low temperature one can perform a Sommerfeld-type expansion of the free energy and it is straightforward to show that (*exercise*) to lowest order in temperature the $\sum_{\vec{k}'} f_{\vec{k}\vec{k}'} \delta n_{\vec{k}'}$ term in Equation (10.39b) is not important and the specific heat is (in 3D) $C = \gamma T$, where

$$\gamma = \frac{k_B^2}{3} \frac{m^* k_F}{\hbar^2} = \frac{m^*}{m} \gamma_0, \tag{10.40}$$

where $\gamma_0 T$ is the corresponding specific heat for the non-interacting Fermi gas. Note that to lowest order in temperature the only effect of interaction on specific heat is to renormalize the electron mass in Fermi liquid theory.

The second example of application we consider is the density–density response function. To determine the density response of the system to an external scalar potential, we solve the (linearized) Boltzmann equation

$$\frac{\partial \delta n_{\vec{k}}(\vec{r}, t)}{\partial t} + \vec{v}_{\vec{k}} \cdot \nabla_{\vec{r}} \delta n_{\vec{k}}(\vec{r}, t) + \delta(\xi^*_{\vec{k}}) \sum_{\vec{k}'} f_{\vec{k}\vec{k}'} \vec{v}_{\vec{k}} . \nabla \delta n_{\vec{k}'}(\vec{r}, t) + \vec{F}^{\text{ext}} . \nabla_{\vec{k}} n^{(0)}_{\vec{k}} = 0, \tag{10.41a}$$

where we have added an external force term in the equation (see Equation (10.30d)). For an external scalar potential, $\vec{F}^{\text{ext}} = e\vec{E}^{\text{ext}} = -e\nabla\Phi^{\text{ext}}$, $\vec{F}^{\text{ext}} . \nabla_{\vec{k}} n^{(0)}_{\vec{k}} \to e\vec{v}_{\vec{k}} . \nabla \Phi^{\text{ext}} \delta(\xi^*_{\vec{k}})$ at $T \to 0$.

Defining $\delta n_{\vec{k}}(\vec{q}, \omega) = \int e^{i(\vec{q}.\vec{r} - \omega t)} \delta n_{\vec{k}}(\vec{r}, t) d^d r dt$, we obtain, after Fourier transforming Equation (10.41a),

$$(\vec{q}.\vec{v}_{\vec{k}} - \omega)\delta n_{\vec{k}} + \vec{q}.\vec{v}_{\vec{k}} \delta(\xi^*_{\vec{k}}) \sum_{\vec{k}'} f_{\vec{k}\vec{k}'} \delta n_{\vec{k}'} = -e\vec{v}_{\vec{k}} . \vec{q} \Phi^{ext} \delta(\xi^*_{\vec{k}}), \tag{10.41b}$$

where we have replaced $\delta n_{\vec{k}}(\vec{q}, \omega) \to \delta n_{\vec{k}}$ for brevity. For rotationally symmetric systems, Equation (10.41b) can be solved by expanding the solutions in terms of spherical harmonics [1]. Here, we shall consider a very simple situation $f_{\vec{k}\vec{k}'} \sim f_0$. In this case, it is easy to see that the solution for $\delta\rho = \sum_{\vec{k}} \delta n_{\vec{k}}$ is

$$\delta\rho(\vec{q},\omega) = \frac{\chi_{0F}(\vec{q},\omega)}{1-f_0\chi_{0F}(\vec{q},\omega)} e\Phi^{\text{ext}}(\vec{q},\omega), \tag{10.42a}$$

where

$$\chi_{0F}(\vec{q},\omega) = \sum_{\vec{k}} \frac{\vec{v}_{\vec{k}}.\vec{q}}{\omega-\vec{v}_{\vec{k}}.\vec{q}} \delta(\xi^*_{\vec{k}}). \tag{10.42b}$$

Note that this is of the same form as the density–density response function deduced from RPA (Equation (10.23c)). More interestingly, it is easy to see that (Homework Q5a), in the small-$\vec{q}$ limit, the Lindhard function $\chi_0(\vec{q},\omega)$ has the same expression as $\chi_{0F}(\vec{q},\omega)$ except that the bare electron mass $m$ is replaced by the effective mass $m^*$, indicating that RPA is in fact an example of Fermi liquid theory with un-renormalized mass $m^* = m$ and interaction $f_0 = V(q \to 0)$. The result (10.42a,b) also suggests that besides quasi-particle–hole-pair excitations, collective excitations can also be produced in Landau's Fermi liquid theory as in RPA. The real limitation of Fermi liquid theory is that it is valid only in the long-wavelength ($q \to 0$) limit where a semi-classical description of the system is valid.

## 10.2.4
## Bosonization Description of Fermi Liquid Theory

Landau's Fermi liquid theory is an effective low-energy theory for fermion systems. It is natural to ask whether the theory can be put into the framework of effective quantum field theory we discussed in Chapter 7. For example, one may imagine that the wavefunction of the $N$-particle fermion system in Fermi liquid theory is written as

$$\psi(\vec{r}_1,\vec{r}_2,\ldots,\vec{r}_N) = \int D\vec{x}\Phi(\vec{x}_1,\ldots,\vec{x}_M)\psi_0([\vec{r}];\vec{x}_1,\ldots,\vec{x}_M), \tag{10.43}$$

where $(\vec{r}_1,\vec{r}_2,\ldots,\vec{r}_N)$ are the coordinates of the bare fermions and $(\vec{x}_1,\vec{x}_2,\ldots,\vec{x}_M)$ are the coordinates of special points in the wavefunction corresponding to positions of the quasi-particles and there is a one-to-one correspondence between $(\vec{x}_1,\vec{x}_2,\ldots,\vec{x}_M)$ and $(\vec{r}_1,\vec{r}_2,\ldots,\vec{r}_N)$ (with $N = M$) in Fermi liquid theory. We may imagine that the low-energy properties of the system are characterized by the wavefunction $\Phi(\vec{x}_1,\vec{x}_2,\ldots,\vec{x}_M)$ such that when we apply the principle of least action to the action

$$S = \int_a^b dt \langle\psi(t)|\left[i\hbar\frac{\partial}{\partial t} - H\right]|\psi(t)\rangle$$

we obtain the Landau transport equation (10.30d).

It turns out that the trial wavefunction approach is not a convenient starting point to understand Fermi liquid theory. The correct starting point is to realize that the ground state of the system is a filled Fermi sea and low-energy excitations in the system can be viewed as (local) deformations in the shape of the Fermi surface (see figure below).

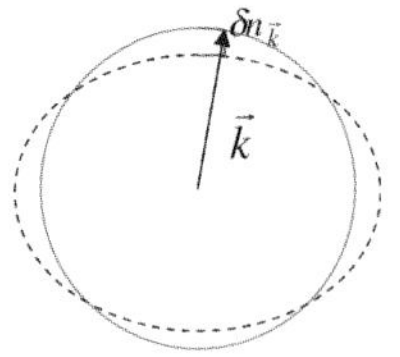


The shape of the Fermi surface can be labeled by a wave vector $\vec{k}(\vec{x}) = (k_F + \delta u(\widehat{n}, \vec{x}))\widehat{n}$, where $\delta u(\widehat{n}, \vec{x})$ labels local small changes in the shape of the Fermi surface in the direction $\widehat{n}$. $\delta u(\widehat{n}, \vec{x})$ can be identified with $\delta n_{\vec{k}}(\vec{x})$ in Fermi liquid theory with $\vec{k} = k_F\widehat{n}$. Note that $\delta u(\widehat{n}, \vec{x})\widehat{n}$ is a real number field and a quantum field theory can be constructed by a canonical quantization scheme if the equation of motion for $\delta u(\widehat{n}, \vec{x})$ is known. This is the spirit of bosonization theory (see Ref. [4]). The Landau energy functional $E[n]$ represents the energy cost for a small distortion in the shape of the Fermi surface in the theory and the Boltzmann equation (10.30d) becomes the quantum equation of motion for the operators $\delta\widehat{u}(\widehat{n}, \vec{x})\widehat{n}$. The (quantum) excitations are bosons representing quantized eigenmodes, which are the solution of the Boltzmann equation representing harmonic oscillatory modes of the Fermi surface. The most non-trivial result is perhaps that there is a one-to-one mapping between the excitation spectrum obtained in bosonization theory and the fermionic excitation spectrum. Interested readers can look at Ref. [4] for an introduction to the bosonization approach.

**Homework Q7 (Lagrangian formulation of Boltzmann equation)**

(a) Introducing $\delta u(\widehat{n}, \vec{x}) = \widehat{n}.\nabla\varphi(\widehat{n}, \vec{x})$, show that the Lagrangian

$$L = \int d^d x \oint d\Omega_k \left( -\widehat{n}.\nabla\varphi(\widehat{n})\partial_t\varphi(\widehat{n}) - \frac{|v_{\vec{k}}|}{2}(\widehat{n}.\nabla\varphi(\widehat{n}))^2 \right)$$

together with the principle of least action produces the equation of motion

$$\frac{\partial\delta n_{\vec{k}}(\vec{r}, t)}{\partial t} + \vec{v}_{\vec{k}} \cdot \nabla_{\vec{r}}\delta n_{\vec{k}}(\vec{r}, t) = 0$$

after identifying $\delta u(\widehat{n}, \vec{x})$ with $\delta n_{\vec{k}}(\vec{x})$. What term should we add to the Lagrangian to obtain the full Landau transport equation? What is the corresponding Hamiltonian? (Note that this is slightly different from $E[n]$.)

### 10.2.5
### Beyond Fermi Liquid Theory?

A New comer to the field may feel uncomfortable with Fermi-liquid theory which seems to base purely on the assumption of adiabaticity and correspondence principle (a classical description becomes valid in the limit $\vec{q}, \omega \to 0$). Is it possible that these assumptions are wrong? Are there other possible fermionic liquid states which are not

Fermi liquids? These are indeed valid questions and physicists have tried very hard to find other plausible fermionic liquid states which are not Fermi liquids. In the absence of a magnetic field or other long-range order the only "confirmed" non-Fermi liquid states physicists found so far is the Luttinger liquid states in one dimension. The problem is hard because there is no "standard procedure" which physicists can apply to find new states of matter. Diagrammatic perturbation theory cannot help because it is also based on adiabaticity. (On the other hand it can be used to justify Fermi liquid theory [1]) You can look at ref. [4] to have a feeling of how difficult it is to find a new states of matter which is also realistic.

## References

1 Pines, D. and Nozieres, P. (1999) *The Theory of Quantum Liquids*, Westview Press.

2 Nagaosa, N. (1999) *Quantum Field Theory in Condensed Matter Physics*, Springer-Verlag Berlin Heidelberg New York.

3 Mahan, G.D. (1990) *Many Particle Physics*, Plenum Press (N. Y. and London).

4 Wen, X.G. (2004) *Theory of Quantum Many-Body Systems*, Oxford University Press.

# 11
# Superconductivity: BCS Theory and Beyond

In this chapter we review the basic features of BCS (Bardeen-Cooper-Schrieffer) theory for superconductivity. We shall see that to understand superconductivity – which is superfluidity in a fermionic system – we need to combine the tools we developed in the previous two chapters. We shall start with the path integral approach – which is an approach that emphasizes more the bosonic (superfluid) aspect of superconductivity. Afterward, we shall consider the Hamiltonian approach – which emphasizes the fermionic aspect more. We shall then derive the important Ginsburg–Landau equation and study *vortices*. The chapter ends with an introduction to superconductor–insulation transitions – an important topic in modern condensed matter field theory.

## 11.1
## BCS Theory for (s-wave) Superconductors: Path Integral Approach

Consider the following Hamiltonian for a fermionic system:

$$H = \int d^d r \hat{\psi}_\sigma^+(\vec{r})\left[-\frac{\hbar^2}{2m}\nabla^2 - \mu\right]\hat{\psi}_\sigma(\vec{r}) - g\int d^d r \hat{\psi}_\uparrow^+(\vec{r})\hat{\psi}_\downarrow^+(\vec{r})\hat{\psi}_\downarrow(\vec{r})\hat{\psi}_\uparrow(\vec{r}) \tag{11.1}$$

where $\psi_\sigma$ and $\psi_\sigma^+$ are the usual electron annihilation and creation operators. The first term represents the usual kinetic energy and the second term represents a short-range attractive interaction ($g>0$) between electrons.

It is now well established that superconductivity is a result of pairing electrons into electron pairs called Cooper pairs and, to describe superconductivity, we need a theory that describes the motion of Cooper pairs. Following Chapter 10, we introduce a Hubbard–Stratonovich (HS) transformation that describes the collective motion of pairs of electrons. This is different from the

*Introduction to Classical and Quantum Field Theory*. Tai-Kai Ng

ISBN: 978-3-527-40726-2

HS transformation introduced in Chapter 10, which describes collective density fluctuations.

We introduce the 'electron-pairing' HS transformation,

$$\exp\left[g\int d^{d+1}x\hat{\psi}_{\uparrow}^{+}(x)\hat{\psi}_{\downarrow}^{+}(x)\hat{\psi}_{\downarrow}(x)\hat{\psi}_{\uparrow}(x)\right]$$
$$=\int D\phi(x)\int D\phi^{*}(x)\exp\left\{\int\left[\frac{1}{g}\phi^{*}(x)\phi(x)-(\bar{\psi}_{\uparrow}(x)\bar{\psi}_{\downarrow}(x)\phi(x)+c.c.)\right]d^{d+1}x\right\}, \tag{11.2}$$

where $x=(\vec{r},\tau)$, $\int d^{d+1}x=\int_0^{\beta}d\tau\int d^{d}x$, and $\phi(x)$ is a complex scalar field which as we shall see represents the wavefunction for Cooper pairs.

**Homework Q1a**

Derive the above HS identity. Show that the HS field must be a complex scalar field. What happens if $g$ is negative?

Therefore, the partition function can be written as

$$Z(\beta)=\int D\phi(x)\int D\phi^{*}(x)\int D\psi(x)\int D\bar{\psi}(x)e^{-S(\psi,\bar{\psi},\phi,\phi^{*})}, \tag{11.3a}$$

where $S=S_{\mathrm{el}}+S_{\mathrm{HS}}$, with

$$S_{\mathrm{el}}=\int d^{d+1}x\left\{\bar{\psi}_{\sigma}(x)\left[\partial_{\tau}-\frac{\hbar^{2}\nabla^{2}}{2m}-\mu\right]\psi_{\sigma}(x)-\phi^{*}(x)\psi_{\uparrow}(x)\psi_{\downarrow}(x)\right.$$
$$\left.-\phi(x)\bar{\psi}_{\downarrow}(x)\bar{\psi}_{\uparrow}(x)\right\} \tag{11.3b}$$

and

$$S_{\mathrm{HS}}=\int d^{d+1}x\left\{\frac{1}{g}\phi^{*}(x)\phi(x)\right\}. \tag{11.3c}$$

To proceed further, we follow Chapter 10 and derive an effective action $S_{\mathrm{eff}}$ for the $\phi$ ($\phi^{*}$) field by integrating out the fermions, that is

$$Z(\beta)\rightarrow\int D\phi(x)\int D\phi^{*}(x)e^{-S_{\mathrm{eff}}(\phi,\phi^{*})}.$$

To derive $S_{\mathrm{eff}}$, we first Fourier transform $S_{\mathrm{el}}$ and write

$$\frac{S_{\mathrm{el}}}{\beta V}=\int d^{d+1}kd^{d+1}k'\bar{\psi}(k)M_{kk'}(\phi,\phi*)\psi(k'), \tag{11.4a}$$

where

$$\bar{\psi}(k) = (\bar{\psi}_\uparrow(k), \quad \psi_\downarrow(-k))$$
$$\psi(k) = \begin{pmatrix} \psi_\uparrow(k) \\ \bar{\psi}_\downarrow(-k) \end{pmatrix} \tag{11.4b}$$

and

$$M_{kk'}(\phi,\phi^*) = \begin{bmatrix} -\delta_{kk'}(i\omega_n - \varepsilon_k + \mu) & -\phi(k-k') \\ -\phi(k'-k)^* & -\delta_{kk'}(i\omega_n + \varepsilon_k - \mu) \end{bmatrix}, \tag{11.4c}$$

where $k = (\vec{k}, i\omega_n)$, $k' = (\vec{k}', i\omega'_n)$, and $\int d^{d+1}k = \frac{1}{\beta V}\sum_{i\omega_n}\int d^d k$.

Using the result

$$[\det M(\phi,\phi^*)] = \int D\psi \int D\bar{\psi} e^{-S_{el}}, \text{ we obtain}$$

$$\frac{1}{\beta V} S_{eff}(\phi,\phi^*) = -\mathrm{Tr}(\ln M(\phi,\phi^*)) + \frac{|\phi|^2}{g}. \tag{11.5}$$

**Homework Q1b**

Derive Equations (11.4a)–(11.5).

To treat this complicated action, we employ the semiclassical approximation for bosons we discussed in Chapters 9 and 10. First, we look for the classical minimum in the action described by

$$\left.\frac{\delta S_{eff}(\phi,\phi^*)}{\delta\phi}\right|_{\phi=\phi^*=\Delta_0} = \left.\frac{\delta S_{eff}(\phi,\phi^*)}{\delta\phi^*}\right|_{\phi=\phi^*=\Delta_0} = 0.$$

As in Chapters 9 and 10, we look for a solution $\phi(\vec{r},\tau) = \Delta_0$ that is translationally invariant in both space and time. In this case, $M_{kk'}(\phi,\phi^*) \to M_{0k}(\Delta_0,\Delta_0^*)\delta_{kk'}$, where

$$M_{0k}(\Delta_0,\Delta_0^*) = -\begin{bmatrix} i\omega_n - \varepsilon_k + \mu & \Delta_0 \\ \Delta_0^* & i\omega_n + \varepsilon_k - \mu \end{bmatrix} \tag{11.6a}$$

and

$$\frac{1}{\beta V} S_{eff}(\Delta_0,\Delta_0^*) = -\frac{1}{\beta V}\sum_{k,i\omega_n} \ln(\det M_{0k}(\Delta_0,\Delta_0^*)) + \frac{|\Delta_0|^2}{g}. \tag{11.6b}$$

Minimizing $S_{eff}(\Delta_0,\Delta_0^*)$ with respect to $\Delta_0$, we obtain the mean-field equation

$$\Delta_0 = g\frac{1}{\beta V}\sum_{\vec{k},i\omega_n} \frac{-\Delta_0}{(i\omega_n - \xi_k)(i\omega_n + \xi_k) - |\Delta_0|^2}, \tag{11.6c}$$

where $\xi_k = \varepsilon_k - \mu$; $\Delta_0$ is the BCS energy gap or the superconductivity order parameter. The equation can be further simplified by performing the summation over frequency.

We obtain the gap equation

$$1 = g\frac{1}{V}\sum_{\vec{k}}\frac{1}{2E_{\vec{k}}}\tanh\left(\frac{\beta E_{\vec{k}}}{2}\right), \tag{11.6d}$$

where $E_{\vec{k}} = \sqrt{\xi_{\vec{k}}^{\,2} + \Delta_0^2}$.

**Homework Q1c**
Derive Equations (11.6c) and (11.6d).

Note that only the energy $E_{\vec{k}}$ appears in the integral equation; therefore, we can introduce the density of states $N(\xi) = \frac{1}{V}\sum_{\vec{k}}\delta(\xi - \xi_{\vec{k}})$ and Equation (11.6d) can be rewritten as

$$1 = g\int N(\xi)\mathrm{d}\xi\frac{1}{2E}\tanh\left(\frac{\beta E}{2}\right), \tag{11.7a}$$

where $E = \sqrt{\xi^2 + \Delta_0^2}$.

In the usual phonon-induced superconductivity the attraction between electrons arises from electrons exchanging phonons and the resulting interaction is attractive only at energy $|\omega| < \omega_D$, where $\omega_D$ is the Debye frequency. In this case, Equation (11.7a) can be replaced by

$$1 = gN(0)\int_{-\omega_D}^{\omega_D}\mathrm{d}\xi\frac{1}{2E}\tanh\left(\frac{\beta E}{2}\right), \tag{11.7b}$$

where we have approximated $N(\xi) \sim N(0)$. The approximation is valid when $\omega_D \ll E_F/\hbar$. We first consider zero temperature ($\beta \to \infty$). In this case, $\tanh\left(\frac{\beta E}{2}\right) \to 1$ and Equation (11.7b) becomes

$$1 = gN(0)\int_{-\omega_D}^{\omega_D}\frac{\mathrm{d}\xi}{2\sqrt{\xi^2 + \Delta_0^2}} = gN(0)\ln\left[\xi + \sqrt{\xi^2 + \Delta_0^2}\right]_0^{\omega_D}$$
$$\sim gN(0)\ln\left[\frac{2\omega_D}{\Delta_0}\right] \tag{11.7c}$$

for $\Delta_0 \ll \hbar\omega_D$. The equation can be solved easily to obtain the BCS result

$$\Delta_0 \sim 2\hbar\omega_D e^{-\frac{1}{gN(0)}}. \tag{11.7d}$$

As temperature rises $\Delta_0$ decreases and vanishes at $T = T_c$. $T_c$ is determined by

$$1 = gN(0)\int_{-\omega_D}^{\omega_D}\mathrm{d}\xi\frac{1}{2\xi}\tanh\left(\frac{\xi}{2k_B T_c}\right). \tag{11.7e}$$

The equation can be solved numerically to yield the result

$$k_B T_c = 1.14\hbar\omega_D e^{-\frac{1}{gN(0)}}. \tag{11.7f}$$

$S_{\text{eff}}(\Delta_0, \Delta_0^*)$ and the solution for $\Delta_0(T)$ are shown schematically in the following figures.

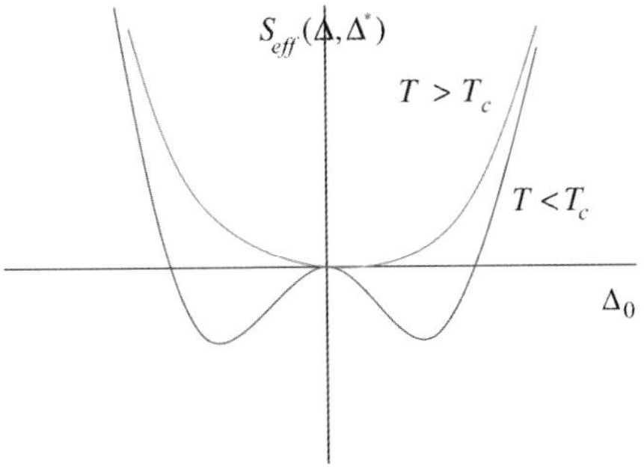


$S_{eff}(\Delta_0,\Delta_0^*)$ as function of $\Delta_0$ for $T > T_c$ and $T < T_c$

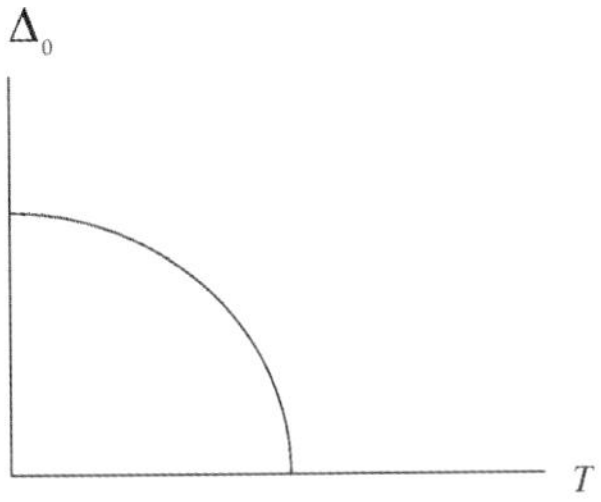


$\Delta_0$ as function of $T$

## 11.1.1 Semiclassical (Gaussian) Theory

To study the superfluid dynamics, we can construct a Gaussian theory by expanding $S_{\text{eff}}$ to second order in $\delta\phi$, $\delta\phi^*$ around the classical minimum, as we have done in Chapter 9 for interacting bosons. In other words, we can construct an effective quadratic action for $\delta\phi$, $\delta\phi^*$ fields that describes the dynamics of Cooper pairs.

Mathematically, we write $M_{kk'} = M_{0k}\delta_{kk'} + M_{1kk'}$ and expand $\ln\det M = \text{Tr}(\ln M)$ in powers of $M_1$.

First, we consider $M_0$ and $G_0 = -M_0^{-1}$. We have

$$M_{0kk'} = \delta_{kk'} \begin{bmatrix} -(i\omega_n - \xi_k) & -\Delta_0 \\ -\Delta_0^* & -(i\omega_n + \xi_k) \end{bmatrix} \tag{11.8a}$$

and

$$G_{0kk'} = -M_{0kk'}{}^{-1} = \frac{\delta_{kk'}}{(i\omega_n)^2 - E_k^2} \begin{bmatrix} (i\omega_n + \xi_k) & -\Delta_0 \\ -\Delta_0^* & (i\omega_n - \xi_k) \end{bmatrix}, \tag{11.8b}$$

where $E_k = \sqrt{\xi_k^2 + |\Delta_0|^2}$ and

$$M_{1kk'}(\phi, \phi^*) = \begin{bmatrix} 0 & -\delta\phi(k-k') \\ -\delta\phi^*(k'-k) & 0 \end{bmatrix}. \tag{11.8c}$$

Expanding

$$-\mathrm{Tr}(\ln M) \sim -\mathrm{Tr}(\ln M_0) + \mathrm{Tr}(G_0 M_1) - \frac{1}{2}\mathrm{Tr}(G_0 M_1 G_0 M_1) + \ldots,$$

we obtain, to quadratic order,

$$S_{\mathrm{eff}}(\delta\phi, \delta\phi^*) \sim -\frac{1}{2}\mathrm{Tr}(G_0 M_1 G_0 M_1) + \frac{|\delta\phi|^2}{g}. \tag{11.9a}$$

**Homework Q2a**

Evaluate $\frac{1}{2}\mathrm{Tr}(G_0 M_1 G_0 M_1)$. Show that it has the form

$$\frac{1}{2}(\pi_1(q, i\omega)\delta A(q, i\omega)\delta A(-q, -i\omega) + \pi_2(q, i\omega)\delta\vartheta(q, i\omega)\delta\vartheta(-q, -i\omega)),$$

where $\phi + \delta\phi = (\Delta_0 + \delta A)e^{i\delta\vartheta}$ and $\delta\phi(q) = i\Delta_0\delta\vartheta(q) + \delta A(q)$ for small $\delta A$ and $\delta\vartheta$. Write down the expressions for $\pi_1(q, i\omega)$ and $\pi_2(q, i\omega)$. Notice that $\mathrm{Tr}(G_0 M_1)$ also contributes to $\pi_2$.

In the continuum limit (small $q = (\vec{q}, i\omega)$), it can be shown after some algebra that at low temperatures $T \ll \Delta_0$ the action can be written as

$$S_{\mathrm{eff}} \sim \int \mathrm{d}^d r \int \mathrm{d}\tau \left\{ \alpha\Delta_0^2(\partial_\tau \delta\vartheta)^2 + \beta\Delta_0^2(\nabla\delta\vartheta)^2 + \varsigma(\delta A)^2 \right\} \\ (\alpha, \beta, \varsigma > 0), \tag{11.9b}$$

which, aside from the time-derivative term, is the same as the effective action for interacting bosons in the semiclassical approximation (see also Ref. [1]). Note that the form of (11.9b) can be deduced from the stability of the action to small fluctuations in $\delta\phi$ and from the invariance of the action with respect to a global U(1) gauge transformation $\phi(x) \to \phi(x)e^{i\theta}$.

**Homework Q2b**

Based on local gauge invariance, generalize Equation (11.9b) for $S_{\mathrm{eff}}$ to include coupling of electrons to an electromagnetic field. Show that the boson field has charge $2e$ and represents pairs of electrons.

Physically, the boson field $\phi$ represents the wavefunction of Cooper pairs, and we can interpret $S_{\mathrm{eff}}$ as describing the dynamics of a liquid of Cooper pairs. In particular, the mean-field solution $\langle\phi\rangle \sim \Delta_0 \neq 0$ implies that the Cooper pairs form a Bose condensate. The similarity of $S_{\mathrm{eff}}$ to the effective action for superfluid bosons suggests that Cooper pairs in superconductors have similar dynamical behavior to ordinary bosonic superfluids. When electromagnetic interaction is included (Homework Q2b), it can be shown that the Cooper pairs have similar dynamical behavior to charged bosonic superfluids (Chapter 9) and superconductivity results.

**Homework Q2c**

By comparing with results in Chapter 9, derive an expression for the supercurrent in terms of $\phi$.

## 11.2 BCS Theory for (s-Wave) Superconductors: Fermion Excitations and Hamiltonian Approach

The Hamiltonian formulation of BCS theory starts with the fermionic Hamiltonian

$$H=\int d^d r\hat{\psi}_\sigma^+(\vec{r})\left[-\frac{\hbar^2\nabla^2}{2m}-\mu\right]\hat{\psi}_\sigma(\vec{r})-g\int d^d r\hat{\psi}_\uparrow^+(\vec{r})\hat{\psi}_\downarrow^+(\vec{r})\hat{\psi}_\downarrow(\vec{r})\hat{\psi}_\uparrow(\vec{r}). \tag{11.10}$$

The interaction between electrons is treated approximately by introducing the Cooper decoupling for the interaction term,

$$\hat{\psi}_\uparrow^+(\vec{r})\hat{\psi}_\downarrow^+(\vec{r})\hat{\psi}_\downarrow(\vec{r})\hat{\psi}_\uparrow(\vec{r})\sim<\hat{\psi}_\uparrow^+(\vec{r})\hat{\psi}_\downarrow^+(\vec{r})>\hat{\psi}_\downarrow(\vec{r})\hat{\psi}_\uparrow(\vec{r})$$
$$+\hat{\psi}_\uparrow^+(\vec{r})\hat{\psi}_\downarrow^+(\vec{r})<\hat{\psi}_\downarrow(\vec{r})\hat{\psi}_\uparrow(\vec{r})>-<\hat{\psi}_\uparrow^+(\vec{r})\hat{\psi}_\downarrow^+(\vec{r})><\hat{\psi}_\downarrow(\vec{r})\hat{\psi}_\uparrow(\vec{r})>, \tag{11.11a}$$

where $\langle O\rangle_{H_{\rm mf}}$ is the thermodynamic average of the operator $O$, the average being performed with the mean-field Hamiltonian $H_{\rm mf}$,

$$H_{\rm mf}=\int d\vec{r}\hat{\psi}_\sigma^+(\vec{r})\left[-\frac{\hbar^2}{2m}\nabla^2-\mu\right]\hat{\psi}_\sigma(\vec{r})$$
$$-g\int d\vec{r}\left[\Delta_0^*\hat{\psi}_\downarrow(\vec{r})\hat{\psi}_\uparrow(\vec{r})+\hat{\psi}_\uparrow^+(\vec{r})\hat{\psi}_\downarrow^+(\vec{r})\Delta_0-|\Delta_0|^2\right], \tag{11.11b}$$

where

$$\Delta_0=<\hat{\psi}_\downarrow(\vec{r})\hat{\psi}_\uparrow(\vec{r})>. \tag{11.11c}$$

Fourier transforming, we obtain the mean-field Hamiltonian in momentum space

$$H_{mf}=\sum_{\vec{k}\sigma}\xi_k\hat{c}_{\vec{k}\sigma}^+\hat{c}_{\vec{k}\sigma}-g\sum_k(\Delta_0^*\hat{c}_{\vec{k}\sigma}\hat{c}_{-\vec{k}-\sigma}+\Delta_0\hat{c}_{-\vec{k}-\sigma}^+\hat{c}_{\vec{k}\sigma}^+). \tag{11.11d}$$

where $\hat{c}_{\vec{k}\sigma}^+(\hat{c}_{\vec{k}\sigma})$ are electron creation (annihilation) operators with momentum $\vec{k}$ and spin $\sigma$. We note that the mean-field Hamiltonian does not conserve particle number. This is a reflection of broken (global) gauge symmetry $\psi(\vec{x})\rightarrow e^{i\vartheta}\psi(\vec{x})$, since the conservation law associated with global U(1) gauge symmetry is the conservation of particle number. A similar situation occurred in Chapter 9 for interacting bosons.

The Hamiltonian can be diagonalized with the Bogoliubov transformation for fermions,

$$\hat{\gamma}_{\vec{k}\sigma}=u_k\hat{c}_{\vec{k}\sigma}+v_k\hat{c}_{-\vec{k}-\sigma}^+,\quad \hat{\gamma}_{-\vec{k}-\sigma}^+=v_k\hat{c}_{\vec{k}\sigma}-u_k\hat{c}_{-\vec{k}-\sigma}^+, \tag{11.12a}$$

where $\hat{\gamma}_{\vec{k}\sigma}^+(\hat{\gamma}_{\vec{k}\sigma})$ are fermion creation and annihilation operators. $u_k$, $v_k$ are numbers determined by the condition that the operators $\hat{\gamma}_{k\sigma}^+(\hat{\gamma}_{k\sigma})$ satisfy standard fermion commutation relations $\{\hat{\gamma}_{\vec{k}'\sigma'},\hat{\gamma}_{\vec{k}\sigma}^+\}=\delta_{\sigma\sigma'}\delta(\vec{k}-\vec{k}')$ and so on, and the Hamiltonian is diagonal when written in terms of the $\hat{\gamma}_{\vec{k}\sigma}^+(\hat{\gamma}_{\vec{k}\sigma})$ operators. After some algebra, we obtain

$$H_{\mathrm{mf}} = \sum_{k\sigma} E_k \widehat{\gamma}^{+}_{\vec{k}\sigma} \widehat{\gamma}_{\vec{k}\sigma} + E_{\mathrm{G}}, \tag{11.12b}$$

where $E_{\mathrm{G}}$ is the ground-state energy and $E_k = \sqrt{\xi_k^2 + |\Delta_0|^2}$ is the fermion excitation energy. It can be shown that the same mean-field equation (11.6d) is obtained for $\Delta_0 = <\hat{\psi}_\downarrow(\vec{r})\widehat{\psi}_\uparrow(\vec{r})>$, indicating that the path integral and Hamiltonian approaches represent the same mean-field theory.

**Homework Q3a**

Derive $u_k$, $v_k$, and the ground-state energy $E_{\mathrm{G}}$ in terms of $\xi_k$ and $\Delta_0$ with the (fermion) Bogoliubov transformation for the BCS mean-field Hamiltonian.

Step (1): Show that $\{\widehat{\gamma}_{\vec{k}'\sigma'}, \widehat{\gamma}^{+}_{\vec{k}\sigma}\} = \delta_{\sigma\sigma'}\delta(\vec{k}-\vec{k}')$ implies the normalization condition $|u_k|^2 + |v_k|^2 = 1$.

Step (2): Show that the Hamiltonian is diagonalized by the matrix equation

$$E_k \begin{pmatrix} u_k \\ v_k \end{pmatrix} = \begin{pmatrix} \xi_k & \Delta_0 \\ \Delta_0^* & -\xi_k \end{pmatrix} \begin{pmatrix} u_k \\ v_k \end{pmatrix}$$

and thus obtain $E_k$.

Step (3): Express $\Delta_0 = <\hat{\psi}_\downarrow(\vec{r})\widehat{\psi}_\uparrow(\vec{r})>$ in terms of expectation value are the $\widehat{\gamma}^{+}_{\vec{k}\sigma}(\widehat{\gamma}_{\vec{k}\sigma})$ operators and thus show that the mean-field equation (11.6d) is obtained.

Note that the fermion excitation spectrum $E_k$ is gapped in the superconducting state. The dynamics of a superconductor is characterized by two types of excitations, a bosonic branch that describes the dynamics of the superfluid collective motion, and the fermionic branch which we derive here. This is similar to bosonic superfluids where the dynamics is governed by a two-fluid model. The main difference is that the 'normal' branch of excitations is fermionic and is gapped for superconductors. The dynamics of boson superfluid is described by the Gross–Pitaevskii equation but it is described by the Ginsburg–Landau (GL) equation (which we shall study in the following) for superconductors.

As in the case of bosonic superfluids, the two branches of excitations interact with each other in general. To study fermion–superfluid interaction, we consider the action for the fermions in the general pairing field $\phi(\vec{x}, t)$:

$$S_{\mathrm{el}} = \int \mathrm{d}^{d+1}x \left\{ \begin{array}{l} \bar{\psi}_\sigma(x)\left[\partial_\tau - \dfrac{\hbar^2\nabla^2}{2m} - \mu\right]\psi_\sigma(x) \\ -|\phi(x)|e^{-2i\vartheta(x)}\psi_\uparrow(x)\psi_\downarrow(x) - |\phi(x)|e^{2i\vartheta(x)}\bar{\psi}_\uparrow(x)\bar{\psi}_\downarrow(x) \end{array} \right\}. \tag{11.13a}$$

The phase fluctuations of the $\phi$ field can be absorbed into the fermion field by the gauge transformation $\psi_\sigma(x) \to \psi_\sigma(x)e^{i\vartheta(x)}$. The action after gauge transformation is

$$S_{\mathrm{el}} = \int \mathrm{d}^{d+1}x \left\{ \begin{array}{l} \bar{\psi}_\sigma(x) \left[ \partial_\tau - \hbar^2 \dfrac{(\nabla + i\vec{A})^2}{2m} - \mu + iA_0 \right] \psi_\sigma(x) \\ -|\phi(x)|\psi_\uparrow(x)\psi_\downarrow(x) - |\phi(x)|\bar{\psi}_\uparrow(x)\bar{\psi}_\downarrow(x) \end{array} \right\}, \tag{11.13b}$$

where $A_\mu \sim 2\partial_\mu \theta$. Recall that $E_k \sim \sqrt{\xi_k^2 + |\phi|^2}$ and local fluctuations in the amplitude of the order parameter field $\phi$ effectively introduce local fluctuations of the fermion energy gap. More interestingly, since the supercurrent is given by $\vec{j}_s \sim |\phi|\nabla\vartheta$, we find that supercurrent flow produces an effective U(1) gauge field acting on the fermions. There are many interesting phenomena in superconductors associated with phase fluctuations of the order parameter, in particular when the phase field has singularities (vortices). You can find many examples in Tinkham's book [2].

## 11.2.1 Variational Wavefunction in BCS Theory

Just like Feyhman's variational wavefunction for interacting boson liquids, BCS theory can be formulated in terms of a variational wavefunction. The variational wavefunction is the ground state of the BCS Hamiltonian (11.11d)

$$|\psi_{\mathrm{BCS}}\rangle = \prod_k (u_k + v_k c_{k\uparrow}^+ c_{-k\downarrow}^+)|0\rangle, \tag{11.14}$$

where $|0\rangle = $ vacuum.

The wavefunction represents a product of wavefunctions of electron pairs with momentum and spin $\vec{k}\sigma$ and $-\vec{k}-\sigma$. $|u_k|^2$ represents the probability that the pairing state is empty, and $|v_k|^2$ represents the probability that the pairing state is occupied. Note that the number of electrons is not fixed in the wavefunction (broken global U(1) gauge symmetry). The BCS mean-field equation can be obtained variationally by minimizing the energy of the system with the wavefunction (11.14), and $u_k$, $v_k$ are determined by the same equation as shown in Homework Q3a.

**Homework Q3b**

Show that $|\psi_{\mathrm{BCS}}\rangle$ is the ground state for the $\hat{\gamma}_{\vec{k}\sigma}^+(\hat{\gamma}_{\vec{k}\sigma})$ operators, that is $\hat{\gamma}_{\vec{k}\sigma}|\psi_{\mathrm{BCS}}\rangle = 0$.

Let us study the exchange symmetry of the BCS wavefunction in more detail. For fermions, the wavefunction should change sign under exchange of two particles. For a pair of electrons interacting via a spin-independent interaction, the wavefunction can be written as a product of spatial and spin parts, $\phi(\vec{x}_1\sigma_1;\vec{x}_2\sigma_2) = \phi_s(\vec{x}_1,\vec{x}_2)\varphi(\sigma_1,\sigma_2)$, where $\sigma = \pm\frac{1}{2}$. $\varphi(\sigma_1,\sigma_2)$ can be symmetric with respect to spin exchange (spin triplet) or anti-symmetric with respect to spin exchange (spin singlet). The corresponding spatial wavefunction should be anti-symmetric (p-wave, f-wave, etc.) for the spin-triplet state, and symmetric (s-wave, d-wave, etc.) for the spin-singlet state.

The different possibilities of pairing of electrons give rise to different pairing states of superconductors. An s-wave superconductor is a coherent state of electron pairs in an s-wave, spin-singlet state, and a p-wave superconductor is a coherent state of electron pairs in a p-wave, spin-triplet state. The pairing states of the electrons are reflected in the coherent factors $u_k$, $v_k$. For the simple point-contact interaction we discussed, $u_k$, $v_k$ are direction independent, corresponding to an s-wave, spin-singlet superconductor. You can look at Ref. [3,4] for an introduction.

### 11.2.2
### GL Equation and Vortex Solution

Historically, the Ginsburg–Landau theory was first written down as a phenomenological theory for superconductors. The equation was derived as an application of Landau's idea for second-order phase transitions (Chapter 7). Later, it was shown that the equation can be derived microscopically by expanding $S_{\text{eff}}$ around $\phi = 0$ to fourth order in $\phi$ at temperatures close to $T_c$ (see Ref. [1] for details), that is

$$S_{\text{eff}}(\phi, \phi^*) \approx S_{\text{eff}}(0) + a(\vec{q}, \omega)\phi^*\phi + b(\vec{q}, \omega)(\phi^*\phi)^2 + \dots. \tag{11.15a}$$

To fourth order and in the continuum limit (gradient expansion keeping only lowest-order derivative terms), we obtain

$$S_{\text{eff}} \sim \int d^d r \int d\tau \left\{ i\gamma\phi^*(\vec{r}, \tau)\partial_\tau\phi(\vec{r}, \tau) + \frac{\hbar^2}{2m^*}\left|\left(\nabla - i\frac{e^*\vec{A}}{\hbar c}\right)\phi\right|^2 + a(T)|\phi|^2 + \frac{b}{2}|\phi|^4 \right\}, \tag{11.15b}$$

where $m^*$, $a$, $b$ are parameters depending on the exact form of $S_{\text{eff}}(\phi, \phi^*)$. $a(T) = a'(T - T_c) < 0$ if $T \leq T_c$. For $T_c \ll E_F$, it can be shown that $a' \sim T_c/E_F$, $b \sim N(0)^{-1}(T_c/E_F)^2$, and $\gamma \sim T_c/E_F$, where $N(0)$ and $E_F$ are the density of states on the Fermi surface and the Fermi energy for the normal-state metal, respectively [1]. The term $i\gamma\phi^*(\vec{r}, \tau)\partial_\tau\phi(\vec{r}, \tau)$ represents dissipation, which is important at $T \sim T_c$ when the superconducting gap is small and lots of quasi-particles are excited across the gap. We have also included the external electromagnetic vector potential $\vec{A}$ in $S_{\text{eff}}$ (see Homework Q2b). $e^* = 2e =$ charge of Cooper pairs. In the original Ginsburg–Landau theory, the time-derivative term was not considered and

$$F_{GL} \sim \int d^d r \left\{ \frac{\hbar^2}{2m^*}\left|\left(\nabla - i\frac{e^*\vec{A}}{\hbar c}\right)\phi\right|^2 + a(T)|\phi|^2 + \frac{b}{2}|\phi|^4 + \frac{B^2}{8\pi} \right\} \tag{11.15c}$$

was regarded as a free energy which is minimized with respect to $\phi$ to determine the thermodynamic behavior of the system. $B^2/8\pi(\vec{B} = \nabla \times \vec{A})$ is the vacuum magnetic field energy.

Minimizing $S_{\text{eff}}$ with respect to $\phi^*$, we obtain the time-dependent GL equation

$$i\gamma\partial_\tau\phi(\vec{r}, \tau) - \frac{\hbar^2}{2m^*}\left(\nabla - i\frac{e^*\vec{A}}{\hbar c}\right)^2\phi + a(T)\phi + b|\phi|^2\phi = 0. \tag{11.16a}$$

We first study steady-state (time-independent) solutions in the following. It is easy to show that a time-independent solution $|\phi|^2 = (-a(T)/b) \sim |\Delta_0|^2$ is obtained at $T < T_c$, corresponding to an approximate solution of the mean-field equation (11.7b) at temperatures close to but less than $T_c$.

Minimizing $S_{\rm eff}$ with respect to $\vec{A}$, we also obtain

$$\frac{1}{4\pi}\nabla\times\vec{B}-\frac{\hbar e^*}{2m^*ci}[\phi^*\nabla\phi-\phi\nabla\phi^*]+\frac{(e^*)^2}{m^*c^2}|\phi|^2\vec{A}(\vec{x})=0. \tag{11.16b}$$

Comparing with Ampere's law $\nabla\times\vec{B}=\frac{4\pi\vec{j}}{c}$, we identify the current density in GL theory,

$$\vec{j}=\frac{\hbar e^*}{2m^*i}[\phi^*\nabla\phi-\phi\nabla\phi^*]-\frac{(e^*)^2}{m^*c}|\phi|^2\vec{A}(\vec{x}). \tag{11.16c}$$

In the absence of $\vec{A}$, a persistent-current solution is obtained by setting $\phi=\tilde{\phi}_0 e^{i\vec{q}.\vec{x}}$, corresponding to a supercurrent $\vec{j}=\frac{\hbar e^*}{m^*}|\tilde{\phi}_0|^2\vec{q}$. Putting this back into Equation (11.16a), it is easy to see that $\tilde{\phi}_0<\phi_0$, that is the magnitude of the order parameter is reduced in the presence of a supercurrent.

**Homework Q4a**

Using Equations (11.16a)–(11.16c), find the maximum current density (critical current) a superconductor can sustain.

Because of its simple form, the Ginsburg–Landau model is often used as an approximated version of $S_{\rm eff}(\phi,\phi^*,\vec{A})$ to study properties of superconductors even at $T\ll T_{\rm c}$. Indeed, many important properties of superconductors we know nowadays were derived first from the Ginsburg–Landau model. It remains one of the most effective tools in understanding properties of exotic superconductors (high-$T_{\rm c}$ cuprates, p-wave superconductors, and superconductors that break parity/time reversal symmetry) nowadays. In the following, we shall apply the equation to study a very important kind of excitation in superconductors – vortices.

First, we show that there are two important length scales in the GL equation. The first length scale can be seen in the steady-state ($\dot{\phi}=0$) situation with no current or magnetic field. In this case, we can choose $\phi$ to be real and the GL equation becomes

$$-\frac{\hbar^2}{2m^*}\nabla^2\phi+a\phi+b\phi^3=0.$$

Assuming that $a<0$ ($T<T_{\rm c}$) and writing $\phi=f\phi_0$, where $\phi_0^2=-\frac{a}{b}$, we can rewrite the equation as

$$-\xi^2\nabla^2 f-f+f^3=0, \tag{11.17a}$$

where

$$\frac{\hbar^2}{2m^*|a|}=\xi^2. \tag{11.17b}$$

$\xi$ is called the coherence length of the superconductor. It is the characteristic length scale where the superconductor order parameter $\phi$ varies. For example, $f(x)\sim f_0 e^{-\frac{|x|}{\xi}}$ near a superconductor–normal-metal boundary, where $|x|=$ distance away from the boundary into the normal metal. Physically it represents roughly the size of a Cooper pair.

The other length scale is the penetration depth and is the length scale where a magnetic field can penetrate inside a superconductor (Meissner effect). We have studied this length scale in Chapter 9 for charged bosons and it is given by

$$\lambda^{-2} = \frac{4\pi e^{*2}}{\hbar^2 c^2} |\phi_0|^2.$$

**Homework Q4b**
Derive the above expression for $\lambda$.

A superconductor with $\xi > (<) \lambda$ is called a type-I (II) superconductor. The two types of superconductors have very different electromagnetic properties [2]. Vortex excitations exist and are important in type-II superconductors.

To understand vortices, first we study the phenomenon of flux quantization.

#### 11.2.2.1 Flux Quantization

To understand flux quantization, it is helpful to re-examine the physics of angular momentum quantization.

Let us consider a general one-particle quantum wavefunction on a 2D ring. In cylindrical coordinates,

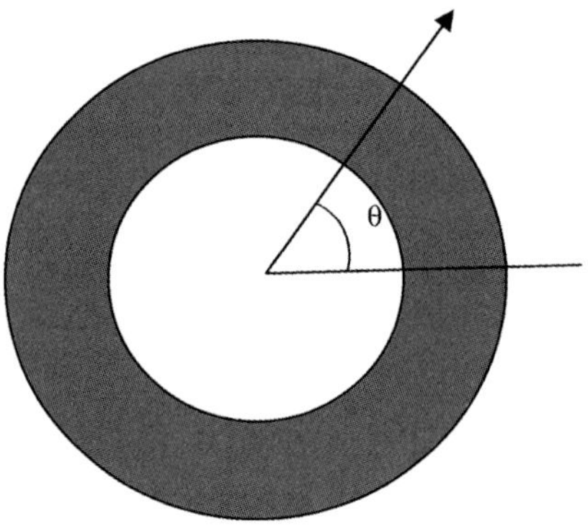


$\psi(\vec{x}) = \sqrt{\rho(r,\vartheta)}e^{i\varphi(r,\vartheta)}$. See above figure.

The single-valuedness of the wavefunction requires that

$\psi(r, \vartheta) = \psi(r, \vartheta + 2\pi)$ and, correspondingly,

$\varphi(r, \vartheta + 2\pi) = \varphi(r, \vartheta) + 2m\pi.$

An example of a wavefunction that satisfies this criterion is $\psi(r,\vartheta) = \sqrt{\rho(r)}e^{im\vartheta}$.

For a single particle, this wavefunction describes a quantum state with angular momentum $L_z = m\hbar$. We note that

single-valuedness of wavefunction⇒angular momentum quantization.

Note that this wavefunction is in general not well defined at $r \to 0$ where all different values of the angle $\vartheta$ collapse to the same point. To make sure that the wavefunction is well defined, we must have $\psi(r \to 0, \vartheta) \to 0$ for $m \neq 0$, that is the amplitude of the wavefunction must vanish at the origin. Note that for a wavefunction on a ring with inner radius $a$, $\psi(r<a, \vartheta) = 0$ and $\psi(r \to 0, \vartheta) \to 0$ is automatically satisfied.

Now, let us examine what happens if the wavefunction $\psi(r, \vartheta)$ describes not a single electron, but is the wavefunction in GL theory describing collective motion of Bose-condensed Cooper pairs.

From Equation (11.16c), the current in the presence of an external electromagnetic field $\vec{A}$ is given by

$$\vec{j} = \frac{\hbar e^*}{2m^* i}[\psi^*\nabla\psi - \psi\nabla\psi^*] - \frac{(e^*)^2}{m^* c}|\psi|^2\vec{A}(\vec{x}) \sim \frac{\hbar e^* n}{m^*}\left[\nabla\varphi(\vec{x}) - \frac{e^*}{\hbar c}\vec{A}(\vec{x})\right] \tag{11.18}$$

at the ring region where $\psi(r,\vartheta) = \sqrt{\rho(r)}e^{im\vartheta} \sim \sqrt{n}e^{im\vartheta}$, $\nabla\psi(\vec{x}) \approx \sqrt{n}\nabla\varphi \sim \sqrt{n}\frac{m}{r}\hat{\vartheta}$.

Therefore, in the absence of a vector field $\vec{A}$, $m \neq 0$ implies that there exists a macroscopic current circulating around the ring or a vortex! The kinetic energy of this macroscopic quantum state is proportional to $\oint d\vec{x}.|\vec{j}_s(\vec{x})|^2$ and is large in general!

For rings without exact cylindrical symmetry, the magnitude of the circulating current can be examined by considering the integral

$$I = \oint d\vec{l}.\vec{j}_s(\vec{x}) = \oint d\vec{l}.\frac{\hbar e^* n}{m^*}\left[\nabla\varphi(\vec{x}) - \frac{e^*\vec{A}(\vec{x})}{\hbar c}\right]$$

around a contour surrounding the ring (arrows in the figure).

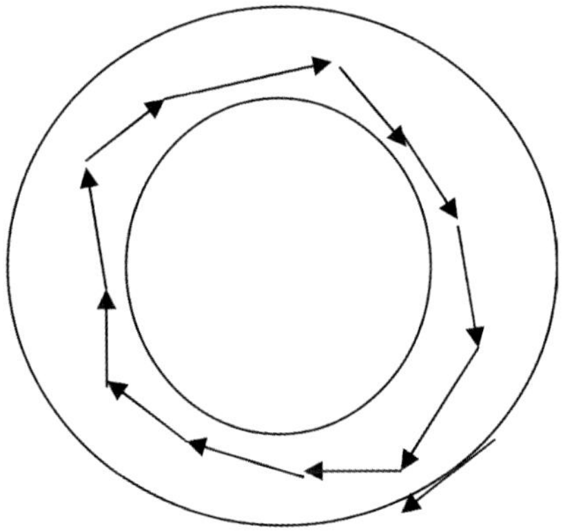

Using Stokes' theorem, we obtain

$$I = \frac{he^* n}{m*}\left[m - \frac{\Phi}{\Phi_0}\right], \tag{11.19}$$

where $\Phi = \iint_S \vec{B}\cdot d\vec{S}$ is the total magnetic flux enclosed in the contour and $\Phi_0 = \frac{hc}{e^*} = \frac{hc}{2e}$ is the magnetic flux quantum associated with charges $e^*$ (see Chapter 6). We note that, first of all, $I$ is quantized in the absence of a magnetic field as a result of single-valuedness of the wavefunction. When a magnetic field is present we note that the net circulating current $I$ is zero if $\Phi = m\Phi_0$, that is when the total magnetic flux trapped inside the ring is a multiple of the fundamental magnetic flux quantum $\Phi_0 = \frac{hc}{2e}$.

Now, let us ask the following question. Imagine that we can change the amount of magnetic flux $\Phi$ trapped in a superconducting ring continuously. How would the energy $E(\Phi)$ of the system change as the flux changes?

Our above analysis suggests that $E(\Phi)$ has a periodic structure, and has minima at $\Phi = m\Phi_0$ (see figure below) because, at these values, the net supercurrent circulating around the ring is zero and the kinetic energy of the Cooper-pair wavefunction is minimized. This result can be confirmed easily in the case of cylindrical rings by solving the Ginsburg–Landau equation directly and will be left as an exercise for readers.

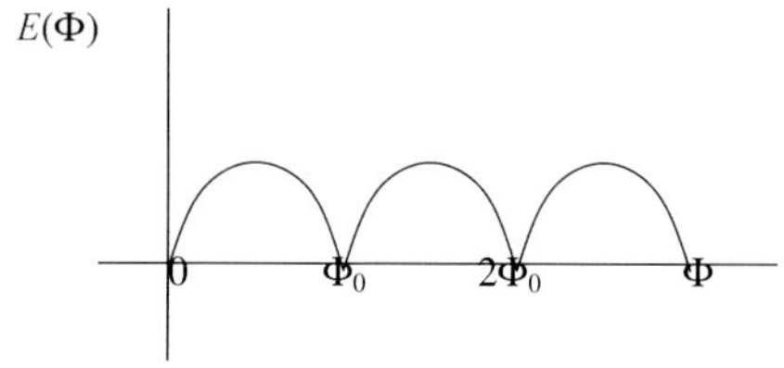


It also implies that the system is stable ($E(\Phi)$ is minimized) when the magnetic flux trapped in the ring is equal to $m\Phi_0$. The magnetic flux trapped in a superconducting ring cannot dissipate to zero by itself because of the high energy barrier. *This phenomenon is called flux quantization and is indeed observed experimentally in superconductors.*

#### 11.2.2.2 **Vortices**

The phenomenon of flux quantization suggests another interesting possibility – the existence of vortex-like topological excitations in superconductors which trap quantized values of magnetic flux.

The idea is simple. We have just seen that a superconducting ring can trap integral values of the magnetic flux quantum. This is permitted because the condition $\psi(r \to 0, \vartheta) \to 0$ is automatically satisfied in ring geometry. However, even in the absence of a ring, we may envision the possibility of finding a solution of the Ginsburg–Landau equation where the wavefunction $\psi(\vec{x})$ vanishes at some spatial point $\vec{x}_0$, with the phase of the wavefunction increasing by integer multiples of $2\pi$ when going around this point, i.e., some kind of self-generated ring structure (see figure next page). If this is possible, a corresponding quantized magnetic flux will be trapped around $\vec{x}_0$:

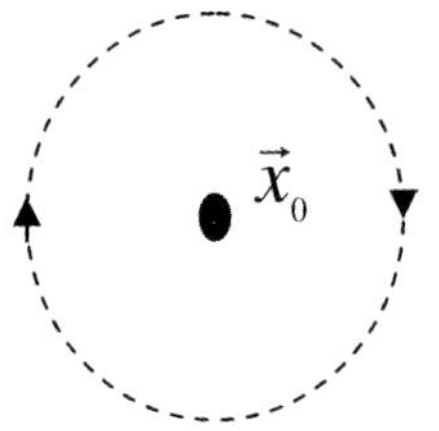


Indeed, such kind of solution was first predicted to exist as a solution of the Ginsburg–Landau equation and was later found experimentally when a type-II superconductor was put under a magnetic field.

Mathematically, a single-vortex solution is given in cylindrical coordinates by

$$\psi(r,\phi,z) = f(r)\phi_0 e^{in\phi}, \tag{11.20a}$$

with $\vec{A}(r,\phi,z) = A(r)\,\widehat{\phi}$, corresponding to a magnetic field in the $z$ direction.

In this case, the time-independent GL equation becomes in cylindrical coordinates

$$-\xi^2\left[\frac{1}{r}\frac{\partial}{\partial r}\left(r\frac{\partial}{\partial r}\right)-\left(\frac{n}{r}-\frac{e^*A(r)}{hc}\right)^2\right]f - f + f^3 = 0, \tag{11.20b}$$

where $\xi$ is the coherence length we discussed before. We also require that

$$\frac{e^*A(r)}{hc} \sim \frac{n}{r} \tag{11.20c}$$

when $r \to \infty$ so that $\oint \vec{A}.dl = n\Phi_0$ (flux quantization) and $\frac{e^*A(r\to 0)}{hc} \to$ constant, so that a finite magnetic field is obtained at $r \to 0$. Therefore, $\left(\frac{n}{r} - \frac{e^*A(r)}{hc}\right) \to 0$ as $r \to \infty$ and $f(r) \to 1$. It is also easy to show that $f(r) \to r^n$ as $r \to 0$, that is a well-behaved solution $f(r)$ is expected to exist although Equation (11.20b) cannot be solved analytically.

Qualitatively, we expect that $f(r) \to r^n$ at small $r$, and it rises gradually to $f(r) \to 1$ at a distance $r \geq \xi$ (dotted line in figure below). Correspondingly, a magnetic flux is trapped at the center of the vortex with $B(r) \sim \frac{1}{r^{1/2}} e^{-\frac{r}{\lambda}}$ (black line in figure below) at large $r$. Note the different length scales that control $f(r)$ and $B(r)$. Interested readers can look at Ref. [2] for details.

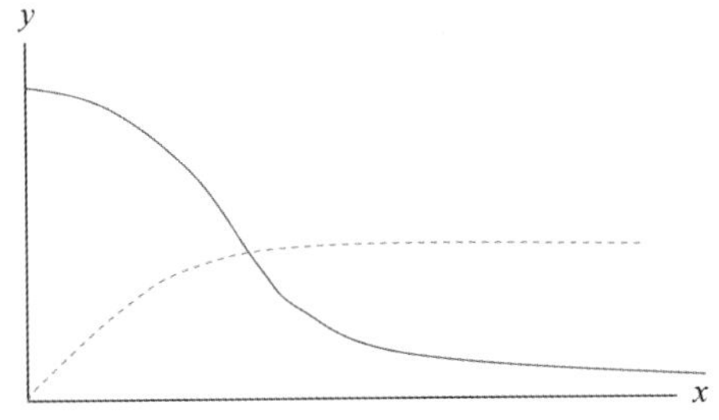


$f(r)$ and $B(r)$ in a vortex solution for $n = 1$ with $\lambda > \xi$.

#### 11.2.2.3 Vortices in Neutral Superfluid and KT Transition

It is interesting to ask whether a corresponding vortex solution exists in a neutral superfluid. This question can be answered by considering a GL equation with $e^* \equiv 0$ and asking whether a vortex solution with $\psi(r,\phi) = f(r)\phi_0 e^{in\phi}$ still exists, that is a vortex solution with no trapped magnetic flux. The corresponding GL equation is

$$-\xi^2\left[\frac{1}{r}\frac{\partial}{\partial r}\left(r\frac{\partial}{\partial r}\right)-\left(\frac{n}{r}\right)^2\right]f - f + f^3 = 0. \tag{11.21}$$

It is not difficult to see that a solution exists with $f(r) \to r^n$ as $r \to 0$ and $f(r) \to 1 - \frac{n^2}{2}\left(\frac{\xi}{r}\right)^2$ as $r \to \infty$. The solution is qualitatively similar to the vortex solution with trapped magnetic flux except that $f(r)$ approaches 1 as $r \to \infty$ more slowly in the present case. In fact, whether a magnetic flux is being trapped or not, vortex solutions are topologically stable because the creation and destruction of one vortex modify the macroscopic condensate wavefunction of the superconductor by changing the value of $n$ in *its boundary condition* $\varphi(r, \phi + 2\pi) = \varphi(r, \phi) + 2n\pi$. Thus, all the Cooper pairs in the condensate are affected when a vortex is created or destroyed. The situation is the same as vortices in the *X–Y* model (Chapter 8), where involvement of all particles in the condensate makes it difficult to create or destroy a single vortex.

The major difference between vortices in a superconductor and vortices in a neutral superfluid can be seen from the energy of a single vortex. At $r \to \infty$,

$$\frac{\hbar^2}{2m^*}\left|\left(\nabla - i\frac{e^*\vec{A}}{\hbar c}\right)\phi\right|^2 + a|\phi|^2 + \frac{b}{2}|\phi|^4 \to \frac{\hbar^2|\phi_0|^2}{2m^*}\left(\frac{n}{r} - \frac{e^*A(r)}{hc}\right)^2 + a|\phi_0|^2 + \frac{b}{2}|\phi_0|^4 \tag{11.22a}$$

and, using Equation (11.15c), it is easy to see that a vortex contributes an additional energy

$$F_V \sim W\int d^2r\left\{\frac{\hbar^2}{2m^*}|\phi_0|^2\left(\frac{n}{r} - \frac{e^*A(r)}{hc}\right)^2\right\} \sim \begin{cases} \text{constant} & \left(A(r) \sim \dfrac{nhc}{e^*r}\right), \\ \dfrac{\hbar^2 n^2}{2m^*}|\phi_0|^2 \ln\left(\dfrac{L}{\xi}\right) & (A(r) = 0), \end{cases} \tag{11.22b}$$

where $W$ and $L$ are the height ($z$ direction) and width ($x$–$y$ plane) of the system, respectively.

We see that the energy diverges logarithmically for vortices in a neutral superfluid as in the case of vortices in the *X–Y* model. However, such a divergence is absent in the corresponding charged superfluid with trapped magnetic flux. As a result, we expect that a Kosterlitz–Thouless (KT)-type phase transition will exist for thin films of neutral superfluids and superconductors with large penetration depth $\lambda$ (see Ref. [1]). Other interesting consequences of vortices in superconductors can be found in Ref. [2].

## 11.3 Superconductor–Insulator Transition

Borrowing results from Chapter 8, we expect that a disordered phase of superconductor/superfluid will exist in two dimensions where superfluidity is lost because of thermal excitation of vortices and the system becomes metallic (KT transition). This new phase of matter was indeed observed in superfluid $He^4$ and superconductor films. A natural question is thus: can a similar phenomenon also exist in other dimensions?

Physically, vortices represent singularities in the phase of the superconductor/superfluid order parameter, and a disordered phase of vortices is a state where the phase of the superconductor or superfluid order parameter is randomized but with the amplitude of the order parameter remaining finite. The question is whether we may obtain similar states in other dimensions because of thermal fluctuations.

We also learnt in Chapter 8 that a quantum system in $D$ dimensions is very often identical to a classical system in $D+1$ dimensions. Thus, we may even ask whether a disordered phase as described above may exist at zero temperature – the mechanism which disordered the phase of the order parameter being not thermal fluctuation but quantum fluctuation.

We may try to guess the nature of such a zero-temperature phase, if it exists. In Chapter 9 we have learnt that the phase and amplitude of a boson field $\psi = \sqrt{\rho}e^{i\vartheta}$ are conjugate variables obeying the commutation rule $[\rho, \vartheta] = i\hbar$. Therefore, a corresponding uncertainty relation exists where $\langle\Delta\rho\rangle\langle\Delta\vartheta\rangle > \hbar$, where $\langle\ldots\rangle$ = ground-state expectation value. For a superconductor where phase is fixed throughout the system, $\langle\Delta\vartheta\rangle \to 0$ and we have $\langle\Delta\rho\rangle \to \infty$ to respect the uncertainty relation, corresponding to a system of broken global gauge symmetry and uncertain particle number. On the other hand, for a phase-disordered state, we expect that $\langle\Delta\vartheta\rangle \to \infty$ and $\langle\Delta\rho\rangle \to 0$, corresponding to a state where the total particle number cannot change at all, that is a solid (insulator). Therefore, we expect that a phase-disordered superconductor/superfluid is an insulator, and a quantum phase transition from a phase-ordered to a phase-disordered state, if it exists, is a superconductor to insulator transition.

### 11.3.1 Rotor Model

To illustrate the plausible existence of a phase-disordered (insulator) quantum state, we assume that the amplitude of the superfluid order parameter is more or less fixed and consider a model that focuses on the phase fluctuations of the superfluid. We consider the phase Lagrangian

$$L[\vartheta] = \frac{1}{2g}\int d^d r(\dot{\vartheta}^2 - c^2(\nabla\vartheta)^2), \qquad (11.23)$$

where $\vartheta$ is the phase of a boson/superconductor field. The model was derived for interacting bosons in Chapter 9. A similar model can be applied to study superconductors if we assume that there is no magnetic field in the system and $\vec{A} = 0$. We shall study some basic properties of this important model here.

The first thing we have to do is to remedy a strong defect of this model. The defect is that the model should be periodic in $\vartheta$, with a period $2\pi$. The periodicity in $\vartheta$ is a consequence of the fact that physically the $\vartheta$ field represents the phase of a complex scalar field with the fundamental property $e^{i\vartheta} = e^{i(\vartheta + 2n\pi)}$, implying that $\vartheta$ and $\vartheta + 2n\pi$ should represent the same point in Hilbert space. However, this information is lost in the Lagrangian $L[\vartheta]$.

The importance of keeping the periodicity of the phase field has been pointed out in Chapter 8, where we saw that it led to the existence of vortices in the X–Y model and quantization of angular momentum in the rotor model. To keep the periodicity of $\vartheta$ here, we first introduce a lattice version of the phase model, where $\vartheta(\vec{x}) \to \vartheta_i$ lives only on lattice points. The derivative in space is replaced by a difference, that is

$$L[\vartheta] \to \frac{1}{2g}\sum_i \dot{\vartheta}_i^2 - \frac{c^2}{2g}\sum_{<i,j>} (\vartheta_i - \vartheta_j)^2, \tag{11.24a}$$

where $\langle i, j\rangle$ are nearest-neighbor sites on a cubic (3D) or square (2D) lattice.

Now, we can see that the periodic phase structure is lost because the Lagrangian is not invariant under the transformation $\vartheta_i \to \vartheta_i + 2n\pi$. To remedy this defect, we replace $(\vartheta_i - \vartheta_j)^2$ by $2(1 - \cos(\vartheta_i - \vartheta_j))$ so that, in the continuum limit, where $\vartheta_i$ is slowly varying in space, $2(1 - \cos(\vartheta_i - \vartheta_i)) \to (\vartheta_i - \vartheta_i)^2$. With this replacement the Lagrangian will be invariant under the transformation $\vartheta_i \to \vartheta_i + 2n\pi$.

Therefore, we end up with the (lattice) rotor model

$$L[\vartheta] = \frac{1}{2g}\sum_i \dot{\vartheta}_i^2 - \frac{c^2}{g}\sum_{<i,j>} [1 - \cos(\vartheta_i - \vartheta_j)]. \tag{11.24b}$$

The physics behind the rotor model can be seen if you imagine the angles $\vartheta_i$ as representing the locations of plane rotors – a particle confined by a rod to move only on a circle (see figure).

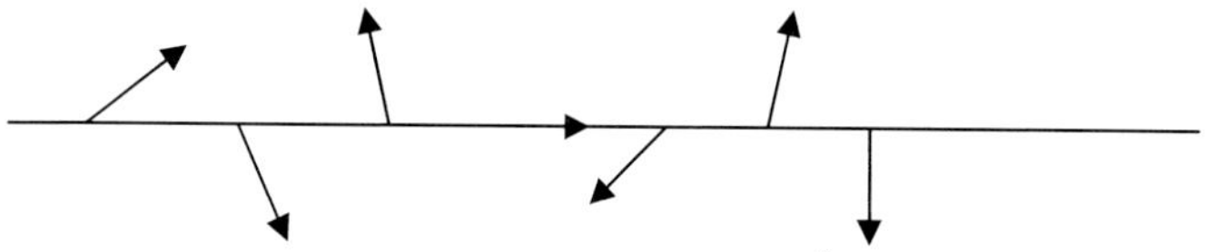

The kinetic energy of the rotors is $T = \sum_i \frac{p_i^2}{2m}$, where $p_i \sim ma\dot{\vartheta}_i$, with $a$ = radius of the rotor. The rotors interact with each other through a potential energy term

$$V = -\alpha \sum_{<i,j>} \vec{r}_i \cdot \vec{r}_j \sim -\alpha a^2 \sum_{<i,j>} \cos(\vartheta_i - \vartheta_j), \tag{11.24c}$$

which restricts the motion of the rotors.

Now, let us quantize the Lagrangian. The conjugate momentum of $\vartheta_i$ is

$$l_i = \frac{\partial L}{\partial \dot{\vartheta}_i} = \frac{\dot{\vartheta}_i}{g} \text{ (or } l_i = ma^2\dot{\vartheta}_i = \text{angular momentum of the rotor).}$$

Therefore, the Hamiltonian is

$$H[\vartheta] = \frac{g}{2}\sum_i l_i^2 + \frac{c^2}{g}\sum_{<i,j>} [1 - \cos(\vartheta_i - \vartheta_j)]. \tag{11.25a}$$

The corresponding (quantum) Hamiltonian is, in the Schrödinger representation,

$$H[\vartheta] = \frac{-\hbar^2 g}{2}\sum_i \frac{\partial^2}{\partial\vartheta_i^2} + \frac{c^2}{g}\sum_{<i,j>}[1-\cos(\vartheta_i-\vartheta_j)]. \tag{11.25b}$$

There is a very important difference between the original phase model and the rotor model here. The original phase model is quadratic and can be exactly solved. The solution represents non-interacting sound waves. The rotor model here is non-quadratic and is not exactly solvable, meaning that interaction has been introduced between the sound waves of the original phase model. *The interaction between sound waves is* not *derived from the original microscopic interacting boson or superconductor Hamiltonian, but is deduced from the requirement of periodicity in the $\vartheta$ field, which is a general requirement independent of the microscopic details of the system.* The philosophy behind this approach is that general geometrical or topological requirements are often more important than the detailed microscopic Hamiltonian in determining the qualitative properties of a system.

### 11.3.2
### Strong- and Weak-Coupling Expansions

The quantum rotor Hamiltonian cannot be solved exactly in general. To see qualitatively what are the plausible eigenstates described by the model, we consider the strong ($g$ large)- and weak ($g$ small)-coupling expansions here. We first consider the strong-coupling expansion (see also Ref. [5]).

**Strong-Coupling Expansion** This is the limit where $g \gg 1$ and the kinetic energy term dominates the Hamiltonian (extreme quantum limit).

To zeroth order, we may neglect the interaction term and obtain

$$H[\vartheta] \sim H_0[\vartheta] = \frac{-\hbar^2 g}{2}\sum_i \frac{\partial^2}{\partial\vartheta_i^2}. \tag{11.26}$$

Obviously, the eigenfunctions are

$$\Psi[\{m\}] = \prod_i (2\pi)^{-1/2}\exp(im_i\vartheta_i), \tag{11.27a}$$

where $m_i$ are integers. The energy of the state $\Psi[\{m\}]$ is

$$E[m] = \sum_i \frac{\hbar^2 g m_i^2}{2}. \tag{11.27b}$$

In particular, the ground state is $\Psi[\{0\}]$, where $m_i = 0$ for all $i$ and is non-degenerate.

The first excited state $\Psi[\{0, \ldots, 0, \pm 1, 0, \ldots, 0\}]$ has $m_i = 0$ for all $i$ except for one particular value of $i = j$. The state is $2N$-fold degenerate ($N$ = number of lattice sites) with excitation energy $E_1 = \frac{\hbar^2 g}{2}$.

**Homework Q5a**

What are the wavefunctions, excitation energies, and degeneracies for the second excited states?

Next, we can study the effect of interaction by (first-order) perturbation theory. First, we consider the ground state. The correction in energy is given by

$$\Delta E_0 = \frac{c^2}{g}\sum_{<i,j>}\left\langle\left[1-\cos(\theta_i-\theta_j)\right]\right\rangle = \frac{c^2}{g}\sum_{<i,j>}\frac{1}{2\pi}\int_0^{2\pi}\mathrm{d}\vartheta_i\frac{1}{2\pi}\int_0^{2\pi}\mathrm{d}\vartheta_j[1-\cos(\vartheta_i-\vartheta_j)]$$
$$= \frac{c^2}{g}\frac{ND}{2}, \tag{11.28}$$

where $D$ = number of nearest-neighbor sites.

The application of first-order perturbation theory to the first excited states is more complicated. We have to apply degenerate perturbation theory because of degeneracy.

**Homework Q5b**

Apply first-order (degenerate) perturbation theory to compute the first excited state spectrum. Show that the degenerate first excited state spectrum is modified into an energy band in first-order perturbation theory. (Hint: the perturbation theory can be simplified by noting that $\frac{1}{2\pi}\int_0^{2\pi} e^{-in\vartheta}e^{\pm i\vartheta}e^{im\vartheta}\mathrm{d}\vartheta = \delta_{m\pm1,n}$ and therefore $e^{\pm i\vartheta_i} = \widehat{a}_i^{\pm}$ can be regarded as a raising (lowering operator) which changes $\Psi[\{n_1, n_2, n_i, n_{i+1}, \ldots, n_N\}]$ into $\Psi[\{n_1, n_2, n_i \pm 1, n_{i+1}, \ldots, n_N\}]$ and $\frac{c^2}{g}\sum_{<i,j>}[1-\cos(\vartheta_i-\vartheta_j)] \rightarrow \frac{c^2}{g}\sum_{<i,j>}\left[1-\left(\frac{a_i^+a_j^- + a_i^-a_j^+}{2}\right)\right]$ is like a hopping term. (Note that the $e^{\pm i\vartheta_i} = \widehat{a}_i^{\pm}$ are, strictly speaking, not boson creation and annihilation operators, as can be checked from their commutation relation.)

Irrespective of the details, the most important result is that, for large g, there is a gap in the excitation spectrum between the ground and excited states, that is the system behaves like an insulator and is no longer a superfluid.

**Weak-Coupling Expansion** Next, we consider the limit $g \rightarrow 0$. In this case, the interaction (or potential energy) dominates the Hamiltonian (semiclassical limit).

Therefore, to zeroth order,

$$H[\vartheta] \sim H_0 = \frac{c^2}{g}\sum_{<i,j>}[1-\cos(\vartheta_i-\vartheta_j)] \tag{11.29}$$

and the ground-state wavefunction is

$$\Psi[\vartheta] = \prod_i \delta(\vartheta_i-\vartheta_0), \tag{11.30}$$

corresponding to a state with all rotors pointing in the same direction $\vartheta_0$. This is a state with *long-range order and broken (rotational) symmetry*.

To solve for the excited states, we have to include the kinetic energy term. We may expect that the low-energy excitation spectrum is dominated by angles $\vartheta_i$ not far away from $\vartheta_0$. If this is true, we can forget about the periodicity requirement of $\vartheta_i$ and expand the interaction term for small $\vartheta_i - \vartheta_j$ to obtain

$$H[\vartheta] \sim \frac{g}{2}\sum_i \dot{\vartheta}_i^2 + \frac{c^2}{2g}\sum_{<i,j>} (\vartheta_i - \vartheta_j)^2,$$

which is the original quadratic phase model on the lattice (Equations (11.23) and (11.24a)). In this case, the Hamiltonian can be diagonalized exactly and we obtain as in Chapter 9 the excited-state spectrum $E_k \sim c|\vec{k}|/g$ at small $\vec{k}$, which is the Goldstone mode associated with the broken global U(1) gauge symmetry in the ground state.

The strong- and weak-coupling expansions suggest that there are at least two phases in the model in the ground state associated with different values of the coupling constant $g$: the ordered state ($g$ small, potential energy dominated) corresponding to a superconductor/superfluid state and the disordered state ($g$ large, kinetic energy dominated) corresponding to an insulator state. A quantum superconductor/superfluid–insulator transition occurs if we can tune the parameter $g$ continuously. Starting from the insulator phase, we expect that the gap of the excitation spectrum $E_g$ decreases as $g$ decreases and an insulator-to-superfluid transition occurs when $E_g \to 0$ and Bose condensation occurs. It is interesting to note that superconductor–insulator transitions were indeed observed in superconducting thin films under magnetic field or when the thickness of the films was gradually reduced (Ref. 6).

The story of the rotor model and the superfluid–insulator transition does not end here. Note that we have assumed that the fluctuation of the $\vartheta$ field is small and have neglected the periodicity in the $\vartheta$ field again in deriving the weak-coupling expansion. Is this approximation reliable?

We can check this result in one dimension if we note that in the continuum limit the one-dimensional quantum rotor model can be mapped into the two-dimensional classical XY model in evaluating the partition function (Chapter 8). In particular, we expect that two different (zero-temperature) quantum phases exist in the 1D quantum rotor model, a KT-like phase and a disordered phase. How does the quantum rotor model behave in these two phases?

**Homework Q6**

Derive the mapping between the partition function of the one-dimensional quantum rotor model and the two-dimensional classical XY model in the continuum limit. Based on this mapping, discuss the properties of the one-dimensional quantum rotor model at zero temperature.

In two dimensions, the quantum rotor model also has non-trivial properties. We shall come back to this problem in Chapter 12, after we discuss lattice gauge field theory.

## References


1 Nagaosa, N. (1999) *Quantum Field Theory in Condensed Matter Physics*, Springer-Verlag, Berlin Heidelberg New York.

2 Tinkham, M. (1996) *Introduction to Superconductivity*, 2nd Ed., McGraw Hill, New York.

**3** Schrieffer, J.R. (1964) *Theory of Superconductivity*, Benjamin/Cummings, (Revised 3rd Printing, 1983).
**4** Mahan, G.D. (1990) *Many-Particle Physics*, Plenum Press - New York & London.
**5** Polyakov, A.M. (1987) *Gauge Fields and Strings*, Harwood Academic Publishers.
**6** Y. Dubi, Y. Meir & Y. Avishai (2007) *Nature* 449,876 and references therein.

# 12
# Introduction to Lattice Gauge Theories

In this chapter we shall examine gauge theories more carefully. Using a simple example of $Z_2$ lattice gauge theory we first show that a hidden non-local structure exists in gauge theories. We then explore the mathematical structure behind the U(1) lattice gauge theory. We shall show that the non-trivial topological structure behind the theory leads to charge quantization and the existence of confinement and plasma phases. The important concept of duality in modern quantum field theory is then introduced using U(1) lattice gauge theory and the superfluid state in $2+1$ dimensions as examples.

## 12.1
## Introduction: U(1) and $Z_2$ Lattice Gauge Theories

Recall that gauge theories are theories with redundancies because of freedom in choosing a gauge. The first thing we want to point out is that in physics there are many examples where a quantum state can be labeled in more than one way, and often we have to fix a representation (or fix a gauge) for the state in order to tackle the problem consistently.

A simple example is particles in a tight-binding model, $H = -t\sum_{<i,j>} c_i^+ c_j$, where $\langle i, j\rangle$ denotes nearest-neighbor sites on a lattice. The eigenstates of this Hamiltonian are plane waves $\phi_{\vec{k}}(\vec{x}) \sim e^{i\vec{k}.\vec{x}}$, where $\vec{x} = a(l, n, m)$ ($l$, $n$, $m$ are arbitrary integers) for a cubic lattice, with $a =$ lattice spacing. We note that two plane-wave states $\phi_{\vec{k}}(\vec{x}) \sim e^{i\vec{k}.\vec{x}}$ and $\phi_{\vec{p}}(\vec{x}) \sim e^{i\vec{p}.\vec{x}}$ represent the same state if $\vec{p} = \vec{k} + \frac{2\pi}{a}(i, j, k)$, since $e^{2\pi ni} = 1$ for arbitrary integer $n$. The physical reason behind this redundancy can be seen more clearly in real space, where the physical wavefunction $\psi(\vec{x})$ is characterized only by its values at $\vec{x} = a(l, n, m)$. In particular, two wavefunctions $\psi_{\vec{p}}(\vec{x})$ and $\psi_{\vec{k}}(\vec{x})$ are considered to be the same as long as they have identical values on the lattice points $\vec{x} = a(l, n, m)$, although they may take different values at other points in (continuous) space. In condensed matter physics this duplication is usually avoided by restricting the wave vector $\vec{k}$ to lie within the first Brillouin zone, $K_1$, or we choose functions $\psi(\vec{x})$ which can be

*Introduction to Classical and Quantum Field Theory.* Tai-Kai Ng

ISBN: 978-3-527-40726-2

written as

$$\psi(\vec{x}) = \sum_{\vec{k}\in K_1} \phi_{\vec{k}} e^{i\vec{k}.\vec{x}}.$$

This is an example where no non-trivial mathematical structure arises in the process of removing the redundancy.

### 12.1.1 Formulating Lattice Gauge Theories

Gauge theories have in general more complicated mathematical structures and the removal of redundancy is much more non-trivial. To understand the mathematical structures associated with gauge theories better, we study gauge theories on lattices in the following. We shall first consider how a U(1) gauge theory can be formulated on a lattice and then discuss a simple version of lattice gauge theory – the $Z_2$ gauge theory. We shall see that structures exist non-trivial mathematical in gauge theories.

To formulate lattice gauge theory, we first consider gauge invariance in a lattice quantum-mechanical problem. To be concrete, we consider a tight-binding quantum-mechanical model on a two-dimensional square lattice. The movement of charged particles from site $i$ to site $j$ is described by the tight-binding Hamiltonian

$$H = -\sum_{<i,j>} t_{ij} a_j^+ a_i + h.c., \tag{12.1a}$$

where $a_i^+$ and $a_i$ are the creation and annihilation operators of the particles on site $i$, respectively, and $\langle i, j\rangle$ denotes nearest-neighbor sites. Let us require that the theory is invariant when the wavefunction $\psi_i$ is multiplied by an arbitrary phase $e^{i\theta_i}$ at every point $i$, that is $\psi_i \to \psi_i e^{i\theta_i}$. The phase transformation can be generated by a transformation $a_i \to a_i e^{-i\theta_i}$, $a_i^+ \to a_i^+ e^{i\theta_i}$ in the Hamiltonian.

The Hamiltonian is invariant if $t_{ij}$ transforms as $t_{ij} \to t_{ij} e^{i(\theta_i - \theta_j)}$ together with the phase change. If $t_{ij}$ is a simple parameter, no such transformation would emerge. It is therefore necessary to think of $t_{ij}$ itself as a field which couples to the phase transformation. This field is nothing but the (lattice) electromagnetic (EM) field $\vec{A}(\vec{r}, t)$.

Let us replace $t_{ij}$ by the expression

$$t_{ij} \to t_{ij} e^{ieA_{ij}} = t_{ij} \exp\left[i\frac{\vec{r}_j - \vec{r}_i}{a} \cdot e\vec{A}\left(\frac{(\vec{r}_i + \vec{r}_j)}{2}\right)\right], \tag{12.1b}$$

where $a = |\vec{r}_i - \vec{r}_{i+\delta}|$ is the lattice spacing and $e$ is the electric charge. The space coordinate of $\vec{A}$ is chosen to be the midpoint between sites $i$ and $j$, that is $\vec{A}(\vec{r})$ is defined on the *links* between lattice points. The requirement that the Hamiltonian $H$ is Hermitian $((t_{ij}e^{ieA_{ij}})^* = t_{ji}e^{ieA_{ji}})$ implies that $A_{ij} = -A_{ji}$, that is $\vec{A}(\vec{r})$ is a vector field. (see figure next page)

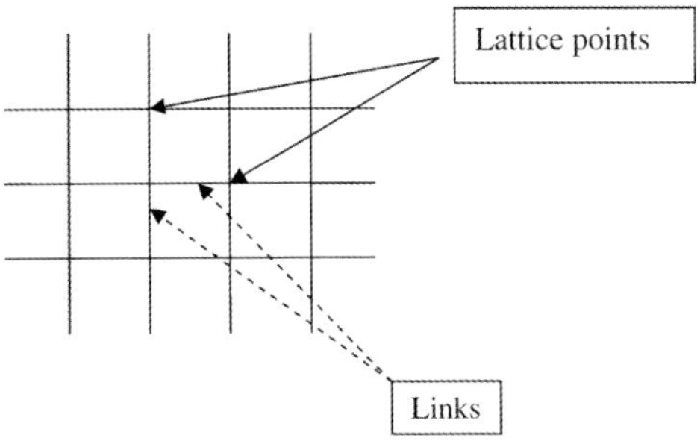


With $A_{ij}$, the transformation $t_{ij} \to t_{ij}e^{i(\theta_i - \theta_j)}$ is replaced by $t_{ij}e^{ieA_{ij}} \to t_{ij}e^{i(eA_{ij} + \theta_i - \theta_j)}$, or

$$eA_{ij} \to eA_{ij} + \theta_i - \theta_j = e\left(A_{ij} + \frac{(\theta_i - \theta_j)}{e}\right), \tag{12.2}$$

which corresponds to the usual gauge transformation $\vec{A} \to \vec{A} + (1/e)a\nabla\theta$ in the continuum limit.

We shall see in the following that it is more important to consider the transformation properties of

$$U_{ij} = e^{ieA_{ij}}, \quad U_{ij} \to e^{i\theta_i} U_{ij} e^{-i\theta_j}, \tag{12.3}$$

since this is the entity that enters a lattice theory. A very important property of $U_{ij}$ is that it goes back to itself when $A_{ij} \to A_{ij} + 2n\pi/e$, that is it is periodic in $A_{ij}$. This is a property which arises naturally in lattice theory but is absent in the continuum theory. In group theory, when group elements are physically different only within a finite interval of the parameter space characterizing the element, the group is said to be compact. The continuous group with elements $\exp[-ieA_{ij}]$ is called a compact U(1) group. The compactness of the lattice U(1) group makes an essential difference between lattice and continuum gauge theories, as we shall see later.

Note that in our discussion a discrete lattice is introduced only to the space coordinate and affects only the spatial components $\vec{A}$ of the vector potential $A_\mu$. We have not 'discretized' time here although it is mathematically possible. As a result, $A_{\mu=0}$ remains a continuous, non-compact variable in our treatment.

We next discuss the dynamics of a gauge field. In the continuum limit, the dynamics is described by the action

$$\begin{aligned} S &= \sum_{\mu,\nu} \frac{1}{16\pi} \int d^4x F_{\mu\nu}^2(x) \\ &= \sum_{\mu,\nu} \frac{1}{16\pi} \int d^4x (\partial_\mu A_\nu(x) - \partial_\nu A_\mu(x))^2. \end{aligned} \tag{12.4a}$$

The question is how the above expression should be modified in lattice gauge theory. We may start by replacing the differentials in $F_{\mu\nu} = \partial_\mu A_\nu - \partial_\nu A_\mu$ by finite differences in spatial coordinates. We obtain (see figure next page)

$$F_{\mu\nu}(i) \to (A_{i+\hat{\mu},i+\hat{\mu}+\hat{\nu}} - A_{i,i+\hat{\nu}} + A_{i,i+\hat{\mu}} - A_{i+\hat{\nu},i+\hat{\mu}+\hat{\nu}})/a, \tag{12.4b}$$

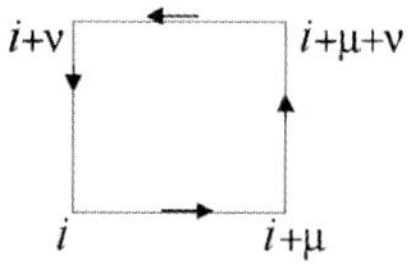


and the corresponding Lagrangian is obviously

$$L \sim \frac{1}{a^2} \sum_{i,\hat{\mu},\hat{\nu}} (A_{i+\hat{\mu},i+\hat{\mu}+\hat{\nu}} - A_{i,i+\hat{\nu}} + A_{i,i+\hat{\mu}} - A_{i+\hat{\nu},i+\hat{\mu}+\hat{\nu}})^2. \tag{12.4c}$$

However, $F_{\mu\nu}(i)$ is not invariant under the transformation $A_{ij} \to A_{ij} + 2\pi/e$ as required in a lattice gauge theory. The periodicity of $A_{ij}$ in the action is recovered if we write the Lagrangian in terms of the more fundamental quantities $U_{ij}$,

$$\begin{aligned} F_{\mu\nu}(i) &\to \exp[ie(A_{i+\hat{\mu},i+\hat{\mu}+\hat{\nu}} - A_{i,i+\hat{\nu}} + A_{i,i+\hat{\mu}} - A_{i+\hat{\nu},i+\hat{\mu}+\hat{\nu}})] \\ &= U_{i,i+\hat{\mu}} U_{i+\hat{\mu},i+\hat{\mu}+\hat{\nu}} U_{i+\hat{\mu}+\hat{\nu},i+\hat{\nu}} U_{i+\hat{\nu},i}, \end{aligned} \tag{12.5a}$$

and

$$\begin{aligned} [F_{\mu\nu}(i)]^2 &\to 1 - \frac{1}{2} (U_{i,i+\hat{\mu}} U_{i+\hat{\mu},i+\hat{\mu}+\hat{\nu}} U_{i+\hat{\mu}+\hat{\nu},i+\hat{\nu}} U_{i+\hat{\nu},i} \\ &\quad + (U_{i,i+\hat{\mu}} U_{i+\hat{\mu},i+\hat{\mu}+\hat{\nu}} U_{i+\hat{\mu}+\hat{\nu},i+\hat{\nu}} U_{i+\hat{\nu},i})^*) \\ &\to 1 - \cos(e(A_{i+\hat{\mu},i+\hat{\mu}+\hat{\nu}} - A_{i,i+\hat{\nu}} + A_{i,i+\hat{\mu}} - A_{i+\hat{\nu},i+\hat{\mu}+\hat{\nu}})). \end{aligned} \tag{12.5b}$$

The smallest unit square is called a plaquette.

The Lagrangian constructed with this procedure is invariant under the transformation $A_{ij} \to A_{ij} + 2\pi n/e$. Putting this together with the time component, the Lagrangian that matches $(1/16\pi)F_{\mu\nu}^2$ in the continuum limit is $(A_0 \to \phi)$

$$\begin{aligned} L &\sim \frac{1}{2g} e^2 \sum_{i,\mu} (\dot{A}_{i,i+\mu} - \phi_{i+\mu} + \phi_i)^2 \\ &- \frac{1}{2g} \sum_i \sum_{\mu\nu} [1 - \cos(e(A_{i+\hat{\mu},i+\hat{\mu}+\hat{\nu}} - A_{i,i+\hat{\nu}} + A_{i,i+\hat{\mu}} - A_{i+\hat{\nu},i+\hat{\mu}+\hat{\nu}})], \end{aligned} \tag{12.6}$$

where we have introduced a coupling constant $g$ in the theory. The action $S$ is invariant with respect to (gauge) transformations

$$eA_{ij} \to eA_{ij} + \theta_i - \theta_j \quad \text{and} \quad e\phi_i \to e\phi_i - \dot{\theta}_i$$

and is also compact in $A_{ij}$. Note that the requirement that the action is written in terms of $U_{ij}$ automatically implies that the action is not quadratic in $A_{ij}$, that is interaction between 'photons' is automatically introduced in lattice gauge theory.

The Hamiltonian can be derived following a standard canonical approach. The momentum conjugate to $A_{i,i+\mu}$ is

$$E_{i,i+\mu} = \frac{\partial L}{\partial \dot{A}_{i,i+\mu}} = \frac{e^2}{g} (\dot{A}_{i,i+\mu} + \phi_i - \phi_{i+\mu}), \tag{12.7}$$

and the conjugate momentum to $\phi$ vanishes $(\frac{\partial L}{\partial \dot{\phi}_i} = 0)$.

The corresponding Hamiltonian is

$$H=\sum_{i,\mu} E_{i,i+\mu}\dot{A}_{i,i+\mu}-L$$

$$=\frac{g}{2e^2}\sum_{i,\mu}E_{i,i+\mu}^2+\frac{1}{2g}\sum_i\sum_{\mu\nu}[1-\cos(e(A_{i+\hat{\mu},i+\hat{\mu}+\hat{\nu}}-A_{i,i+\hat{\nu}}+A_{i,i+\hat{\mu}}$$

$$-A_{i+\hat{\nu},i+\hat{\mu}+\hat{\nu}})]-\tfrac{g}{e^2}\sum_{i,\mu}\phi_i(E_{i,i+\mu}-E_{i-\mu,\mu}). \qquad (12.8)$$

As in continuum U(1) gauge theory, there is no dynamics (time derivatives) associated with the $\phi$ field and the equation of motion with respect to $\phi$ is a constraint condition (Gauss' law) (see Chapter 6)

$$\Gamma_i=\sum_\mu(E_{i,i+\mu}-E_{i-\mu,\mu})=0, \qquad (12.9)$$

which corresponds to $\nabla.\vec{E}=0$ in the continuum limit. (see figure below)

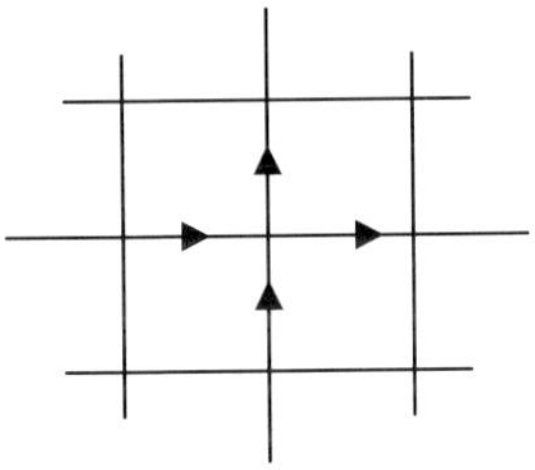

The corresponding quantum Hamiltonian is (subject to the constraint equation (12.9))

$$H=-\frac{\hbar^2 g}{2e^2}\sum_{i,\mu}\frac{\partial^2}{\partial A_{i,i+\mu}^2}$$

$$+\frac{1}{2g}\sum_i\sum_{\mu\nu}[1-\cos(e(A_{i+\hat{\mu},i+\hat{\mu}+\hat{\nu}}-A_{i,i+\hat{\nu}}+A_{i,i+\hat{\mu}}-A_{i+\hat{\nu},i+\hat{\mu}+\hat{\nu}})],$$

$$(12.10)$$

which is not quadratic and cannot be diagonalized exactly as in the continuum U(1) gauge theory. The (local) observables in the theory are electric fields defined on links

$$E_{i,i+\mu}=\frac{e^2}{g}(\dot{A}_{i,i+\mu}+\phi_i-\phi_{i+\mu}) \qquad (12.11a)$$

and magnetic fluxes through plaquettes (mod $2\pi$)

$$B_i a^2=A_{i+\hat{\mu},i+\hat{\mu}+\hat{\nu}}-A_{i,i+\hat{\nu}}+A_{i,i+\hat{\mu}}-A_{i+\hat{\nu},i+\hat{\mu}+\hat{\nu}}. \qquad (12.11b)$$

### 12.1.2
### $Z_2$ Gauge Theory

To obtain some feeling of the non-trivial mathematical structure associated with gauge transformations, we consider a simple statistical model with Hamiltonian

$$H = -g \sum_{\text{plaquette}} \bar{U}_{i,i+\hat{\mu}} \bar{U}_{i+\hat{\mu},i+\hat{\mu}+\hat{\nu}} \bar{U}_{i+\hat{\mu}+\hat{\nu},i+\hat{\nu}} \bar{U}_{i+\hat{\nu},i}, \tag{12.12}$$

where $\bar{U}_{ij} = \pm 1$ are link variables defined around plaquettes. The model can be considered as a simplified model of U(1) lattice gauge theory where (i) there is no dynamics and (ii) the general (continuous) link variable $U_{ij} = e^{iA_{ij}}$ is replaced by a discrete link variable $\bar{U}_{ij} = \pm 1$. As a result, the magnetic flux passing through a plaquette is restricted to $Ba^2 = (0, \pi)^*$ (mod $2\pi$). Note that the Hamiltonian is invariant with respect to a 'discrete' gauge transformation

$$\bar{U}_{ij} \to s_i \bar{U}_{ij} s_j^{-1}, \tag{12.13}$$

where $s_i = \pm 1$ are site variables. The model is called $Z_2$ gauge theory because the continuous U(1) gauge transformation is broken down to a discrete gauge transformation with only two elements (see also Refs. [1, 2]). To see the non-trivial mathematical structure behind gauge transformation, we study the phase-space structure of this model.

To begin with, we first consider the model on a single plaquette and count the total number of physically distinct states in this case. (see figure below)

There are four link variables $U_{12}$, $U_{23}$, $U_{34}$, and $U_{41}$ in this case, and four independent site variables $s_i$ ($i = 1$–$4$) which generate the gauge transformation. Without the site variables, the total number of independent states is $2^4 = 16$. With the site variables, all configurations $\{U_{ij}\}$ which can be transformed to each other via a gauge transformation are equivalent (gauge invariance). The total number of independent gauge transformations is equal to the number of independent products $(s_1 s_2)$, $(s_2 s_3)$, $(s_3 s_4)$, and $(s_4 s_1)$ and is $2^3 = 8$, since fixing $(s_1 s_2)$, $(s_2 s_3)$, and $(s_3 s_4)$ determines $(s_4 s_1)$. The total number of physically independent states is $\frac{16}{8} = 2$ in this model, corresponding to two possible values of 'magnetic flux' $= 0, \pi$ through the plaquette.

Next, we consider a ladder as shown in the figure below. We also impose a periodic boundary condition such that site $1 =$ site $N + 1$ and site $1' =$ site $N' + 1$.

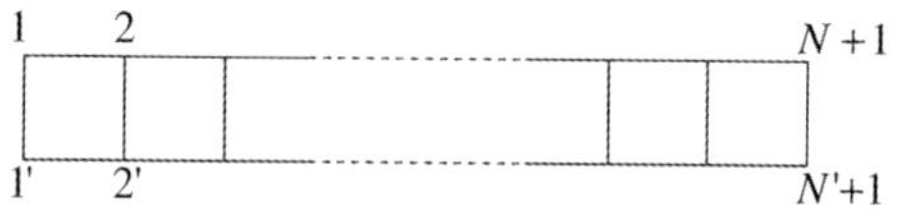


The total number of link variables is $2N + N = 3N$ (note the periodic boundary condition) and the total number of independent gauge transformations is $2N - 1$.

**Homework Q1a**
Show that the total number of independent gauge transformations is $2N - 1$ in the above ladder $Z_2$ model.

Therefore, the total number of states is

$$N_S = \frac{2^{3N}}{2^{2N-1}} = 2 \times 2^N. \tag{12.14a}$$

Note a 'mystery' here. The total number of plaquettes in the model is $N$, and therefore we expect that the total number of physically distinct states to be $2^N$ = total number of ways to distribute magnetic fluxes in the plaquettes. Why is the number of states 'doubled' in Equation [12.14a]?

A hint to the origin of the 'doubling of states' can be seen by studying the model with open boundary condition, that is the sites $1$, $1'$, $N+1$, and $N+1'$ are independent. In this case, we find that the total number of independent gauge transformation is $2N$ and the total number of independent states is

$$N_S = \frac{2^{3N}}{2^{2N}} = 2^N, \tag{12.14b}$$

in agreement with our naïve expectation.

**Homework Q1b**
Show that the total number of independent gauge transformations becomes $2N$ if the periodic boundary condition is replaced by a fixed boundary condition.

Note that the total number of independent plaquettes is *identical* (= $N$) for both boundary conditions and the extra degree of freedom is a result of the periodic boundary condition.

To see the origin of the doubling of states, we consider a ladder with $N=4$ and look at the following two configurations: (see figure below)

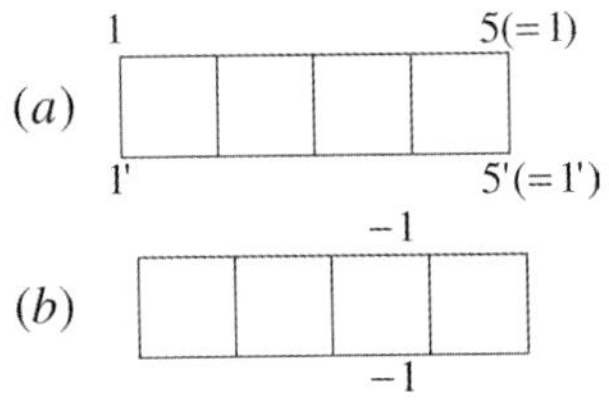


In configuration (a), all the $U_{i,i+1}$ and $U_{i,i'}$ are equal to 1, whereas, in configuration (b), all the $U_{i,i+1}$ except for $U_{34}$ and $U_{3'4'}$ are equal to 1 and all the $U_{i,i'}$ are equal to 1. Both configurations have zero flux passing through all plaquettes. The question is: are these two configurations equivalent? To answer the question, let us try bringing one configuration into the other by performing gauge transformations. Let us assume that originally $s_i = 1$ for all $i$.

It is easy to see that for a fixed boundary condition we can bring configuration (b) into configuration (a) by simply changing $s_4, s_{4'}, s_5, s_{5'}$ from 1 to $-1$.

However, this cannot be done for a periodic boundary condition because changing $s_5, s_{5'}$ would also change $s_1, s_{1'}$. The impossibility of bringing configuration (b) into configuration (a) through a gauge transformation can be seen from the objects

$$\Phi_C = U_{12}U_{23}U_{34}U_{45(1)} \quad \text{and} \quad \Phi'_C = U_{1'2'}U_{2'3'}U_{3'4'}U_{4'5'(1')}, \tag{12.15}$$

which are gauge dependent for an open boundary condition but becomes gauge invariant ($U_{45} = U_{41}$, etc.) under a periodic boundary condition. Configurations (a) and (b) have different values of $\Phi_C, \Phi'_C$ under a periodic boundary condition and thus represent different states of the system.

$\Phi_C, \Phi'_C$ are not independent of each other. It can be seen that $\Phi_C = \Phi'_C$ when there are $2n\pi$ fluxes (including zero flux) in the system and $\Phi_C = -\Phi'_C$ when there are $(2n+1)\pi$ fluxes in the system. The doubling of states under a periodic boundary condition is associated with the two possible values of $\Phi_C, \Phi'_C$ for each given magnetic flux configuration in the system. For Hamiltonians that are local in space, the energy of the system should be independent of $\Phi_C, \Phi'_C$, since these are 'global' properties of the system depending on the boundary condition. As a result, all states are (at least) doubly degenerate in ladder $Z_2$ models under a periodic boundary condition.

The above analysis can be extended to $Z_2$ models on two-dimensional square lattices (see Ref. [1]), where it can be seen that, under a periodic boundary condition, all states are four-fold degenerate with local Hamiltonians.

**Question**

What happens when the above $Z_2$ gauge theory model is replaced by a U(1) gauge theory model where $U_{ij} \to e^{iA_{ij}}$ and $s_i \to e^{i\vartheta_i}$? And what would happen to a particle field coupled to the gauge field as in Equation (12.1a,b)? Will the wavefunctions of the particles differ in the different degenerate gauge field configurations?

The sensitive dependence of phase space structure on boundary condition for $Z_2$ gauge theory suggests that Hilbert space structures of gauge theories are intrinsically non-local. *The Hilbert space is divided into topologically distinct classes (different boundary conditions) which do not communicate with each other through local processes.* This is a rather surprising conclusion. We shall learn more about the non-trivial Hilbert space structure of U(1) gauge theory in the following.

As in the study of the rotor (quantum *X–Y*) model, we shall study U(1) lattice gauge theory using the strong- and weak-coupling expansions, corresponding to the limits $g \to \infty$ and $g \to 0$ in Equation (12.10) (see also Ref. [2]). We hope that we can understand the physics of lattice gauge theory through systematic expansion in powers of $g^{-1}$ or $g$ in the two limits.

## 12.2
## Strong- and Weak-Coupling Expansions in U(1) Lattice Gauge Theory

(i) First, we consider weak-coupling expansion ($g \to 0$).

In this limit, the zeroth-order Hamiltonian is

$$H \to H_0 = \frac{1}{2g}\sum_i \sum_{\mu\nu}[1-\cos(e(A_{i+\hat{\mu},i+\hat{\mu}+\hat{\nu}}-A_{i,i+\hat{\nu}}+A_{i,i+\hat{\mu}}-A_{i+\hat{\nu},i+\hat{\mu}+\hat{\nu}})] \tag{12.16a}$$

and the ground state of the system can be obtained by minimizing $H_0$ with respect to the $A$'s. It is obvious that the ground state corresponds to a state with zero magnetic flux (mod $2\pi$) through all plaquettes, that is

$$B_i = A_{i+\hat{\mu},i+\hat{\mu}+\hat{\nu}}-A_{i,i+\hat{\nu}}+A_{i,i+\hat{\mu}}-A_{i+\hat{\nu},i+\hat{\mu}+\hat{\nu}} = 0.$$

Note that a weak-coupling expansion can be performed only in 2D and 3D (or a 1D ladder) since we need the existence of 'plaquettes' to define the Hamiltonian $H_0$. We shall consider the case of a one-dimensional chain when we discuss strong-coupling expansion.

As in the case of the quantum $X$–$Y$ model, excitations of the system are described by the semiclassical approximation representing small fluctuations of the $\vec{A}$ field about the ground state, that is we approximate

$$H \sim -\frac{\hbar^2 g}{2e^2}\sum_{i,\mu}\frac{\partial^2}{\partial A_{i,i+\mu}^2} + \frac{e^2 a^2}{4g}\sum_i (A_{i+\hat{\mu},i+\hat{\mu}+\hat{\nu}}-A_{i,i+\hat{\nu}}+A_{i,i+\hat{\mu}}-A_{i+\hat{\nu},i+\hat{\mu}+\hat{\nu}})^2 \tag{12.16b}$$

with constraint

$$\Gamma_i = \sum_\mu (E_{i,i+\mu}-E_{i-\mu,i}) = -i\hbar\sum_\mu\left(\frac{\partial}{\partial A_{i,i+\mu}}-\frac{\partial}{\partial A_{i,i-\mu}}\right) = 0, \tag{12.16c}$$

which reduces to the usual electromagnetism in the small-$\vec{q}$ (continuum) limit. The weak-coupling phase is also called the plasma phase of U(1) lattice gauge theory. Note that the Lagrangian is no longer invariant with respect to the transformation $A_{ij} \to A_{ij} + 2\pi n/e$ in the semiclassical approximation and the importance of loss of compactness is not clear. We shall return to this question in the next section.

**Homework Q2**

Show that in weak-coupling expansion the theory reduces to the usual continuum U (1) gauge theory in the long-wavelength limit. What is the photon spectrum in the short-wavelength regime?

(ii) Next, we consider the strong-coupling limit ($g \to \infty$), which is a much more interesting limit because the invariance of the Lagrangian with respect to the transformation $A_{ij} \to A_{ij} + 2\pi n/e$ is kept explicitly. To zeroth order, we have to diagonalize the Hamiltonian

$$H \to H_0 \sim -\frac{\hbar^2 g}{2e^2}\sum_{i,\mu}\frac{\partial^2}{\partial A_{i,i+\mu}^2}. \tag{12.17}$$

It is obvious that, in the absence of constraint $\Gamma_i = \sum_\mu (E_{i,i+\mu} - E_{i-\mu,\mu}) = 0$ (Gauss' law), the eigenstates are given by

$$\Psi[\{k_{i,i+\mu}\}] = \prod_{i,\mu} \exp[ik_{i,i+\mu}A_{i,i+\mu}], \tag{12.18a}$$

where each link in the lattice is associated with a factor $\exp(ik_{i,i+\mu}A_{i,i+\mu})$. The energy of the system is

$$E[k] = \sum_{i,\mu} \frac{\hbar^2 g k^2_{i,i+\mu}}{2e^2}. \tag{12.18b}$$

The 'physical' eigenstates must also obey the constraint (Gauss' law)

$$\Gamma_i = -i\hbar \sum_\mu \left( \frac{\partial}{\partial A_{i,i+\mu}} - \frac{\partial}{\partial A_{i-\mu,i}} \right) \Psi[A] = 0, \tag{12.18c}$$

or $\sum_\mu (k_{i,i+\mu} - k_{i-\mu,i}) = 0$.

### 12.2.1 Compactness of Gauge Field and Charge Quantization

Next, we examine the consequences of the compactness condition that the transformation $A_{ij} \to A_{ij} + 2\pi n/e$ must leave the system invariant. The requirement means that the quantum-mechanical wavefunction must satisfy

$$\Psi[A] = \prod_{i,\mu} \exp(ik_{i,i+\mu}A_{i,i+\mu}) = \prod_{i,\mu} \exp(ik_{i,i+\mu}(A_{i,i+\mu} + 2\pi n/e)), \tag{12.19a}$$

which gives rise to a quantization condition

$$k_{i,i+\mu} = n_{i,i+\mu} e, \tag{12.19b}$$

where $n_{i,i+\mu} = -n_{i+\mu,i}$ are arbitrary integers.

The energy of the system becomes quantized as a result. The ground state corresponds to $n_{i,i+\mu} = 0$ for all $i$, $\mu$ (Equation (12.18b)). The excited states are states with electric flux $E_{i,i+\mu} = (n_{i,i+\mu}e)$ linking between two sites $i$ and $i + \mu$ in the system and are separated by energy gaps $\sim \frac{\hbar^2 g}{2}$ from the ground state. The constraint condition $\Gamma_i = \sum_\mu (E_{i,i+\mu} - E_{i-\mu,\mu}) = 0$ implies that there is a constraint on the allowed configurations of the electric field. The sum of electric fields entering (or leaving) any lattice site must be zero (Gauss' law). The lowest-energy excitation on a (2D) square lattice is $E_1 = \sum_{i,\mu} \frac{4\hbar^2 g}{2}$, corresponding to electric fields circulating around a plaquette as shown in the following figure.

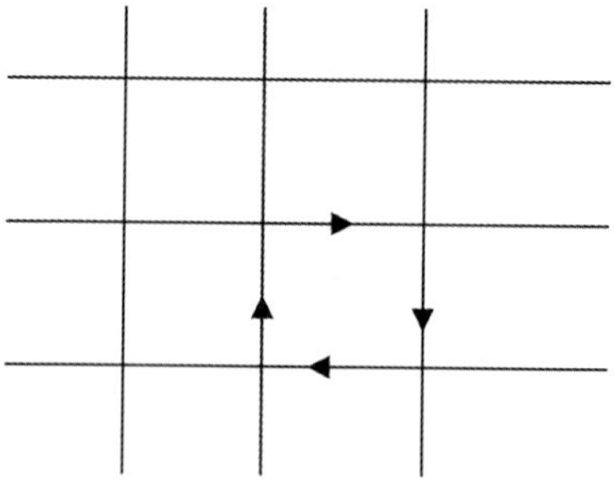

This unusual phase of lattice gauge theory is called the strong-coupling phase, or confining phase, for reasons we shall see in the following.

Now, let us introduce charges in the system by generalizing the constraint condition to

$$Q_i = \sum_{\mu} (E_{i,i+\mu} - E_{i-\mu,\mu}), \tag{12.20}$$

where $Q_i$ is the charge located at site $i$. (Note that charges are located on lattice sites but not on links between two sites.) Equation (12.20) is the analogue of Gauss' law $\nabla.\vec{E} = \rho$ in the continuum limit. The compactness condition (12.19b) implies that $Q_i = e\sum_{\mu}(n_{i,i+\mu} - n_{i-\mu,\mu})$, meaning that charges are quantized in integer units of $e$.

We see that the *compactness of the gauge field automatically implies charge quantization*. This is another example of quantization from topology. In this case, the topology comes from the compactness of U(1) gauge group.

### 12.2.2 Charge Confinement

The strong-coupling phase has the important property of charge confinement.

Imagine that we have two charges $\pm e$ separated by a distance $d$ on the square lattice. What is the ground-state energy of the system in this case? How does it depend on the separation of charges $d$?

To minimize the energy, we may want to have zero electric flux $E_{i,i+\mu} = 0$ for all links in the system. However, it is impossible in this case because of the constraint $Q_i = \sum_{\mu}(E_{i,i+\mu} - E_{i-\mu,\mu}) \neq 0$ on two sites. In fact, it is easy to see that the lowest-energy electric field configuration that is allowed is the configuration below (for $d = 5$):

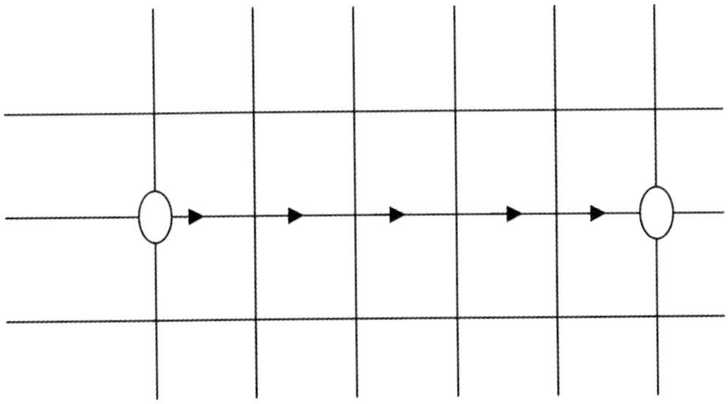

Obviously, for general separation $d$, the energy of the configuration is $E_d = \sum_{i,\mu} \frac{\hbar^2 g}{2} d$, and is proportional to the distance between the two charges, that is *the two charges are bound by a potential energy that is proportional to their distance* and we need infinite energy to separate two charges of opposite sign (or bringing $d \rightarrow \infty$). This is very different from usual electromagnetism (weak-coupling phase), where the electrostatic potential energy goes as $Q^2/d$ (in 3D) between charges. For this reason the strong-coupling phase of lattice U(1) gauge theory is also called the confinement phase.

### 12.2.3
### Finite 1/g Correction, Loop and String Gas

To see whether the above conclusions remain valid when g is finite, we have to examine the effect of interaction

$$V=\frac{1}{2g}\sum_{i}\sum_{\mu\nu}[1-\cos(ea(A_{i+\hat{\mu},i+\hat{\mu}+\hat{\nu}}-A_{i,i+\hat{\nu}}+A_{i,i+\hat{\mu}}-A_{i+\hat{\nu},i+\hat{\mu}+\hat{\nu}})].\tag{12.21a}$$

As in the case of the quantum rotor model, we may introduce raising and lowering operators $\widehat{a}^{\pm}_{i,i+\mu}=e^{\pm ieaA_{i,i+\mu}}$, which change $n_{i,i+\mu}$ to $n_{i,i+\mu}\pm 1$. In this representation,

$$V=\frac{1}{2g}\sum_{i}\sum_{\mu\nu}\left[1-\left(\frac{\widehat{a}^{+}_{i+\hat{\mu},i+\hat{\mu}+\hat{\nu}}\widehat{a}^{-}_{i,i+\hat{\nu}}\widehat{a}^{+}_{i,i+\hat{\mu}}\widehat{a}^{-}_{i+\hat{\nu},i+\hat{\mu}+\hat{\nu}}+\widehat{a}^{-}_{i+\hat{\mu},i+\hat{\mu}+\hat{\nu}}\widehat{a}^{+}_{i,i+\hat{\nu}}\widehat{a}^{-}_{i,i+\hat{\mu}}\widehat{a}^{+}_{i+\hat{\nu},i+\hat{\mu}+\hat{\nu}}}{2}\right)\right]\tag{12.21b}$$

and the effects of $V$ are to create (destroy) loops of electric field around a plaquette, pictorially,

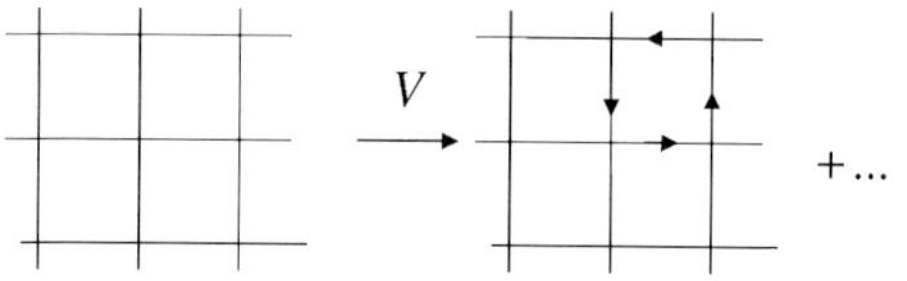


or to deform a loop,

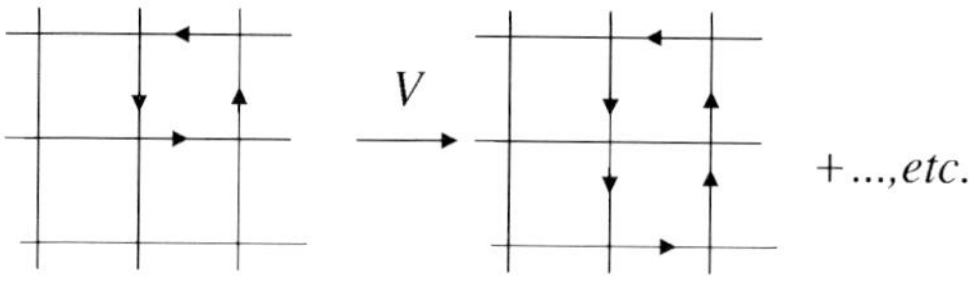


Thus, $V$ introduces dynamics to the electric field loops. The electric field loops can be created and/or destroyed, change in shape, and can move in space (in second-order perturbation theory). The true ground state is a superposition of loops of electric fields. In the strong-coupling limit, the potential energy for creating loops is high and the probability of finding loops in the ground state is small. Therefore, the system can be described as a dilute gas of electric field loops.

In the presence of charges, the consideration is similar except that open strings can also exist in the system. The end points of the strings represent charges by Gauss' law and the strings can deform and move (if the charges can move) in the presence of the interaction $V$.

Note that, in the strong-coupling expansion, a U(1) lattice gauge theory can be viewed as a theory of electric field loops and strings, which is a picture very different from the weak-coupling limit where the electric field is represented by a continuous vector variable in space. It is interesting to ask what a plasma phase looks like when described in the language of loops and strings. Why are loops and strings absent in the weak-coupling phase? How do they 'disappear'?

### 12.2.4 Confinement-to-Plasma Phase Transition and String–Net Condensation

Wen [1] proposed that the transition from strong-coupling (confinement) phase to weak-coupling (plasma) phase can be viewed as a kind of insulator–superfluid transition we discussed in chapter 11. The electric field loops and strings 'Bose condense' when g is small enough, giving rise to the plasma phase. To understand this Bose-condensation picture, we note that to zeroth-order approximation the ground state $|\psi_P\rangle$ in the weak-coupling expansion is the ground state of the Hamiltonian (12.16a) (or (12.21b)) with energy zero, that is

$$\sum_i \sum_{\mu\nu} \left[ 1 - \left( \frac{\widehat{a}^{+}_{i+\hat{\mu},i+\hat{\mu}+\hat{\nu}} \widehat{a}^{-}_{i,i+\hat{\nu}} \widehat{a}^{+}_{i,i+\hat{\mu}} \widehat{a}^{-}_{i+\hat{\nu},i+\hat{\mu}+\hat{\nu}} + \widehat{a}^{-}_{i+\hat{\mu},i+\hat{\mu}+\hat{\nu}} \widehat{a}^{+}_{i,i+\hat{\nu}} \widehat{a}^{-}_{i,i+\hat{\mu}} \widehat{a}^{+}_{i+\hat{\nu},i+\hat{\mu}+\hat{\nu}}}{2} \right) \right]$$

$$|\psi_P\rangle = 0 |\psi_P\rangle$$

and is a coherent state of the $\widehat{a}^{\pm}_{i,i+\mu} = e^{\pm iea A_{i,i+\mu}}$ operators with $\widehat{a}^{\pm}_{i,i+\mu}|\psi_P\rangle = |\psi_P\rangle$, for all links $(i, i+\mu)$, that is

$$|\psi_P\rangle = \sum_{\{n_{i,i+\mu}\}} \prod_{i,\mu} \exp(ien_{i,i+\mu} A_{i,i+\mu}) \tag{12.22}$$

is a superposition of states with all possible configurations $\{n_{i,i+\mu}\}$ which satisfy Gauss' law $\sum_\mu (n_{i,i+\mu} - n_{i-\mu,\mu}) = 0$, which is some kind of Bose-condensed state of electric field loops. You can look at Ref. [1] for more details.

The plasma phase Hamiltonian (12.16b, c) describes Gaussian fluctuation corrections to this 'string–net condensation' state, very much like the semiclassical theory of interacting bosons which describes corrections to the 'pure' Bose-condensed ground state. The Goldstone modes for ordinary boson liquids are sound waves, whereas the Goldstone modes for string–net condensation are photons which are transverse excitations as a result of Gauss' law.

The existence of two very different physical pictures – the conventional field theory picture (weak-coupling expansion) and the string–loop picture (strong-coupling expansion) – to describe the same gauge theory system provides an example of duality where a physical system can be represented in two apparently very different, but yet equivalent, ways. The existence of duality alerts us that physical pictures/ theories which are apparently very different may not be alien to each other after all. They may just represent different starting points to understand the same physical system.

As discussed in Chapter 11, we have to be cautious about employing perturbation theory to illustrate the existence of Coulomb ($g \to 0$) and confinement ($g \to \infty$) phases in lattice gauge theory because of loss of compactness in the weak-coupling expansion. In the following, we shall show that the weak-coupling phase of U(1) lattice gauge theory is not stable for any finite value of g in two dimensions and the only stable phase is the strong-coupling phase where charges are always confined. The plasma phase is stable in 3D for small enough $g$. The instability of the weak-coupling phase is a result of monopole-like instantons in 2 + 1D, as we shall see below.

**Homework Q3**

Show that in 1D U(1) lattice gauge field theory charges are always confined irrespective of the magnitude of $g$. (Hint: write down the gauge theory Hamiltonian in 1D and derive the electrostatic potential between charges assuming Gauss' law.)

## 12.3 Instantons in 2 + 1D U(1) Lattice Gauge Theory

In this section we examine U(1) gauge theory in 2 + 1D. We shall show that magnetic-monopole-like instantons exist in 2 + 1D U(1) lattice gauge theory. The instantons destabilize the plasma phase making the strong-coupling phase the only stable phase in 2 + 1D.

The effect of monopole-like instantons on the plasma phase is very similar to that of vortices on the superfluid phase in 2D (see Chapter 8 on the Kosterlitz–Thouless (KT) transition). The main difference is that the monopoles are always deconfined in U(1) lattice gauge theory, whereas vortices are confined at low temperatures $T < T_{KT}$.

We start with the continuum limit of U(1) gauge theory at imaginary time (see Chapter 6). The action is just

$$S = \frac{1}{2g}\sum_{\mu,\nu}\int d^3x F_{\mu\nu}^2(x) = \frac{1}{2g}\int d^3x(\nabla\times\vec{A})^2 = \frac{1}{2g}\int d^3x\vec{B}^2, \tag{12.23a}$$

where $x = (\tau, x, y)$, $(\nabla\times\vec{A})_0 = \partial_x A_y - \partial_y A_x$, $(\nabla\times\vec{A})_x = \partial_y A_\tau - \partial_\tau A_y$, and $(\nabla\times\vec{A})_y = \partial_\tau A_x - \partial_x A_\tau$. Note that the electric fields in real time become magnetic fields at imaginary time. The classical equation of motion is

$$\nabla\times\vec{B} = 0 \quad \text{and} \quad \nabla.\vec{B} = 0.\ (B = \nabla\times\vec{A}), \tag{12.23b}$$

which can be solved by functions of the form

$$-\nabla^2\vec{A} = 0, \quad \text{with} \quad \nabla.\vec{A} = 0\ (\text{Lorentz gauge}), \tag{12.23c}$$

and the only solution in free space with boundary condition $\vec{A}(|x|\to\infty) = 0$ is $\vec{A}(x) = 0$.

However, as in the case of the $X$–$Y$ model in 2D, solutions with point-like singularity exist. An example of such a solution is

$$\vec{B}(\vec{x}) = \frac{1}{2}\frac{q\vec{x}}{|\vec{x}|^3} - 2\pi q\hat{\tau}\delta(x)\delta(y)\theta(-\tau) \tag{12.24}$$

with $\nabla.\vec{B} = \nabla.(\nabla\times\vec{A}) \sim 2\pi q\delta(\vec{x})$ if we neglect the second term, and represents a magnetic monopole located at the origin $x = 0$. The magnetic field of the monopole comes from a flux tube (solenoid) that connects the monopole to $(-\infty, 0, 0)$ [2, 3]. In real time, Equation (12.24) represents a quantum-tunneling process where a quantum state with a magnetic flux tube (a magnetic flux penetrating through only one plaquette in lattice gauge theory) that exists at time $t<0$ is quantum tunneled to a state when the flux tube vanishes $t>0$. An additional EM field is induced (first term in Equation (12.24)) at both $t>0$ and $t<0$ as a result of this tunneling process. The magnetic flux tube would cause infinite energy in a true continuum theory. However in U(1) lattice gauge theory, the energy cost for $B_\tau$ is actually

$$E = \frac{1}{2g}\sum_i [1-\cos(\Phi_i)],$$

where $\Phi_i \sim B_{\tau i}a^2$ is the total flux passing through plaquette $i$. In particular, there is no energy cost if $\Phi_i \sim 2\pi n$ ($n=$ integer). Therefore, the magnetic flux tube in the monopole solution (12.24) does not cost any energy if $q=$ integer and Equation (12.24) is an allowed solution in lattice gauge theory.

Next, we compute the energy cost of a single monopole at imaginary time. The energy cost is

$$E \sim \frac{1}{2g}\int d^3x \vec{B}^2 = \frac{2\pi}{2g}\int_{r_c}^{L} r dr \int_0^{\beta} d\tau \left(\frac{q}{2(r^2+\tau^2)}\right)^2 \sim \frac{\pi q^2}{2g}\left(\frac{1}{r_c}\right), \tag{12.25}$$

where $L\sim$ size of the (2D) system and $\beta=(k_B T)^{-1}$. We have taken the limit $L, \beta \to \infty$ in the last expression; $r_c \sim a$ is a lattice cutoff. Note a similarity between here and a vortex in the Heisenberg model (Chapter 8), where a vortex has a finite core energy because of a short-distance cutoff $\xi \sim$ lattice spacing in the theory.

There is, however an important difference between vortex and monopole solutions. A single vortex in 2D has energy $E \sim J\ln\left(\frac{L}{\xi}\right)$ and diverges in the limit $L \to \infty$, whereas the energy cost for exciting a monopole is finite. This difference is responsible for the absence of a plasma phase in 2 + 1D U(1) lattice gauge theory, as we shall see below.

### 12.3.1
### Plasma and Confinement Phases

We now consider the contribution of monopoles to the partition function. The magnetic field from a collection of monopoles is

$$\vec{B}(\vec{x}) = \sum_i \left(\frac{q_i}{2}\right)\left[\frac{(\vec{x}-\vec{x}_i)}{|\vec{x}-\vec{x}_i|^3} - 4\pi\hat{\tau}\delta(x-x_i)\delta(y-y_i)\theta(-(\tau-\tau_i))\right], \tag{12.26}$$

where the $\vec{x}_i$ are the locations of the magnetic monopoles and the energy of such a configuration is (borrowing our knowledge in electrostatics)

$$E[\{q_i\}] \sim \frac{1}{2g}\int d^3x \vec{B}^2 = \frac{\pi}{2g}\sum_{i,j}\frac{q_i q_j}{|\vec{x}_i - \vec{x}_j|}. \tag{12.27a}$$

Note that there is no contribution to the energy from the flux tubes. The corresponding contribution to the partition function is

$$Z[\{q_i\}] = \sum_{\{q_i\}} e^{-\frac{\pi}{2g}\sum_{i,j}\frac{q_i q_j}{|\vec{x}_i - \vec{x}_j|}}. \tag{12.27b}$$

It is interesting to note that the partition function (12.27b) is essentially the same as the partition function for vortices (Chapter 8) except that the two-dimensional logarithmic Coulomb interaction between vortices is replaced by the three-dimensional $q^2/r$ Coulomb interaction between monopoles. In particular, the interaction looks the same after Fourier transformation to momentum space. Because of the lattice cutoff (12.25), we shall follow Chapter 8 and replace the self interaction of monopoles by a constant energy in Equation (12.27b), that is

$$E[\{q_i\}] = \frac{\pi}{2g}\sum_{i,j}\frac{q_i q_j}{|\vec{x}_i - \vec{x}_j|} \rightarrow \frac{\pi}{2g}\sum_{i\neq j}\frac{q_i q_j}{|\vec{x}_i - \vec{x}_j|} + \frac{\mu_0}{g}\sum_i q_i^2, \tag{12.28}$$

where $\mu_0 \sim \frac{\pi}{2r_c}$.

As in the case of vortices, we expect that there may be two phases associated with the magnetic monopoles – the low-temperature (small g, weak-coupling) phase where the monopoles are confined in pairs and the high-temperature (large g, strong-coupling) phase where the monopoles are deconfined to form a monopole gas. In the following, we shall show that the strong-coupling phase is the only stable phase in 2 + 1D using a mean-field theory which is essentially the same one we applied to study the KT transition.

Following Chapter 8, we introduce an auxiliary vector field $\vec{j}$ (duality transformation) and rewrite the partition function as

$$Z[\vec{A}] = \int D\vec{A}e^{-\frac{1}{2g}\int d^3x(\nabla\times(\vec{A}+\vec{A}_M))^2}$$

$$= Z\left[\vec{A},\vec{j}\right] = \int D\vec{A}\int D\vec{A}_M\int D\vec{j}e^{-\frac{1}{2}\int d^3x[2i\vec{j}.(\nabla\times(\vec{A}+\vec{A}_M))+g(\vec{j})^2]}, \tag{12.29}$$

where $\vec{A}$ is the vector potential associated with the usual (regular) magnetic field and $\vec{A}_M$ is a singular vector potential associated with magnetic monopoles. Integrating by parts, we obtain

$$\int d^3x\vec{j}.(\nabla\times\vec{A}) = -\int d^3x\vec{A}.(\nabla\times\vec{j})$$

and integrating over the $\vec{A}$ fields results in the constraint $\nabla\times\vec{j} = 0$. Therefore, we can write $\vec{j} = \nabla\phi$, where $\phi$ is an auxiliary scalar potential. The corresponding partition function is

$$Z\left[\vec{A},\vec{j}\right] \rightarrow Z\left[\vec{A}_M,\phi\right] = \int D\phi\int D\vec{A}_M e^{-\frac{1}{2}\int d^3x[2i\nabla\phi.(\nabla\times\vec{A}_M)+g(\nabla\phi)^2]}. \tag{12.30a}$$

Integrating by parts again, we obtain

$$\int d^3x \nabla\phi.(\nabla \times \vec{A}_M) = -\int d^3x \phi\nabla.(\nabla \times \vec{A}_M) = -\int d^3x \phi\rho,$$

where $\rho(\vec{x}) = \left[2\pi \sum_i q_i \delta^{(3)}(\vec{x}-\vec{x}_i)\right]$ is the magnetic monopole density. The partition function is therefore

$$Z\left[\vec{A}_M, \phi\right] \to Z[\{q_i\}, \phi] = \sum_{\{q_i\}} \int d\phi e^{-\frac{1}{2}\int d^2x[-2i\phi\rho + g(\nabla\phi)^2]}, \tag{12.30b}$$

where we have replaced the integration over $\vec{A}_M$ by summing over the monopole configurations. It is straightforward to show that integrating over the $\phi$ field produces the partition function (12.27b).

As in Chapter 8, we now replace the partition function (12.30b) by a mean-field partition function where the monopole density is given by

$$\langle n_\pm(\vec{x})\rangle = a^{-3} e^{-\frac{\mu^\pm_{\text{eff}}(\vec{x})}{g}}, \tag{12.31a}$$

where $n_\pm(\vec{x}) = \sum_i \delta(q_i \mp 1)\delta^{(2)}(\vec{r}-\vec{r}_i)$ is the density of monopoles with charge $q$. We assume here that $q = \pm 1$ for simplicity. $\mu^\pm_{\text{eff}}(\vec{x}) = \mu_{\text{eff}} \mp ig\phi(\vec{x})$ is an effective monopole energy to be determined self consistently.

To determine $\mu_{\text{eff}}$ and $\phi(\vec{x})$, we consider the increase in energy when an additional monopole $Q(\vec{x}) = \pm Q\delta(\vec{x})$ is put into the system. The increase in energy is

$$\mu_{\text{eff}}(Q) = \int d^3x\left(-ig\phi(\vec{x})(\delta\langle\rho(\vec{x})\rangle + Q\delta(\vec{x})) + \frac{g^2}{2}(\nabla\phi(\vec{x}))^2\right), \tag{12.31b}$$

where $\delta\rho(\vec{x}) = n_+(\vec{x}) - n_-(\vec{x})$ and $\phi(\vec{x})$ are the induced change in monopole density and auxiliary potential by $Q$, respectively. $\delta\rho(\vec{x})$ can be eliminated using Equation (12.31a), where

$$\delta\langle\rho(\vec{x})\rangle = a^{-3} e^{-\frac{\mu^+_{\text{eff}}(\vec{x})}{g}} - a^{-3} e^{-\frac{\mu^-_{\text{eff}}(\vec{x})}{g}} = a^{-3} e^{-\frac{\mu_{\text{eff}}}{g}}\left(e^{i\phi(\vec{x})} - e^{-i\phi(\vec{x})}\right) \sim i n_0 \phi(\vec{x}), \tag{12.31c}$$

where $n_0 = <n_+(\vec{x})> + <n_-(\vec{x})>$

$$= 2a^{-3} e^{-\frac{\mu_{\text{eff}}}{g}}. \tag{12.32a}$$

is the average monopole density in the system. Putting Equation (12.31c) back into Equation (12.31b) and minimizing the energy with respect to $\phi$, we obtain after some simple algebra (see also Chapter 8)

$$\mu_{\text{eff}}(Q = \pm 1) = \mu_0 + \frac{g}{2(2\pi)^3}\int \frac{d^3q}{(gq^2 + 2n_0)}. \tag{12.32b}$$

The equation is essentially the same as the equation determining the vortex densities in the 2D Coulomb gas problem except that the integral is carried out now in 3D.

Despite the mathematical similarly between monopole and vortex gases the conclusion coming out from the equation is very different because of the *difference in dimensions.* The integral (12.32b) is finite in the limit $n_0 \to 0$ in 3D but diverges logarithmically in 2D. As a result, there exists no self-consistent solution with $n_0 = 0$ in 3D although such a solution exists in 2D at $T < T_{KT}$.

**Homework Q4**

Solve the above mean-field equation to show that there exists no solution with $n_0 = 0$ for arbitrary values of the coupling constant g. What is $\mu_{eff}$ in the small-g limit?

Thus, unlike the corresponding Coulomb gas problem in 2D, there is only one phase with deconfined monopoles in 2 + 1D U(1) lattice gauge theory. We expect that the phase corresponds to the strong-coupling phase where electric charges are confined.

### 12.3.2 Wilson Loop

To show that the phase confines electric charge, we compute the energy of two electric charges $\pm e$ located at positions $(x, y) = (R, 0)$ and $(x, y) = (0, 0)$ separated by a distance $R$. With the electric charges the imaginary-time action becomes

$$H = \frac{1}{2g}\int_0^\beta d\tau \int dxdy\vec{B}^2 + \int_0^\beta d\tau \int dxdy\vec{A}.\vec{j}_e$$
$$= \frac{1}{2g}\int_0^\beta d\tau \int dxdy\vec{B}^2 + e\int_0^\beta (A_\tau(R, o, \tau) - A_\tau(0, 0, \tau))d\tau, \tag{12.33a}$$

where $\vec{j}_e = e\hat{\tau}(\delta(x-R)\delta(y) - \delta(x)\delta(y))$ is the 3-current associated with the static charge. The corresponding free energy is

$$F = -\frac{1}{\beta}\ln(\mathrm{Tr}(e^{-\beta H})) = F_0 + F_e,$$

where $F_0$ is the energy of the pure electromagnetic field and $F_e$ is the correction in energy coming from the charges. In particular, we expect from the previous section that $F_e \sim e^2 R$ for large $R$ if the monopole-disordered phase confines electric charge.

In practical calculations, it is more convenient to replace the two static charges by a dynamical process corresponding to creation of a pair of charges from the vacuum at $\tau \sim 0$, separating by a distance $R$ quickly, and coming back and annihilating at time $\tau \sim \beta$ (see figure next page).

The process can be represented by a space–time loop called a Wilson loop. The free energy should be the same as the free energy from two static charges to leading order in $R$ and $\beta$ as long as the time taken for separating the charges is small compared with $\beta$.

The corresponding action for the Wilson loop is

$$H = \frac{1}{2g}\int_0^\beta d\tau \int dxdy\vec{B}^2 + e\oint_C \vec{A}.d\vec{l} = \frac{1}{2g}\int_0^\beta d\tau \int dxdy\vec{B}^2 + e\oint_S \vec{B}.d\vec{S}, \tag{12.33b}$$

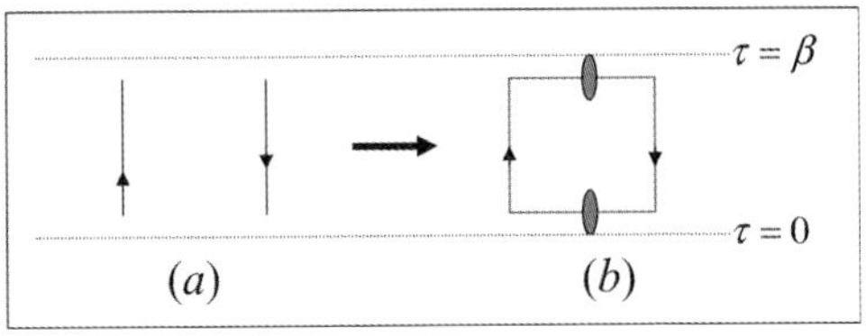


Schematic figure showing (a) two static charges of opposite sign separated by distance $R$; (b) a pair of opposite charges is created in the vacuum at $\tau \sim 0$, separating by the distance $R$ quickly, and coming back and annihilating at time $\tau \sim \beta$.

where $C$ represents the loop and $S$ is an area enclosed by the loop. To proceed further, we again write $\vec{B} = \nabla \times (\vec{A} + \vec{A}_M)$ and introduce the auxiliary field $\vec{j}$ as before. Going through similar algebra, we obtain

$$\nabla \times (\vec{j} - ie\vec{l}) = 0,$$

where $\int \vec{l}\mathrm{d}^3x = \oint_S 1\mathrm{d}\vec{S}$ is the contribution from the current loop. Therefore, we can write $\vec{j} = e\vec{j}_0 + \nabla\phi$, where $\vec{l} = i\vec{j}_0$ and

$$H \to \frac{1}{2}\int_0^\beta \mathrm{d}\tau \int \mathrm{d}x\mathrm{d}y[-2i\nabla\phi.(\nabla \times \vec{A}_M) + g(e\vec{j}_0 + \nabla\phi)^2]$$
$$= \frac{1}{2}\int_0^\beta \mathrm{d}\tau \int \mathrm{d}x\mathrm{d}y[2i\phi\rho + g(e\vec{j}_0 + \nabla\phi)^2]. \tag{12.34a}$$

Compared with Equation (12.30b), we see that the presence of external charge introduces an additional term $\sim e\vec{j}_0$ in the action. The additional energy introduced by the $e\vec{j}_0$ term can be computed in the mean-field approximation, where $\rho \to \langle\rho\rangle \sim in_0\phi$ and

$$H \to \frac{1}{2}\int_0^\beta \mathrm{d}\tau \int \mathrm{d}x\mathrm{d}y[2n_0\phi^2 + g(e\vec{j}_0 + \nabla\phi)^2], \tag{12.34b}$$

which is a quadratic action for the $\phi$ field and the additional energy associated with the $e\vec{j}_0$ term is

$$H_0 = \frac{1}{2}\int_0^\beta \mathrm{d}\tau \int \mathrm{d}x\mathrm{d}y[2n_0\phi_0^2 + g(e\vec{j}_0 + \nabla\phi_0)^2], \tag{12.35a}$$

where $\phi_0$ is the solution of the classical equation of motion

$$\frac{\delta H}{\delta\phi} = 2n_0\phi - g\nabla.(e\vec{j}_0 + \nabla\phi) = 0. \tag{12.35b}$$

**Exercise**

Show that Equation (12.35a, b) is correct even quantum mechanically.

Careful readers may have detected a suspicious 'error' in our derivation. Expanding Equation (12.35a), the coupling term between $\nabla\phi$ and $\vec{j}_0$ is

$$g\int d^3x\nabla\phi.\vec{j}_0 = g\oint_S \nabla\phi.d\vec{S} = 0.$$

Physically, the surface integral actually originates from a line integral (Wilson loop) through $\oint_C \vec{A}.d\vec{l} = \oint_S \vec{B}.d\vec{S}$, meaning that we should be able to define a function $\vec{A}_\phi$ by $\oint_S \nabla\phi.d\vec{S} = \oint \vec{A}_\phi.d\vec{l}$, or $\nabla\phi = \nabla\times\vec{A}_\phi$ if the integral is non-zero.

This is impossible for regular scalar functions $\phi$ and $g\int d^3x\nabla\phi.\vec{j}_0 = 0$, meaning that there should be no coupling between $\nabla\phi$ and $\vec{j}_0$ for regular $\phi$'s! Notice, however, that $\nabla\phi \sim \vec{B}_M$ represents a monopole magnetic field, which is a transverse field when the magnetic flux tube is taken into account. We have replaced the singular vector potential from a magnetic monopole by a scalar potential in our derivation when we ignore the magnetic flux tube. The two 'errors' in our derivation cancel each other, making it possible to couple $\nabla\phi$ and $\vec{j}_0$. Note that, for this reason, in a state where monopoles are absent and $\phi$ is regular, there should be *no* coupling between $\nabla\phi$ and $\vec{j}_0$ and the solution of the classical equation of motion (12.35b) is simply $\phi_0 = 0$. In this case, the energy associated with the charges is

$$E_{NM} = \frac{g}{2}\int_0^\beta d\tau\int dxdy\left[e\vec{j}_0\right]^2, \tag{12.36}$$

which leads to the usual logarithmic interaction $V_0(R) \sim e^2\ln(R/a)$ between charges.

**Homework Q5**

Show that Equation (12.36) leads to the usual logarithmic interaction between charges.

In the presence of monopoles, we have to solve Equation (12.35b). Choosing a flat surface on the $x-\tau$ plane between $x = 0$ and $x = R$ as '$S$', we obtain

$$\vec{j}_0(x,y,\tau) = \theta(R-x)\theta(x)\theta(\beta-\tau)\theta(\tau)\delta(y)\,\widehat{y}$$

and

$$2n_0\phi - g\nabla^2\phi = ge\frac{d\delta(y)}{dy}$$

for $x$, $\tau$ far away from the surface boundary. The solution in this case is simply

$$\phi(x,y,\tau) \sim -\mathrm{sgn}(y)(e)e^{-\frac{|y|}{\lambda}}, \tag{12.37a}$$

where $\lambda^{-2} = \frac{2n_0}{g}$.

Putting this back into Equation (12.35a) and integrating over d$y$, we obtain

$$H_0 \sim \int_0^{\beta} d\tau \int dx(2n_0\lambda) = 2n_0\lambda A = \sqrt{\frac{g}{2n_0}}(R\beta), \tag{12.37b}$$

which is proportional to the area of the loop $A = R\beta$. The error in the calculation is mainly coming from the boundary region and is of order $2n_0\lambda(\lambda \times 2(R+\beta))$, which is much smaller than $\sqrt{\frac{g}{2n_0}}(R\beta)$ in the limit $R, \beta \gg \lambda$. The corresponding electrostatic potential between charges is $V(R) \sim \sqrt{\frac{g}{2n_0}}R$ and increases linearly with distance between charges.

The physical meaning of the classical solution can be understood in real time where $B_y \sim -\frac{d\phi_0}{dy} \rightarrow E_x$. The solution represents a string of electric field passing between the two charges, in agreement with the confinement picture we discussed in the last section. The only difference is that the string of electric field is distributed in space with finite width of order $\lambda$ in the instanton theory.

## 12.4
## Duality Between a Neutral Superliquid and U(1) Gauge Theory Coupled to Charged Bosons

We have learned that a U(1) lattice gauge theory can be understood in dual pictures, the strong-coupling picture where strings and loops become the fundamental objects in the theory, and the weak-coupling picture corresponding to the usual field theory approach. We shall show in this section that duality is not limited to the strong- and weak-coupling expansions of lattice gauge theory, and other forms of duality are also possible. Using the duality transformation technique we used so often in the last section, we shall show that a duality picture also exists between an ordinary (neutral) boson liquid and a model of charged bosons interacting with a U(1) gauge field in 2 + 1 dimensions.

We start with the continuum phase model of the boson liquid. The partition function is expressed in terms of an imaginary-time path integral with Lagrangian

$$\begin{aligned} L_\theta &= \frac{1}{2}\int d^2r\left\{\frac{1}{g}[\partial_\tau\theta(\vec{r})]^2 + \rho_s[\nabla\theta(\vec{r})]^2\right\} \\ &= \frac{1}{2g}\int d^2r \sum_{\mu=0,1,2}\left[\partial_\mu\theta(\vec{r})\right]^2, \end{aligned} \tag{12.38a}$$

where $\theta$ is the boson phase field, $g \sim$ interaction between bosons, and $\rho_s = N_0/m$ is the 'bare' boson superfluid density. $\partial_0 = \partial_\tau$ and $\partial_{1(2)} = c\partial_{x(y)}$, where $c = \sqrt{g\rho_s}$ is the sound-wave velocity. The corresponding partition function is

$$Z[\theta] = \int D\theta \exp\left[-\frac{1}{2g}\sum_\mu \int_0^\beta dx^0 \int d^2x\left(\nabla\theta(x)\right)^2\right], \tag{12.38b}$$

where $\vec{\nabla} = (\partial_0, \partial_1, \partial_2)$.

As in Chapters 8 and 11, we need to include singular field configurations (vortices) in the partition function. To allow for singular field configurations, we let $\nabla\theta \to \nabla\theta + \vec{A}$, where $\vec{A}$ represents the singular part of the $\theta$ field. We then introduce a dual field $j_\mu$ in the partition function,

$$Z[\theta] \to Z[\theta, A, j] = \int D\theta DADj\exp\left[-\frac{1}{2}\sum_\mu \int_0^\beta dx^0 \int d^2x\left[2i\vec{j}.(\nabla\theta + \vec{A}) + g\vec{j}^2\right]\right]. \tag{12.38c}$$

The physical meaning of the dual field $j_\mu$ can be seen most easily by minimizing the action with respect to $j_\mu$, where we obtain $\vec{j} \sim i(\nabla\theta + \vec{A})$, indicating that $j_\mu$ represents the superfluid 3-current in the original interacting boson problem.

Next, we eliminate the $\theta$ field through integration by parts, and obtain

$$\int d^3x\vec{j}.\nabla\theta \to -\int d^3x\theta\nabla.\vec{j}. \tag{12.38d}$$

Integration over the $\theta$ field results in the constraint $\nabla.\vec{j} = 0$, which is just the imaginary-time representation of the continuity equation

$$\frac{\partial}{\partial t}\rho + \frac{\partial}{\partial x}j_x + \frac{\partial}{\partial y}j_y = 0,$$

where $\rho(\vec{j})$ is the charge (current) density of the original bosons. The constraint $\nabla.\vec{j} = 0$ can be satisfied in 3D by writing $\vec{j} = \nabla \times \vec{a}$ and the partition function (12.38c) becomes

$$Z[\theta, A, j] \to \int DADa\exp\left[-\frac{1}{2}\sum_\mu \int_0^\beta dx^0 \int d^2x\left[2i(\nabla \times \vec{a}).\vec{A} + g(\nabla \times \vec{a})^2\right]^2\right] \tag{12.39a}$$

or

$$Z[\theta, A, j] \to \int Dj_\nu Da\exp\left[-\frac{1}{2}\sum_\mu \int_0^\beta dx^0 \int d^2x\left[-2i\vec{a}.\vec{j}_\nu + g(\nabla \times \vec{a})^2\right]^2\right] \tag{12.39b}$$

after integration by parts, where $\vec{j}_\nu = \nabla \times \vec{A}$.

To understand the physical meaning of $\vec{j}_\nu$, first we recall from Chapter 8 that $j_{\nu 0} = \frac{\partial}{\partial x}A_y - \frac{\partial}{\partial y}A_x$ represents the vortex density. Together with the continuity equation $\nabla.\vec{j}_\nu = 0$, we infer that $j_{\nu x}, j_{\nu y}$ represents the corresponding vortex current.

Note that in real time $(\nabla \times \vec{a})^2 \to \frac{1}{2}\sum_{\mu\nu=0,1,2} F_{\mu\nu}^2$ is just the action of a U(1) gauge field in the continuum limit. Therefore, Equation (12.39b) implies that the interaction between vortices and supercurrent flow in the system has the same form as interaction between particles (vortices) with a U(1) gauge field (~supercurrent flow). Vorticity becomes the 'charge' of the vortices and supercurrent flow becomes gauge field coupling to the vortices. Note that the dynamics (kinetic energy term) of vortices is absent in this approach. The dynamics of vortices can be inferred from the following simple argument.

Recall that a finite energy $m \sim c^2/g \sim \rho_s$ is needed to create a (static) vortex in 2D. In 2 + 1D, a vortex will trace out a space–time trajectory (see figure below), which becomes a line defect at 3D (imaginary time).

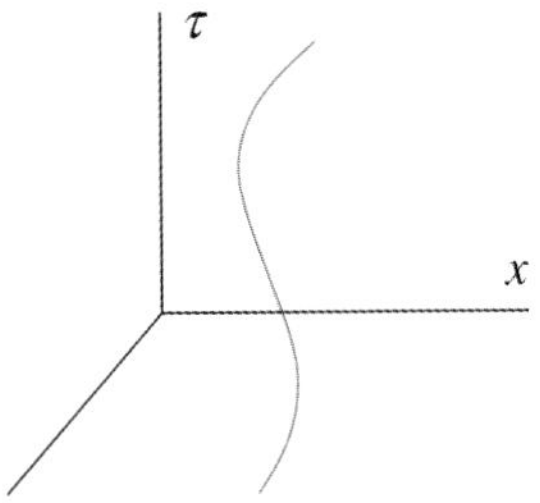


In imaginary-time formulation, the action $S$ is simply the energy of this vortex line, which is simply $S = m \times L$, where $L$ = length of trajectory in space–time $= \int_0^\beta \mathrm{d}\tau \sqrt{1 + \left(\frac{\mathrm{d}\vec{x}}{\mathrm{d}\tau}\right)^2}$, where $\vec{x} = (x, y)$. The corresponding action in real time is $S = -m \int_0^t \mathrm{d}t \sqrt{1 - \left(\frac{\mathrm{d}\vec{x}}{\mathrm{d}t}\right)^2}$, which is the action of a (free) relativistic particle in 2 + 1D, that is vortices behave as free relativistic particles in the absence of interaction with the gauge field $\vec{a}$.

Assuming that exchange of vortices does not introduce non-trivial Berry phases, the corresponding quantum field theory for vortices should be a relativistic field theory of charged bosons interacting with a U(1) gauge field. The simplest possibility of the action is

$$S = \int \mathrm{d}t \int \mathrm{d}^2x (|(\partial_t - i a_0)\phi|^2 - |(\nabla - i\vec{a})\phi|^2 - m^2|\phi|^2) + \frac{g}{2} F_{\mu\nu}^2, \tag{12.40}$$

where $\phi$ represents a vortex field in coherent state representation. The neutral boson or quantum rotor model becomes a model of relativistic charged bosons interacting with a U(1) gauge field when viewed from the angle of vortices.

It is interesting to note that, if the vortices are fermions, the corresponding effective relativistic field theory would be a theory of Dirac fermions coupled to a U(1) gauge field. Indeed, it has been proposed that this situation may be realized in quantum disordered anti-ferromagnets in two dimensions [4].

### 12.4.1
### 'Vortices' in Vortex Liquid

The action (12.40) is interesting because it is a model of bosons in itself. Let us consider the limit $m = 0$ and derive the semiclassical effective action for this model starting from a Bose-condensed state with $\langle \phi \rangle \neq 0$. It is not difficult to see that the effective action for the phase field ($\phi = \sqrt{\rho} e^{i\varphi}$) has the same form as Equation (12.38a), except that $\partial_\mu \theta \to \partial_\mu \varphi - i a_\mu$, and is the same as the phase action for superconductors. In particular, vortices exist in this model as in usual superconductors. The question is: what does a 'vortex' of a vortex liquid represent in terms of the original bosons?

Note that an auxiliary EM field exists in this model and a vortex in this model will bind an auxiliary magnetic flux $\Phi_0$ as in usual superconductors. As explained in the discussion after Equation (12.38a–d), the auxiliary EM field coupling to vortices represents super 3-current of the original boson system, and a magnetic field represents density fluctuation of the original bosons. Therefore, a quantized auxiliary magnetic flux represents a quantized unit of boson density fluctuation. In fact, it represents exactly one boson [1], that is a 'vortex' of the vortex liquid is nothing but the boson of the original boson liquid. Vortices and bosons are 'dual' particles to each other. The superfluid phase of one is the insulator phase of the other, as we shall see in the following.

We can examine the behavior of the system in both representations. From Equation (12.38a), it is easy to see that a superfluid phase for bosons is realized in the small-$g$ limit, whereas from Equation (12.40) it is obvious that $g \to 0$ corresponds to the confinement phase of U(1) gauge theory. In this limit, vortices (= charges coupled to the auxiliary U(1) gauge field) are confined in pairs and there are no free vortices in the system. Therefore, the system is a superfluid phase for bosons but an insulator phase for vortices. Vortices are liberated when $g$ becomes large and $m \to 0$ (plasma phase), corresponding to the insulator phase of the original boson superfluid from our analysis in Chapter 11.

One may deduce from the instanton analysis in the previous section that a U(1) gauge field is confining in 2 + 1D and vortices are always confined. Therefore, there should be only one quantum phase at zero temperature where the system is always a superfluid (of the original bosons). This is not correct because of the presence of a matter field. We shall not go into the details here except to point out that the dynamics of gauge fields is strongly modified by the presence of a matter field and as a result both superfluid and insulator phases may exist in the system.

## References


1 Wen, X.G. (2004) *Theory of Quantum Many-Body Systems*, Oxford University Press.

2 Polyakov, A.M. (1987) *Gauge Fields and Strings*, Harwood Academic Publishers.

3 Sakurai, J.J. (1994) *Modern Quantum Mechanics*, Addison-Wesley.

4 Ng, T.K. (1999) *Physical Review Letters*, **82**, 3504.

# Appendix: One-Particle Green's Function in Second-Order Perturbation Theory

In this appendix we shall evaluate the retarded fermion Green's function

$$G_{nR}(t-t') = i\langle c_n(t)c_n^+(t') + c_n^+(t')c_n(t)\rangle\theta(t-t')$$

to second order in the interaction. We start with Equation (7.27) in Chapter 7, where the fermion Green's function is written as

$$G_{nR}(\omega) = \frac{1}{\omega + i\delta - \varepsilon_n - \Sigma_n(\omega + i\delta) + \mu} \text{ with self energy}$$

$$\Sigma_n(t-t') = \sum_{m,m'} H_{nm'} R_{m'm}(t-t') H_{mn}, \tag{A.1}$$

where $|n\rangle = c_n^+|G\rangle$ ($c_n|G\rangle$) and the $|m\rangle$ are states orthogonal to $|n\rangle$.

Here,

$$R_{m'm}(t) = \langle m'|\left(i\hbar\frac{\partial}{\partial t} - H_{m'm} + \mu\right)^{-1}|m\rangle \tag{A.2}$$

and the sum over $m$, $m'$ is performed over all states orthogonal to $|n\rangle$. Note that we have introduced the chemical potential $\mu$ here since the single-particle Green's function involves states with one more/less particle (see Chapter 5) than the original state. Although the states $|n\rangle$ and $|m\rangle$ are difficult to construct in general, they can be taken as the eigenstates of $H_0$ – the unperturbed Hamiltonian – when we consider second-order correction to the self energy, since the self energy (A.1) is already second order in $H_{mn}$ – the non-diagonal part of the Hamiltonian. Writing $H = H_0 + V$, we can write

$$\Sigma_n^{(2)}(t-t') = \sum_{m,m'} V_{nm'} R^{(0)}_{m'm}(t-t') V_{mn}, \tag{A.3}$$

where

$$R^{(0)}_{m'm}(t-t') = \langle m'|\left(i\hbar\frac{\partial}{\partial t} - H_{0m'm} + \mu\right)^{-1}|m\rangle = \delta_{m'm}\langle m|\left(i\hbar\frac{\partial}{\partial t} - \varepsilon_m + \mu\right)^{-1}|m\rangle,$$

*Introduction to Classical and Quantum Field Theory.* Tai-Kai Ng

ISBN: 978-3-527-40726-2

where $\varepsilon_m$ is the excitation energy for the state $|m\rangle$ in the unperturbed Hamiltonian. Fourier transforming, we obtain

$$\Sigma_{nR}^{(2)}(\omega+i\delta)=\sum_m \frac{|V_{nm}|^2}{\hbar\omega+i\delta-\varepsilon_m+\mu}, \tag{A.4}$$

which is the standard result in second-order perturbation theory, except that the self energy is evaluated at general energy $\hbar\omega+i\delta$, but not at the unperturbed energy $\varepsilon_n$. We shall evaluate Equation (A.4) directly in the following and compare the result with standard diagrammatic perturbation theory.

To evaluate Equation (A.4), we have to determine what are the allowable $\varepsilon_n$ and $V_{nm}$ with an explicit Hamiltonian. We shall consider a spinless fermion gas with

$$H = H_0 + V = \sum_{i=1,\ldots,N,\sigma} (\varepsilon_{\vec{k}}-\mu)c_{\vec{k}}^{+}c_{\vec{k}} + \frac{1}{2V}\sum_{\vec{q},\vec{k},\vec{k}'} V(\vec{q})c_{\vec{k}-\vec{q}}^{+}c_{\vec{k}'+\vec{q}}^{+}c_{\vec{k}'}c_{\vec{k}} \tag{A.5}$$

where $\varepsilon_{\vec{k}}$ is the kinetic energy of the fermion gas, $\mu$ is the chemical potential, and $V(\vec{q})$ is an interaction which conserves total momentum.

$$|n\rangle = c_n^{+}|G\rangle \rightarrow c_{\vec{k}\sigma}^{+}|\mathrm{FS}\rangle(|\vec{k}| > k_\mathrm{F}) \quad \text{for } t > t' \tag{A.6a}$$

and

$$|n\rangle = c_n|G\rangle \rightarrow c_{\vec{k}}|\mathrm{FS}\rangle(|\vec{k}| < k_\mathrm{F}) \quad \text{for } t < t' \tag{A.6b}$$

are states constructed by adding (removing) an electron with momentum $|\vec{k}| > (<)k_\mathrm{F}$ to or from a filled Fermi sea with Fermi momentum $k_\mathrm{F}$ and the states $|m\rangle$ can be any states linked to the states $|n\rangle$ by the interaction $V$.

The possible states $|m\rangle$ can be separated into two groups. For $|\vec{k}| > k_\mathrm{F}$, they are

$$|m^{(\mathrm{I})}\rangle \sim c_{\vec{k}-\vec{q}}^{+}c_{\vec{p}+\vec{q}}^{+}c_{\vec{p}}|\mathrm{FS}\rangle \tag{A.7a}$$

and

$$|m^{(\mathrm{II})}\rangle \sim c_{\vec{p}-\vec{q}}^{+}c_{\vec{p}'+\vec{q}}^{+}c_{\vec{p}'}c_{\vec{p}}c_{\vec{k}}^{+}|\mathrm{FS}\rangle, \tag{A.7b}$$

where $\vec{p},\vec{p}' \neq \vec{k}$ in $|m^{(\mathrm{II})}\rangle$.

Physically, $|m^{(\mathrm{I})}\rangle$ is created by scattering a fermion with momentum $\vec{k}$ to a state with momentum $\vec{k}-\vec{q}$ and creating simultaneously a particle–hole pair with momentum $\vec{q}$ (figure (a)), whereas $|m^{(\mathrm{II})}\rangle$ is created by spontaneous creation of two particle–hole pairs from the filled Fermi sea without involving the fermion state $\vec{k}$ (figure (b)).

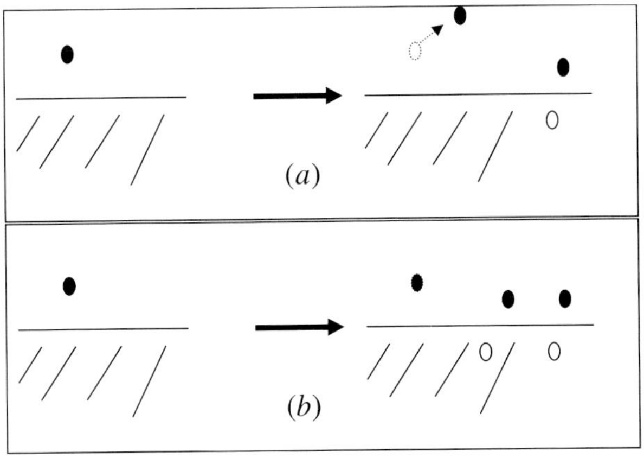


First, we consider the second kind of process.

Using Equation (A.5), we obtain with straightforward algebra

$$\varepsilon_m^{(\mathrm{II})} = \xi_{\vec{p}'+\vec{q}} + \xi_{\vec{p}-\vec{q}} - \xi_{\vec{p}} - \xi_{\vec{p}'} + \xi_{\vec{k}}, \tag{A.8a}$$

$$V_{nm}^{(\mathrm{II})} = [V(\vec{q}) - V(\vec{p}-\vec{p}'-\vec{q})]n_{\vec{p}'}n_{\vec{p}}(1-n_{\vec{p}'+\vec{q}})(1-n_{\vec{p}-\vec{q}}), \tag{A.8b}$$

and

$$\Sigma_{\mathrm{R}}^{(2(\mathrm{II}))}(\vec{k}, \omega + i\delta) = \Sigma'(\omega - \xi_{\vec{k}} + i\delta),$$

where $\xi_k = \varepsilon_k - \mu$,

$$\begin{aligned}\Sigma'(\omega) \sim\ & 2\sum_{\vec{p},\vec{p}',\vec{q}} V(\vec{q})^2 \left(\frac{n_{\vec{p}'}n_{\vec{p}}(1-n_{\vec{p}'+\vec{q}})(1-n_{\vec{p}-\vec{q}})}{\omega - \xi_{\vec{p}'+\vec{q}} - \xi_{\vec{p}-\vec{q}} + \xi_{\vec{p}'} + \xi_{\vec{p}}}\right) \\ & - 2\sum_{\vec{p},\vec{p}',\vec{q}} V(\vec{q})V(\vec{p}-\vec{p}'-\vec{q}) \left(\frac{n_{\vec{p}'}n_{\vec{p}}(1-n_{\vec{p}'+\vec{q}})(1-n_{\vec{p}-\vec{q}})}{\omega - \xi_{\vec{p}'+\vec{q}} - \xi_{\vec{p}-\vec{q}} + \xi_{\vec{p}'} + \xi_{\vec{p}}}\right).\end{aligned} \tag{A.9}$$

The two terms are represented by the following diagrams in diagrammatic perturbation theory:

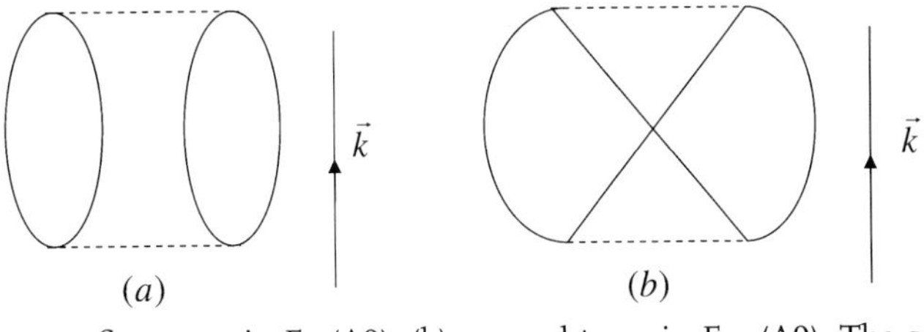


(a)–first term in Eq.(A9); (b)- second term in Eq. (A9). The solid line represents electron Green's functions. The dash line represents interaction.

Note that the electron Green's function $g(\vec{k})$ is separated from the interaction. This is also reflected in the self energy, where $\Sigma_{n\mathrm{R}}^{(2(\mathrm{II}))}(\omega + i\delta) = \Sigma'(\omega - \xi_{\vec{k}} + i\delta)$ depends on $\vec{k}$ only through a shift in energy. In particular, $\Sigma_{n\mathrm{R}}^{(2(\mathrm{II}))}(\omega \to \xi_{\vec{k}})$ is independent of $\vec{k}$. Physically, this process represents a correction to the ground-state energy which is independent of whether an extra electron $\vec{k}$ is present in the system. The self energy $\Sigma^{(2(\mathrm{II}))}_{n\mathrm{R}}(\omega \to \xi_{\vec{k}})$ contributes to an overall shift in chemical potential in the one-electron Green's function.

The first kind of process is more interesting because it affects the electron with momentum $\vec{k}$ directly. In this case,

$$\varepsilon^{(\mathrm{I})}_{m} = \xi_{\vec{p}+\vec{q}} - \xi_{\vec{p}} + \xi_{\vec{k}-\vec{q}}, \tag{A.10a}$$

$$V^{(\mathrm{II})}_{nm} = [V(\vec{q}) - V(\vec{k}-\vec{p}-\vec{q})]n_{\vec{p}}(1-n_{\vec{p}+\vec{q}})(1-n_{\vec{k}-\vec{q}}), \tag{A.10b}$$

and

$$\begin{aligned}\Sigma^{(2(\mathrm{I}))}_{\mathrm{R}}(\vec{k},\omega) \sim\ & 2\sum_{\vec{p},\vec{p}',\vec{q}} V(\vec{q})^2 \left(\frac{n_{\vec{p}}(1-n_{\vec{p}+\vec{q}})(1-n_{\vec{k}-\vec{q}})}{\omega - \xi_{\vec{p}+\vec{q}} - \xi_{\vec{k}-\vec{q}} + \xi_{\vec{p}}}\right) \\ & - 2\sum_{\vec{p},\vec{p}',\vec{q}} V(\vec{q})V(\vec{k}-\vec{p}-\vec{q}) \left(\frac{n_{\vec{p}}(1-n_{\vec{p}+\vec{q}})(1-n_{\vec{k}-\vec{q}})}{\omega - \xi_{\vec{p}+\vec{q}} - \xi_{\vec{k}-\vec{q}} + \xi_{\vec{p}}}\right).\end{aligned} \tag{A.11}$$

The two terms are represented by the following diagrams in diagrammatic perturbation theory:

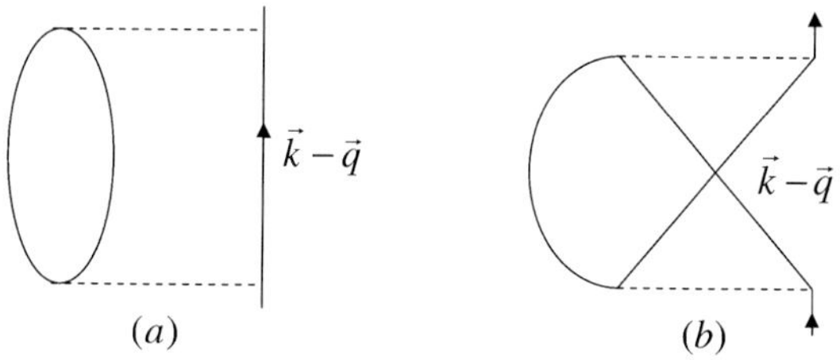


(a)–first term in Eq.(A11); (b)- second term in Eq. (A11).

Next, we examine the imaginary part of $\Sigma^{(2(\mathrm{I}))}_{\mathrm{R}}(\vec{k},\omega)$ for a non-singular interaction $V(\vec{q})$. To see its qualitative behavior, we can replace $\left[V(\vec{q}) - V(\vec{k}-\vec{p}-\vec{q})\right]$ by a constant $\sim U$. In this case, the imaginary part of the self energy is given by the expression

$$\begin{aligned}\mathrm{Im}\Sigma^{(2(\mathrm{I}))}_{\mathrm{R}}(\vec{k},\omega) &\sim -\pi U^2 \sum_{\vec{p},\vec{p}',\vec{q}} (n_{\vec{p}}(1-n_{\vec{p}+\vec{q}})(1-n_{\vec{k}-\vec{q}}))\delta(\omega - \xi_{\vec{p}+\vec{q}} - \xi_{\vec{k}-\vec{q}} + \xi_{\vec{p}}) \\ &= -\pi U^2 \int d\varepsilon_1 \rho(\varepsilon_1) \int d\varepsilon_2 \rho(\varepsilon_2) \int d\varepsilon_3 \rho(\varepsilon_3) n(\varepsilon_1)(1-n(\varepsilon_2))(1-n(\varepsilon_3))\delta(\omega - \varepsilon_3 - \varepsilon_2 + \varepsilon_1) \\ &= -\pi U^2 \int d\varepsilon_2 \rho(\varepsilon_3 + \varepsilon_2 - \omega) \int d\varepsilon_3 \rho(\varepsilon_2)\rho(\varepsilon_3) n(\varepsilon_3 + \varepsilon_2 - \omega)(1-n(\varepsilon_2))(1-n(\varepsilon_3))\end{aligned}$$

where $\rho(\varepsilon)$ is the density of states.

At zero temperature $n(\varepsilon) = \theta(-\varepsilon)$ and we obtain

$$\begin{aligned}\mathrm{Im}\Sigma^{(2(\mathrm{I}))}_{\mathrm{R}}(\vec{k},\omega) &\sim -\pi U^2 \int_0^{\omega} d\varepsilon_2 \int_0^{\omega-\varepsilon_2} d\varepsilon_3 \rho(\varepsilon_2)\rho(\varepsilon_3)\rho(\varepsilon_3 + \varepsilon_2 - \omega) \\ &\sim -\frac{1}{2}\pi U^2 \rho(0)^3 \omega^2\end{aligned} \tag{A.12}$$

at small $\omega$, where $\rho(0)$ is the density of states on the Fermi surface. The result $\mathrm{Im}\Sigma_R(\vec{k},\omega) \sim \omega^2$ for small $\omega$ is a general result for fermion liquids with regular interaction and non-vanishing Fermi surface. The same conclusion can be derived for $|\vec{k}| < k_F$ and will be left as an exercise for readers.

# Index



*Introduction to Classical and Quantum Field Theory.* Tai-Kai Ng

ISBN: 978-3-527-40726-2

**x**



**y**



**z**